Environmental Catalysis in Advanced Oxidation Processes, 2nd Edition

Environmental Catalysis in Advanced Oxidation Processes, 2nd Edition

Guest Editors

Albin Pintar
Gregor Žerjav

MDPI

Basel • Beijing • Wuhan • Barcelona • Belgrade • Novi Sad • Cluj • Manchester

Guest Editors
Albin Pintar
Department of Inorganic Chemistry and Technology
National Institute of Chemistry
Ljubljana
Slovenia

Gregor Žerjav
Department of Inorganic Chemistry and Technology
National Institute of Chemistry
Ljubljana
Slovenia

Editorial Office
MDPI AG
Grosspeteranlage 5
4052 Basel, Switzerland

This is a reprint of the Special Issue, published open access by the journal *Catalysts* (ISSN 2073-4344), freely accessible at: https://www.mdpi.com/journal/catalysts/special_issues/1L0L796L42.

For citation purposes, cite each article independently as indicated on the article page online and as indicated below:

Lastname, A.A.; Lastname, B.B. Article Title. *Journal Name* **Year**, *Volume Number*, Page Range.

ISBN 978-3-7258-6228-3 (Hbk)
ISBN 978-3-7258-6229-0 (PDF)
https://doi.org/10.3390/books978-3-7258-6229-0

Contents

About the Editors

Albin Pintar

Albin Pintar is a Research Professor (Res. Prof. Dr.) and head of the research programme group "Integrated Approach to Water Pollution Prevention" at the Department of Inorganic Chemistry and Technology, National Institute of Chemistry, Ljubljana, Slovenia. He obtained his MSc and PhD degrees from the University of Ljubljana in 1992 and 1996, respectively. He was a visiting scientist at the Tokyo Institute of Technology (Japan) and the Research Institute of Catalysis and Environment (Villeurbanne, France). His research interests include heterogeneous catalysis, environmental catalysis, kinetics and mechanisms of catalytic reactions in multiphase reactors, wastewater treatment, production of hydrogen-rich mixtures from renewables, and process intensification. He is the author of over 200 scientific papers and review articles in peer-reviewed journals, as well as several international patents (h-index: 53). He received a prestigious national award for significant scientific achievements in chemical engineering. Albin Pintar is chairman of the Catalysis Division at the Slovenian Chemical Society, serves as Associate Editor of *RSC Advances*, and is a member of the Editorial Boards of the MDPI journals *Catalysts* and *Molecules*. Since 2018, Albin Pintar has been chairman of the annual meetings of the Slovenian Chemical Society and has chaired various international scientific conferences (CAMURE-5 & ISMR-4, EAAOP-6, 2020 EFCATS Summer School, Ceria 2024, and NPM-8 & PAOT-9).

Gregor Žerjav

Gregor Žerjav is a Research Assistant Professor at the Department of Inorganic Chemistry and Technology, National Institute of Chemistry, Ljubljana, Slovenia. He obtained his PhD in Chemistry in 2015, following several years as a young researcher at the Jožef Stefan Institute, where he developed anti-corrosion and hydrophobic coatings for metallic surfaces. In 2014, he was a visiting PhD student at the University of Udine, Italy. Since then, he has focused on the synthesis and characterisation of heterogeneous catalysts, particularly TiO_2- and g-C_3N_4-based systems, as well as catalytic processes for environmental protection, advanced oxidation, wastewater treatment, and NO_x reduction. In 2016, Dr. Žerjav received the Henkel Prize for his outstanding doctoral dissertation. Between 2018 and 2019, he was an ERC-funded visiting scientist at Chalmers University of Technology in Göteborg, Sweden, where he investigated TiO_2-based plasmonic catalysts for NO_x reduction. In 2023, he also conducted research at Khalifa University in Abu Dhabi, UAE, on photocatalytic air oxidation processes. He has coordinated and collaborated on several international projects, including with Brookhaven National Laboratory (USA), the N. D. Zelinsky Institute of Organic Chemistry (Russia), and CSIR-NEERI (India). He also contributes to the scientific community as a reviewer for leading journals and as a Guest Editor of the MDPI journal *Catalysts*. Furthermore, he has co-organised several international conferences, including EAAOP-6, Ceria 2024, and NPM-8 & PAOT-9. His research makes a significant contribution to the development of catalytic materials and sustainable processes for environmental protection.

catalysts MDPI

Editorial

Environmental Catalysis in Advanced Oxidation Processes, 2nd Edition—Preface to the Special Issue

Albin Pintar * and Gregor Žerjav *

Department of Inorganic Chemistry and Technology, National Institute of Chemistry, Hajdrihova 19, SI-1001 Ljubljana, Slovenia
* Correspondence: albin.pintar@ki.si (A.P.); gregor.zerjav@ki.si (G.Ž.)

This Editorial, "Environmental Catalysis in Advanced Oxidation Processes, 2nd Edition", highlights the expanding role of catalytic and photocatalytic oxidation technologies in addressing today's pressing environmental challenges. Bringing together advances in catalyst design and characterization, innovative reactor engineering, and mechanistic insights into oxidation pathways, this collection reflects the dynamic evolution of advanced oxidation processes (AOPs). Beyond showcasing technical progress, it underscores the importance of AOPs as key tools for achieving cleaner air and water through sustainable and energy-efficient approaches. Motivated by the global pursuit of green and resilient environmental technologies, this Editorial invites scientists, engineers, and students to explore new opportunities and inspire future breakthroughs in catalysis, environmental chemistry, and process engineering.

Featuring a collection of original research articles, this Special Issue begins with a study on the photocatalytic degradation of levofloxacin in aqueous media using metal-free graphitic carbon nitride (g-C_3N_4) under simulated solar irradiation (Contribution 1). The authors demonstrated the catalyst's high stability and identified the key roles of superoxide anion radicals ($^{\bullet}O_2{}^{-}$) and photogenerated holes (h^+) in the oxidative degradation pathway.

In Contribution 2, biochar derived from banana peel, promoted with copper phosphide (Cu_3P), was used to activate peroxydisulfate for the efficient removal of bisphenol S (BPS). The 2% Cu_3P/BPB/SPS system achieved 90% BPS degradation within 10 min, and both radical and non-radical pathways were shown to participate in the transformation, with the authors proposing a detailed degradation mechanism predicated on UHPLC/TOF-MS analysis.

The photo-Fenton degradation of sulfamethoxazole (SMX) was investigated to monitor changes in aromaticity and colour during treatment (Contribution 3). Kinetic modelling revealed that an optimal molar ratio of 1 mol SMX:10 mol H_2O_2 enabled a 71% elimination of aromatic intermediates, yielding colourless water with minimal turbidity and providing a predictive tool for efficient oxidant usage under environmentally friendly conditions.

Contribution 4 explored modified gasification slag as a low-cost catalyst for persulfate activation in the degradation of acetaminophen, highlighting the role of carbonyl and graphitic nitrogen functional groups as active sites. Non-radical oxidation pathways predominated, demonstrating the potential for industrial waste utilization in wastewater treatment.

Mo_2C was evaluated as a persulfate activator for the degradation of losartan in tertiary wastewater treatment in Contribution 5. The Mo_2C/SPS system achieved complete LOS removal and significant *E. coli* inactivation, particularly under simulated solar irradiation, showcasing synergistic effects between photocatalysis and advanced oxidation for both chemical and microbiological remediation.

In Contribution 6, the influence of Au loading and TiO_2 support morphology on the catalytic wet air oxidation of glyphosate was reported. TiO_2+Au catalysts demonstrated enhanced degradation rates, with the Schottky barrier and surface adsorption of degradation products playing critical roles in catalyst activity and longevity.

Catalytic wet air oxidation was further applied for the disposal of excess activated sludge from a coal chemical wastewater treatment process using Cu-Ce/γ-Al_2O_3 catalysts (Contribution 7). High removal rates of volatile suspended solids (93.2%) and chemical oxygen demand (78.3%) were achieved, and the process generated volatile fatty acids that could serve as carbon sources for subsequent biological treatment.

Innovative reactor designs were highlighted in two studies. First, a rotating photodisc reactor (RPR) using ZnO doped with aluminum nanoparticles demonstrated 50–62% pyridine degradation under natural light, emphasizing the importance of catalyst doping and reactor design for enhanced photodegradation (Contribution 8). Second, in Contribution 9, a 3D-printed modular photoelectrochemical (PEC) reactor was developed to couple seawater splitting with solid-state hydrogen storage, showing excellent performance and stability over 300 h across a broad pH range.

This Editorial also features two comprehensive reviews. Contribution 10 provides an in-depth overview of the synthesis and photocatalytic performance of distinct nano-oxides, including Ti, Zn, Cu, Fe, Ag, Sn, and W oxides, illustrating strategies to enhance photocatalytic efficiency through heterostructure formation and material modification. The second review summarizes recent advances in the removal of organic drugs using AOPs, highlighting methods such as UV, ozone, Fenton-based processes, and heterogeneous photocatalysis while discussing their advantages, limitations, and mechanistic insights (Contribution 11).

Taken together, the contributions in this Special Issue provide mechanistic insights, practical knowledge, and technological innovations that advance the field of environmental catalysis and AOPs. They demonstrate how material design, process intensification, and reactor engineering can improve the efficiency of pollutant removal. This Editorial is intended as an up-to-date resource for researchers, engineers, and students, offering a comprehensive perspective on state-of-the-art strategies to address global environmental challenges and promote sustainable approaches for cleaner water and air.

We extend our heartfelt gratitude to all the authors for their invaluable contributions, without which this Special Issue would not have been possible. We hope that the original research papers and review articles included will help advance the understanding and resolution of current challenges in the field. Our sincere thanks also go to MDPI's *Catalysts* journal for granting us the opportunity to serve as Guest Editors and contribute to the ongoing development of environmental catalysis for water remediation. We are particularly grateful to the Assistant Editor, Ms. Patty Ge, for her dedicated efforts and collaboration throughout the publication process. Finally, we express our appreciation to all the reviewers for their essential and thoughtful evaluations, which greatly enriched this Special Issue.

Author Contributions: Conceptualization, A.P. and G.Ž.; writing—original draft preparation, A.P. and G.Ž.; writing—review and editing, A.P. and G.Ž.; funding acquisition, A.P. All authors have read and agreed to the published version of the manuscript.

Acknowledgments: We acknowledge the financial support from the Slovenian Research and Innovation Agency (ARIS) through core funding No. P2-0150.

Conflicts of Interest: The authors declare no conflicts of interest.

List of Contributions:

1. Xuan, T.N.; Thi, D.N.; Thanh, C.L.; Thi, T.M.; Dieu, T.L.; Duc, T.N.; Horváth, O. Assessing the Degradation of Levofloxacin in Aqueous Media by Metal-Free g-C_3N_4 Photocatalyst Under Simulated Solar Light Irradiation. *Catalysts* **2024**, *14*, 837. https://doi.org/10.3390/catal14110837.
2. Ioannidi, A.A.; Logginou, O.; Kouvelis, K.; Petala, A.; Antonopoulou, M.; Mantzavinos, D.; Frontistis, Z. Peroxydisulfate Activation by Biochar from Banana Peel Promoted with Copper Phosphide for Bisphenol S Degradation in Aqueous Media. *Catalysts* **2024**, *14*, 789. https://doi.org/10.3390/catal14110789.
3. Villota, N.; Duoandicoechea, U.; Lombraña, J.I.; De Luis, A.M. Kinetic Modelling of Aromaticity and Colour Changes during the Degradation of Sulfamethoxazole Using Photo-Fenton Technology. *Catalysts* **2024**, *14*, 718. https://doi.org/10.3390/catal14100718.
4. Si, W.; Qi, F.; Wang, K.; Wang, Q.; Zeng, Z.; Niu, Y.; Huang, Z. Modified Gasification-Slag-Driven Persulfate Activation for Highly Efficient Degradation of Acetaminophen: N/O Active Site Regulation and Nonradical Oxidation. *Catalysts* **2023**, *13*, 1512. https://doi.org/10.3390/catal13121512.
5. Ioannidi, A.A.; Vlachodimitropoulou, M.; Frontistis, Z.; Petala, A.; Koutra, E.; Kornaros, M.; Mantzavinos, D. Assessing the Efficacy of a Mo_2C/Peroxydisulfate System for Tertiary Wastewater Treatment: A Study of Losartan Degradation, E. coli Inactivation, and Synergistic Effects. *Catalysts* **2023**, *13*, 1285. https://doi.org/10.3390/catal13091285.
6. Žerjav, G.; Albreht, A.; Pintar, A. The Influence of Au Loading and TiO_2 Support on the Catalytic Wet Air Oxidation of Glyphosate over TiO_2+Au Catalysts. *Catalysts* **2024**, *14*, 448. https://doi.org/10.3390/catal14070448.
7. Wang, Z.; Qin, S.; Zheng, W.; Lou, X.; Zeng, X.; Wu, T. The Catalytic Wet Oxidation of Excess Activated Sludge from a Coal Chemical Wastewater Treatment Process. *Catalysts* **2023**, *13*, 1352. https://doi.org/10.3390/catal13101352.
8. Montalvo, C.; Lemus, E.; Aguilar, C.A.; Cerón, R.M.; Cerón, J.G.; Robles, J.C.; Ruiz, A. Rotating Photo-Disc Reactor (RPR) Used in the Photo-Degradation of Pyridine Using Zinc Oxide as a Catalyst Composited with Aluminum Nanoparticles and Irradiated with Natural Light. *Catalysts* **2025**, *15*, 437. https://doi.org/10.3390/catal15050437.
9. Wyżga, P.; Macyk, J.; Lin, Y.-C.; Jensen, E.H.; Guzik, M.N.; Bieńkowski, K.; Solarska, R.; Macyk, W. 3D-Printed Reactor for Coupling Photoelectrochemical (Sea)Water Splitting with Solid-State H_2 Storage. *Catalysts* **2024**, *14*, 941. https://doi.org/10.3390/catal14120941.
10. Pavlović, J.; Tušar, N.N.; Rajić, N. The Synthesis and Photocatalytic Efficacy of Distinct Nano-Oxides in the Breakdown of Organic Contaminants. *Catalysts* **2024**, *14*, 771. https://doi.org/10.3390/catal14110771.
11. Umair, M.; Kanwal, T.; Loddo, V.; Palmisano, L.; Bellardita, M. Review on Recent Advances in the Removal of Organic Drugs by Advanced Oxidation Processes. *Catalysts* **2023**, *13*, 1440. https://doi.org/10.3390/catal13111440.

catalysts

MDPI

Article

Peroxydisulfate Activation by Biochar from Banana Peel Promoted with Copper Phosphide for Bisphenol S Degradation in Aqueous Media

Alexandra A. Ioannidi [1], Orestia Logginou [1], Konstantinos Kouvelis [1], Athanasia Petala [2], Maria Antonopoulou [3], Dionissios Mantzavinos [1,*] and Zacharias Frontistis [4,*]

1 Department of Chemical Engineering, University of Patras, Caratheodory 1, University Campus, GR-26504 Patras, Greece; alex.ioannidi@chemeng.upatras.gr (A.A.I.); koskouv@chemeng.upatras.gr (K.K.)
2 Department of Environment, Ionian University, GR-29100 Zakynthos, Greece; apetala@ionio.gr
3 Department of Sustainable Agriculture, University of Patras, GR-30131 Agrinio, Greece; mantonop@upatras.gr
4 Department of Chemical Engineering, University of Western Macedonia, GR-50132 Kozani, Greece
* Correspondence: mantzavinos@chemeng.upatras.gr (D.M.); zfrontistis@uowm.gr (Z.F.)

Abstract: In this work, the decomposition of bisphenol S (BPS) by biochar derived from banana peel (BPB) promoted by copper phosphide (Cu_3P) was examined. Different materials with Cu_3P loadings from 0.25 to 4.00 wt.% on biochar were synthesized, characterized using the Brunauer–Emmett–Teller (BET) method, X-ray diffraction (XRD) and a scanning electron microscope (SEM) equipped with an energy dispersive spectrometer (EDS), and evaluated. Nearly all of the synthesized materials exhibited low to moderate adsorption capacity, attributable to their limited surface area (<3.1 m^2/g). However, in the presence of sodium persulfate (SPS), the 2%Cu_3P/BPB/SPS system was capable of removing 90% of 500 μg/L BPS in less than 10 min. The system's performance was enhanced under inherent pH, and the reaction rate followed pseudo-first-order kinetics with respect to BPS and persulfate concentrations. Interestingly, the presence of 250 mg/L of sodium chloride had a negligible effect, while low to moderate inhibition was observed in the presence of bicarbonates and humic acid. In contrast, significant retardation was observed in experiments performed in real matrices, such as secondary effluent (WW) and bottled water (BW). According to scavenging experiments, both radical and non-radical mechanisms participated in the BPS degradation. Four transformation products were identified using the UHPLC/TOF-MS system in negative ionization mode, with two of them having higher molecular weights than BPS, while the other two TBPs involved the ring-opening reaction, and a BPS decomposition pathway was proposed.

Keywords: endocrine disruptor; persulfate; banana; copper phosphide; transformation products

1. Introduction

The modern lifestyle has led to a significant increase in wastewater complexity, making the adoption of more efficient water treatment methods an urgent need. The complexity and difficulty of managing wastewater arises not only due to the higher concentration of conventional pollutants but also due to the occurrence of emerging pollutants (EPs) such as pharmaceuticals and personal care products (PPCPs) as well as endocrine-disrupting chemicals (EDCs) in aqueous media [1]. Among the different technologies that have been proposed in recent years, advanced oxidation processes (AOPs) seem to be gaining ground, not only because of the increased yields recorded, but also due to their "green" nature [2]. The oxidative power of AOPs is attributed to reactive oxygen species, particularly hydroxyl radicals ($^{\bullet}OH$), which are generated in situ. However, other reactive species, such as sulfate radicals $SO_4^{\bullet-}$ can also act either exclusively or as supplementary agents in various AOPs [3,4]. Sulfate radical generation derives from the activation of persulfates or peroxydisulfates by different methods, including thermal [5], base [6], UV radiation [7], ultrasound [8], electrochemical [9], and chemical activation [10]. Considering chemical activation, transition metals such as ferrous iron (Fe^{2+}), cobalt (Co^{2+}) and manganese (Mn^{2+}) are dissolved in aqueous media and react with the persulfate anion initiating reactions to produce the highly reactive sulfate radicals, thus eliminating the need for energy input. However, dissolved metals can lead to secondary pollution and require an additional treatment step for their removal before effluent discharge, increasing both the cost and complexity. Additionally, the pH and composition of aqueous matrices significantly affect transition metal species, often leading to a reduction in process performance [11]. To overcome these challenges, researchers have turned to heterogeneous activation systems using different catalytic materials, mainly metal oxides, as persulfate activators [12]. Specifically, cobalt oxides (CoO_x) have been successfully used in wastewater treatment systems for the activation of persulfates (SPS/CoO_x systems) and the elimination of emerging contaminants, such as pharmaceuticals and personal care products, through the formation of sulfate radicals [13].

In addition, a less toxic and more abundant metal, such as copper (Cu), has been adopted as a heterogeneous persulfate activator in various formulations, such as CuO_x [14] and Cu_3P [15]. Regarding CuO_x, two opposing perspectives on the degradation mechanism of probe compounds have been reported [16,17]. Zhang et al. [16] suggested a non-radical pathway for the decomposition of 2,4-dichlorophenol in the CuO/persulfate system, while Lei et al. [17] attributed the observed phenol degradation in the CuO-Fe_3O_4/persulfate system to sulfate radicals. A subsequent study on the CuO/persulfate system bridged these views, proposing that both radical and non-radical mechanisms may occur depending on the experimental conditions [18]. Apart from metal-based heterogeneous persulfate activators, metal-free materials have also been successfully adopted in EC degradation systems. In this class, carbon-based materials hold the lion's share due to their robust mechanical properties, tunable chemical composition, large surface area, porosity, and flexible physical forms [19]. For example, Zhang et al. used granular activated carbon as a heterogeneous peroxymonosulfate activator for the degradation of azo dye Acid Orange 7 (AO7) and concluded that AO7 was effectively removed through reactions with sulfate radicals [20]. AO7 degradation was also studied by Chen et al. [21] using carbon nanotubes as the peroxymonosulfate activator. According to the authors, although the dominant degradation mechanism is based on radical formation, under specific conditions, such as high chloride concentrations in the aqueous media, the non-radical mechanism also contributes [21]. In contrast, another study reported that single- or multi-walled carbon nanotubes can activate persulfate to effectively degrade phenol without involving the formation of $SO_4^{\bullet -}$ [22]. A novel approach that considers carbon-based materials was presented by Bekris et al. [23], who examined a commercial graphene nano powder as a heterogeneous persulfate activator for the destruction of the endocrine disruptor propylparaben (PP) and found that over 95% of propylparaben (PP) was removed in a very short reaction time (15 min), while both sulfate and hydroxyl radicals with hydroxyl radicals being the dominant species participating in PP degradation [23].

Recently, biochar (BC), a key category of carbon-based heterogeneous catalysts, has been applied in AOPs for the degradation of various micropollutants [24]. Biochars are produced through the thermal decomposition of biomass in an environment with little or no oxygen, a process known as pyrolysis. This process converts biomass into a stable, carbon-rich material that resembles charcoal but has specific properties that make it very useful in a variety of applications. Among BCs desirable characteristics, high surface area, porous structure, thermal stability, and a great amount of surface functional groups are the most crucial for environmental applications [25]. It should be noted, however, that BC properties are highly connected with the synthesis conditions (pyrolysis temperature and duration) and the pristine waste material used as the carbon source (such as crop residues, food and vegetable wastes) [26]. Finally, it should be noted that the use of materials that are residues from other processes, such as biochar, is highly promoted in the context of sustainability and the circular economy for the integrated design of new processes and systems.

Biochar from pomegranate peel was effectively used to activate persulfate for the degradation of Losartan in water showing great stability and reusability properties [27], while BC derived from lemon stalks showed promising efficiency for the destruction of another pharmaceutical agent, sulfamethoxazole, in a biochar (BC)/persulfate (SPS) system [28]. Furthermore, in the study by Grilla et al., the authors pointed out the significant impact of the presence of various inorganic ions in real water matrices on the efficiency of BC derived from spent malt rootlets for the decomposition of the antibiotic trimethoprim [29]. In addition, different modifications, such as anion doping, metal deposition and so on, have been examined to further improve biochar activity. For instance, Zhu et al. [30] enhanced the catalytic performance of biochar by utilizing nitrogen doping, which significantly influenced the BC's surface characteristics, porosity, and graphitic carbon structure. Additionally, Hussain et al. [31] explored the combination of biochar with transition metals, particularly Fe^0, to activate persulfate for the oxidation of nonylphenol. This approach resulted in a degradation rate that was markedly higher than the one observed when persulfate was activated by either Fe^0 or biochar alone.

Recently, transition metal phosphides (TMPs) have gained attention for their application in AOPs, particularly in combination with persulfate, showing high efficiency in removing persistent organic pollutants [32]. TMPs are characterized by a relatively small difference in electronegativity between most transition metals and phosphorus, allowing the M−P bond to exhibit both covalent and ionic characteristics, resulting in partial charges on the metal and phosphorus ions. This facilitates the redox cycle of $M^{(n+m)+}/M^{n+}$, enhancing catalytic activity [33]. Additionally, the strong M−P bond provides TMPs with high stability, even in acidic conditions [34]. The phosphorus lattice, which serves as a scaffold for metal nanoclusters, features a graphene-like network structure that enhances electron transfer and interfacial reactions [35]. Due to these properties, TMPs have been recognized as versatile catalysts in hydrogenation systems, water–gas shift reactions, and energy storage.

Indeed, Wang et al., [36] in their interesting work, examined the combination of copper phosphide with biochar obtained from phytic acid-doped rice brans for the degradation of sulfamethoxazole using activated persulfate. The study revealed promising results including the destruction of 0.25 mg/L of the antibiotic in less than 30 min. Interestingly, the authors used phytic acid as a source of phosphorus in the context of the circular economy; however, a relatively high loading of copper was used (9–13% Cu). Since lower metal loadings are considered less expensive and show higher dispersion while at the same time suffering from lower leaching, the use of significantly lower loadings (up to approximately 4%) was considered in this work. At the same time, banana peel was used as the biomass source since it constitutes a significant part of food waste in several countries.

In this context, this study first synthesized biochar (banana peel biochar, BPB) by pyrolyzing banana peels at 850 °C for 3 h. It was then modified by the addition of different loadings of copper phosphide (Cu_3P, 0.25–4.00% wt.). Subsequently, Cu_3P/BPB catalysts were tested as heterogeneous sulfate persulfate (SPS) activators for the degradation of bisphenol S (BPS) in water. BPS is a chemical compound mainly used as a substitute for the endocrine disruptor Bisphenol A (BPA) in various products, such as plastics and thermal paper. Like BPA, BPS is an organic compound with two phenolic groups; however, instead of methylene, BPS contains a sulfur group. The use of BPS has increased as BPA has been restricted or banned in many countries due to concerns about its health effects. However, there are worries that BPS may have similar negative effects, such as estrogenic activity and high toxicity, while research also questions the investigation of the transformation products derived from BPS degradation.

From this perspective, the influence of Cu_3P loading on the biochar, along with the effect of pH, was examined, as well as the impact of the initial concentrations of BPS, SPS, catalyst, and the water matrix on BPS degradation. Additionally, the reaction mechanism and the transformation products were investigated for the first time with the proposed system.

2. Results

2.1. Physicochemical Characteristics

The specific surface area (SSA) of the synthesized catalysts was determined through nitrogen physisorption at liquid nitrogen temperature (77 K), employing the Brunauer–Emmett–Teller method. It was found that the SSA of the BPB was 1.5 $m^2\ g^{-1}$ and doping with Cu_3P slightly increased the SSA of the BPB. Specifically, for all Cu_3P/BPB samples analyzed, the SSA value was 2.0 $\pm$ 0.5 $m^2\ g^{-1}$ except for 2% Cu_3P/BPB, which had an SSA of 3.1 $m^2\ g^{-1}$ (Table S1). Additionally, the characteristic BET isotherms of BPB, 0.5% Cu_3P/BPB and 2% Cu_3P/BPB are displayed in Figure S1. This substantially low value is likely due to the high temperature (850 °C) during the pyrolysis, which typically leads to the decomposition of the organic compounds and the formation of condensed structures with reduced porosity [28,37]. Additionally, the introduction of Cu_3P onto the biochar's surface did not alter the overall SSA value as anticipated, due to the low loading.

The pristine biochar as well as the optimum copper phosphide-modified catalyst (2% Cu_3P/BPB) were characterized by means of X-ray diffraction (XRD) and the results are summarized in Figure 1. The XRD pattern of the pristine biochar (BPB) indicates that the main crystalline phase is attributed to sylvite (KCl) with characteristic diffraction peaks observed at 28.45°, 40.71°, 50.32°, 58.32°, 66.55° and 73.86°, corresponding to (200), (220), (222), (400), (420), (422) crystallographic planes, respectively (JCPDS No 41-1476) [38]. The presence of KCl crystallites stems from the high potassium content in the banana peel (approximately 80 g/kg), resulting in the formation of potassium-based salts of high crystallinity after pyrolysis [39]. Additionally, no distinct peaks corresponding to carbon were detected, presumably due to the amorphous nature of the carbonaceous species.

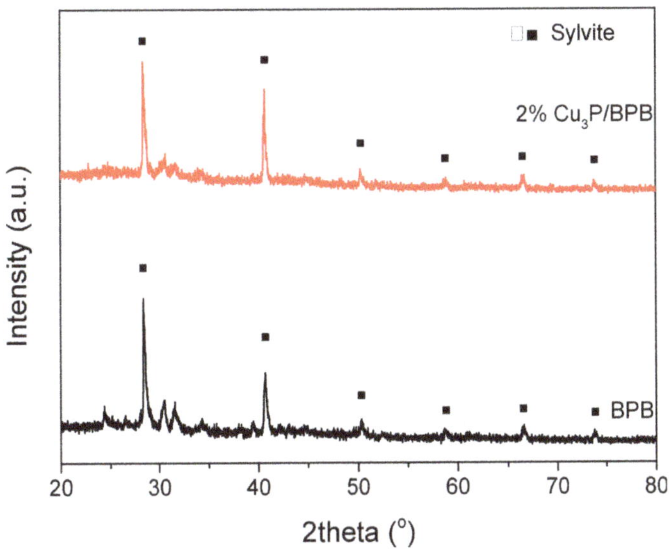

Figure 1. XRD pattern of the pristine biochar and the 2% Cu_3P/BPB composite catalyst.

As expected, the incorporation of Cu_3P onto the biochar surface did not result in any observable diffraction peaks associated with Cu_3P. This absence can be explained by the low Cu_3P content (2%) in the composite catalyst along with the homogeneous distribution of copper phosphide across the biochar surface.

Scanning electron microscopy (SEM) and energy dispersive spectrometer analysis (EDS) were conducted to examine the surface morphology, elemental composition and microstructural features of the synthesized catalysts (Figure 2). As demonstrated in Figure 2A,B, both the pristine biochar and the copper-modified catalyst exhibited a highly crystalline structure, and the morphology appeared to resemble a hollow rod. The similar structure of both catalysts underscores the repeatability and consistency of the employed synthesis method.

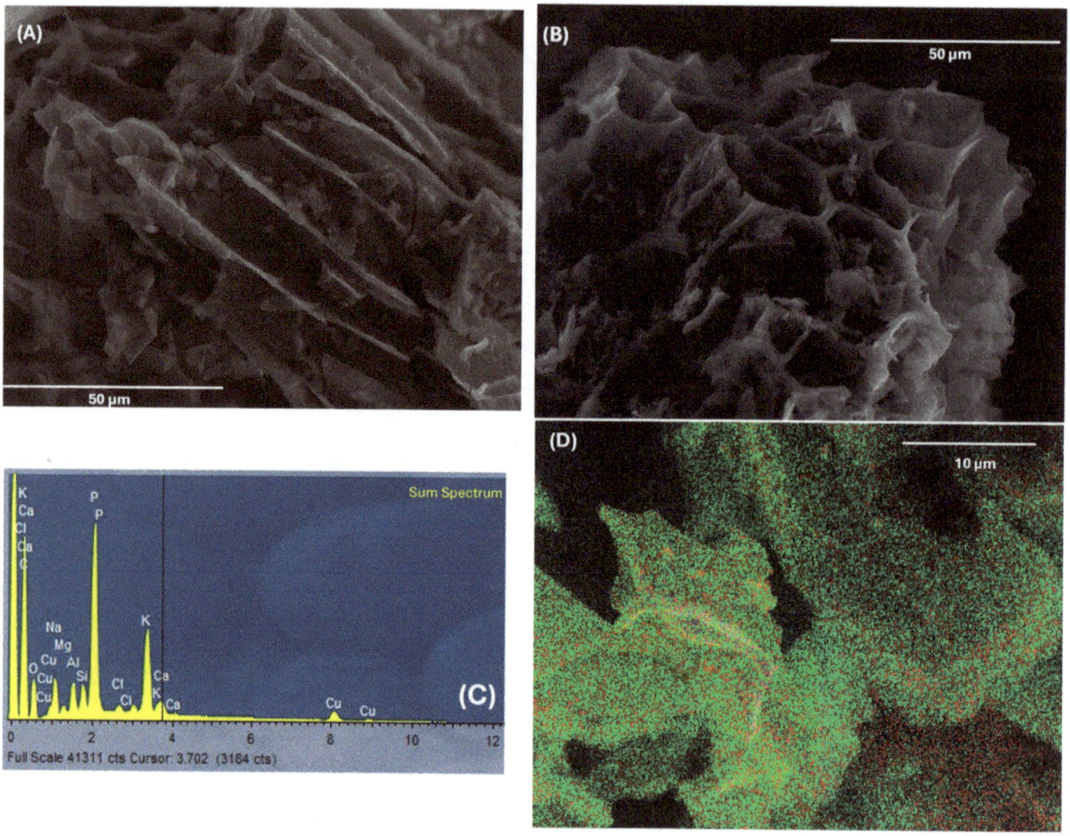

Figure 2. SEM images of (**A**) BPB and (**B**) 2% Cu_3P/BPB, (**C**) EDS spectrum corresponding to 2% Cu_3P/BPB and (**D**) elemental mapping for copper (red dots) and phosphorus (green dots).

The EDS analysis (Figure 2C,D) confirmed the successful formation of Cu_3P with the presence of characteristic peaks for both phosphorus (P) and copper (Cu) and its uniform dispersion. The reinforced intensity of the phosphorus peak can be attributed to the inherent P content in the banana peel. Furthermore, the EDS data revealed that the wt.% of carbon in both samples exceeded 80% and confirmed the potassium-rich nature of the BPB (>5 wt.%).

2.2. Catalytic Results

2.2.1. Catalytic Activity of Cu_3P/BPB

The synthesis conditions are crucial for catalytic performance; therefore, to investigate the catalytic efficiency of BPB and the enhanced Cu_3P/BPB samples (0.25–4.00% wt.) in activating persulfate (SPS), catalytic experiments for BPS (0.5 mg/L) degradation were conducted in ultrapure water (UPW), and the results are presented in Figure 3. As shown in Figure 3A, the biochar derived from banana peel exhibits very low catalytic activity for BPS oxidation. The addition of 0.25% Cu_3P to BPB increased its catalytic activity, resulting in 45% BPS removal within 15 min. Additionally, 0.5% Cu_3P/BPB, 1% Cu_3P/BPB, 2% Cu_3P/BPB, and 4% Cu_3P/BPB achieved 62%, 82%, 95%, and 100% BPS removal in the same time frame, respectively.

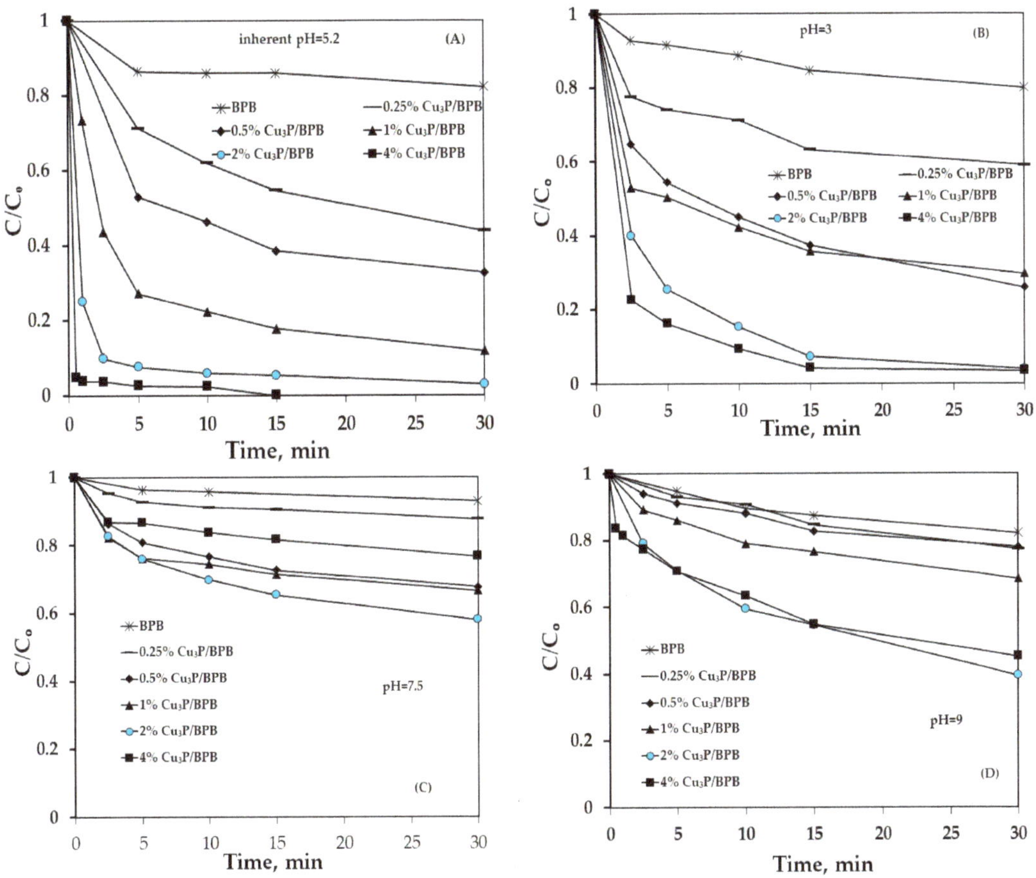

Figure 3. Effect of copper phosphide content (0–4% wt. Cu_3P) on the degradation of BPS over Cu_3P/BPB and SPS at (**A**) inherent pH = 5.1, (**B**) pH_{buffer} = 3, (**C**) pH_{buffer} = 7.5, and (**D**) pH_{buffer} = 9. Experimental conditions: [BPS] = 500 μg/L, [# wt.% Cu_3P/BPB] = 250 mg/L and [SPS] = 500 mg/L.

Recognizing that adsorption is essential for heterogeneous catalysis, adsorption experiments were performed using 250 mg/L of the catalysts at inherent pH in the absence of persulfate and the results are shown in Figure S2. It was observed that low loadings of Cu_3P (0.25%, 0.50%, and 1.00% wt.) did not affect the adsorption efficiency of the biochar, with BPS adsorption being negligible. However, increasing the Cu_3P loading to 2% wt. resulted in noticeable BPS removal (36% within 15 min). Further increasing the Cu_3P content to 4% wt. led to a decrease in BPS removal to 22% within 15 min. This reduction in adsorption may be due to an excess of Cu_3P, which leads to the blocking of active sites on the catalysts [40]. Thus, no higher loadings of Cu_3P on BPB were prepared.

Furthermore, since environmental matrices and wastewaters exhibit pH variability, all the synthesized materials were tested catalytically at various pH levels using buffer solutions to avoid changes in pH during treatment. At pH 3, the order of catalytic efficiency of the Cu_3P/BPB samples did not change, but an inhibition of BPS degradation was observed for all Cu_3P/BPB samples compared to BPS degradation at inherent pH. At pH 7.5 and 9, all the samples showed even lower catalytic performance (Figure 3B,C) for BPS. However, the most notable finding in Figure 3B,C is that the 4% Cu_3P/BPB exhibited a lower catalytic yield than the 2% Cu_3P/BPB.

The efficiency of the proposed system under alkaline conditions (pH 9) is shown in Figure 3D. As shown, the oxidation of BPS was significantly decreased compared to the results at inherent pH and under acidic conditions. A similar behavior observed in different heterogeneous advanced oxidation processes is mainly attributed to the effect of pH on both the generation of reactive species and the adsorption capacity. In addition, in heterogenous systems, metal leaching can contribute to reduced efficiency or lead to homogeneous reactions.

It is well established that sulfate radical formation is more favorable under acidic conditions, whereas under alkaline conditions, $SO_4^{\bullet-}$ reacts with HO^- to produce additional $^{\bullet}OH$. However, the redox potential of hydroxyl radicals diminishes as the solution becomes more alkaline, leading to a hindrance in BPS oxidation [41]. Despite the favorable formation of $SO_4^{\bullet-}$ at pH 3, an excess of these radicals can lead to self-scavenging reactions, significantly decreasing process efficiency

The other factor that can explain the behavior of 4% Cu_3P/BPB and 2% Cu_3P/BPB under neutral and alkaline conditions is adsorption. As previously mentioned, the 4% Cu_3P/BPB material exhibited a lower adsorption capacity than the 2% Cu_3P/BPB at inherent pH, and this capacity decreases even further as the pH increases from 5.2 to 7.5 compared to the 2% Cu_3P/BPB (Figure S2B). These results confirm the crucial role of pH in the interplay between mass transfer (adsorption) and chemical processes (mechanism and production of reactive species), thereby affecting the system's efficacy. Therefore, considering both the synthesis cost and the material's performance in BPS degradation across a wide pH range, the 2% Cu_3P/BPB was selected for further examination.

2.2.2. Effect of Initial Concentration of SPS, 2% Cu_3P/BPB, and BPS

The effect of the concentration of SPS, 2% Cu_3P/BPB, and BPS on BPS degradation in the 2% Cu_3P/BPB/SPS system was examined and the results are depicted in Figure 4. As shown in Figure 4A, the degradation of BPS gradually accelerated with an increase in SPS concentration. When the SPS concentration increased from 0 mg/L to 500 mg/L, the performance of the system was greatly enhanced. The BPS removal increased from 15% to 93% within 5 min. Referring to k_{app} as shown in Table 1, the k_{app} increased 9.5-fold when the SPS dosage was raised from 65 mg/L to 500 mg/L. Nevertheless, the subsequent parameters were examined using 250 mg/L SPS in spite of the great efficiency that the present system showed using 500 mg/L SPS. However, keeping in mind that increased oxidant concentration raises associated costs, thus hindering industrial application, and considering the WHO's suggested limit for sulfate anions, which affects the taste of drinking water [42], the subsequent parameters were examined using 250 mg/L SPS.

Table 1. k_{app} values for 0.5 mg/L BPS degradation with 250 mg/L 2% Cu_3P/BPB for various initial concentrations of SPS in UPW and inherent pH.

[SPS], mg/L	k_{app}, min^{-1}	R^2
0	0.0315 ± 0.0021	0.97
65	0.0620 ± 0.0016	0.98
125	0.0934 ± 0.0018	0.98
250	**0.1295 ± 0.0017**	**0.95**
500	0.5885 ± 0.0033	0.97

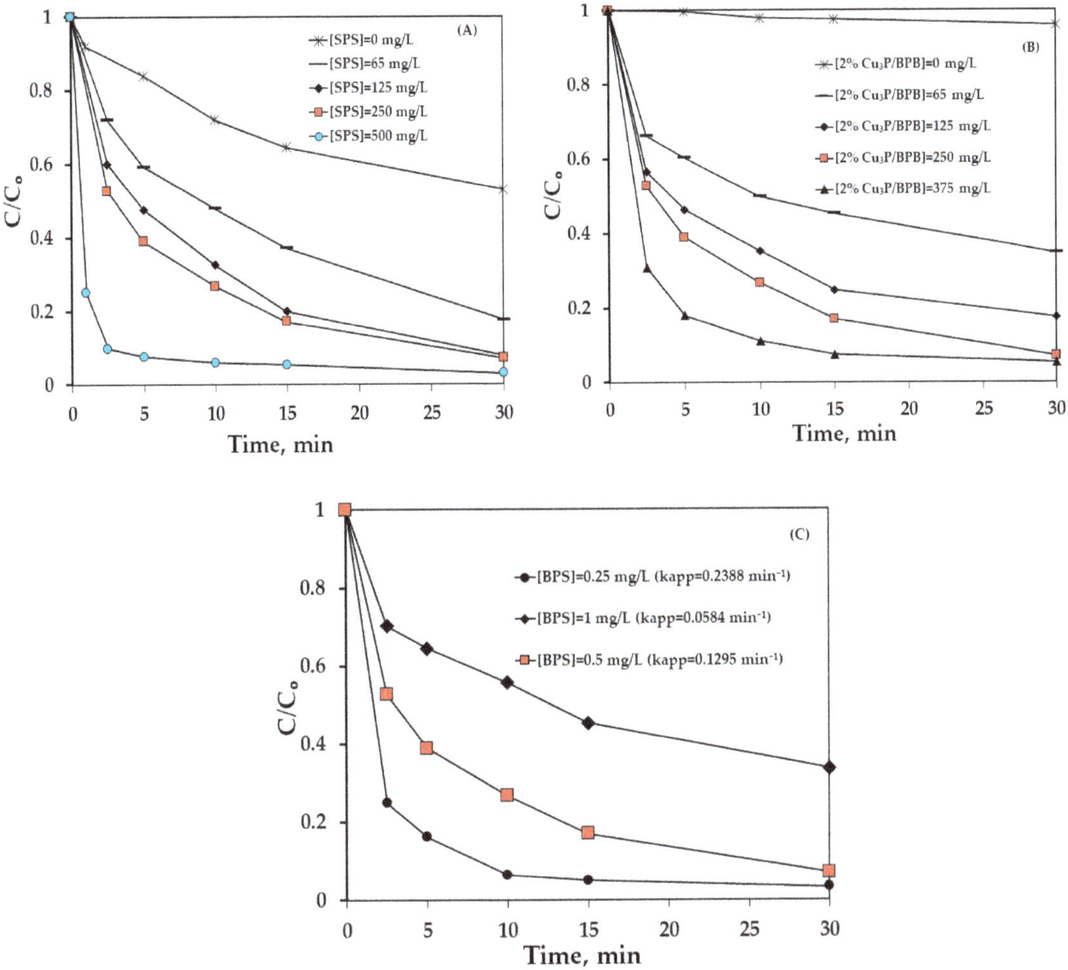

Figure 4. (**A**) Effect of initial concentration of SPS on the degradation of 0.5 mg/L BPS with 250 mg/L 2% Cu_3P/BPB at inherent pH = 5.2. (**B**) Effect of initial concentration of 2% Cu_3P/BPB on the degradation of 0.5 mg/L BPS with 250 mg/L SPS at inherent pH = 5.2 and (**C**) effect of initial concentration of BPS on its degradation with 250 mg/L 2% Cu_3P/BPB and 250 mg/L SPS at inherent pH = 5.2 (numbers in brackets show apparent rate constants).

Furthermore, different concentrations of 2% Cu_3P/BPB were applied to illustrate the effect of the treated biochar dosage using 250 mg/L SPS and 0.5 mg/L BPS (Figure 4B). A higher concentration of 2% Cu_3P/BPB was beneficial for system efficiency, since more active sites were available for the activation and adsorption of SPS and BPS, respectively. BPS removal rates of 2%, 50%, 65%, 74%, and 90% were achieved at 10 min when the biochar concentrations were 0 mg/L, 65 mg/L, 125 mg/L, 250 mg/L, and 375 mg/L, respectively. In addition, the experiment conducted without 2% Cu_3P/BPB demonstrated the inability of SPS alone to degrade BPS, despite being a mild oxidant. Considering the satisfactory efficiency of the system in UPW using 250 mg/L of 2% Cu_3P/BPB, where BPS removal after 30 min of reaction was nearly equivalent to that achieved using 375 mg/L of 2% Cu_3P/BPB, the concentration of 2% Cu_3P/BPB was set at 250 mg/L.

Continuously, knowing that the performance of advanced oxidation processes is strongly dependent on the concentration of micropollutants, the performance of the proposed system was examined using different BPS concentration in the μg/L range [43].

Thus, three experiments were conducted with initial BPS concentrations of 0.25 mg/L, 0.5 mg/L, and 1 mg/L (Figure 4C). As seen in Figure 4C, the 2% Cu_3P/BPB/SPS system showed very promising results, especially at lower BPS concentrations. BPS was almost completely oxidized within 10 min when its initial concentration was 0.25 mg/L. Correspondingly, the apparent kinetic constant for the initial concentration of 0.25 mg/L BPS is nearly double that of the initial concentration of 0.5 mg/L and 4.1-fold higher than that of the initial concentration of 1 mg/L. The reduction in the value of the apparent rate constant indicates that the degradation rate of BPS follows pseudo-first-order kinetics. Similar behavior has been observed in comparable systems (biochar/persulfate systems) as well as in other AOPs, such as photocatalysis and electrochemical oxidation [42,44,45].

2.2.3. Effect of Water Matrix

Figure 5 shows the performance of the 2% Cu_3P/BPB/SPS process on BPS degradation for various synthetic and real water matrices. Table 2 presents the physicochemical characteristics of BW and WW. Interestingly, the presence of NaCl did not affect the performance of the system, whereas the presence of bicarbonates and humic acid had a negative impact on BPS degradation. This could be attributed to the hindrance of BPS adsorption caused by HCO_3^- and humic acid (HA), as well as the competition between BPS and both inorganic and organic matter for the produced reactive species and the catalyst surface.

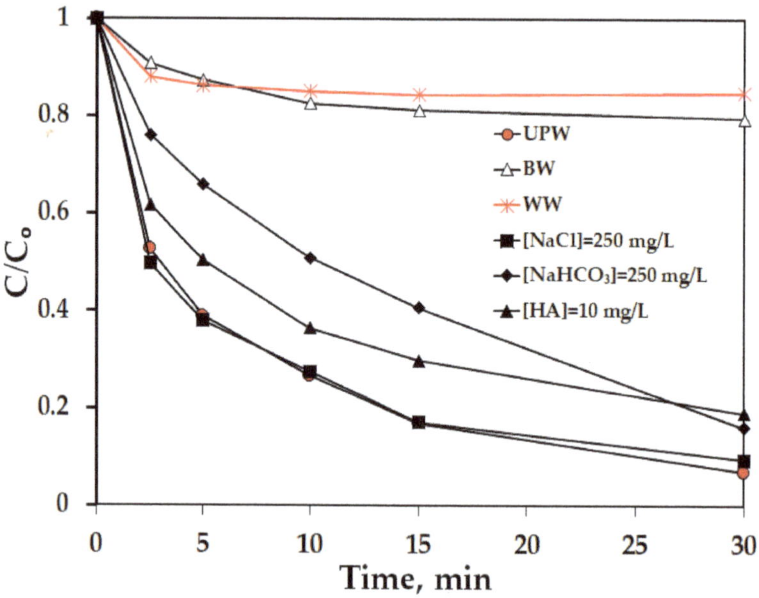

Figure 5. Effect of the water matrix on 0.5 mg/L BPS degradation with 250 mg/L 2% Cu_3P/BPB and 250 mg/L SPS.

The negligible and negative effects of NaCl and HCO_3^-, respectively, have also been reported by Zhong et al. [46], who studied the elimination of tetracycline using persulfate activated by N and Cu codoped biochar. Additionally, based on the results regarding the effect of pH, the increase in pH to alkaline conditions may contribute to the inhibition of BPS in the presence of bicarbonate [46]. Furthermore, since it is well known that Cl^- and HCO_3^- react with $SO_4^{\bullet-}$ and $HO^{\bullet}$, the observation of negligible and moderate inhibition likely suggests that both radical and non-radical mechanisms contribute to the destruction of BPS.

Table 2. Physicochemical characteristics of the water matrices used in this work.

Parameter	Wastewater (WW)	Bottled Water (BW)
pH	8.5	7.4
Conductivity (20 °C), [μS/cm]	934	513
Total dissolved solids (TDS), [mg/L]	654	312
Total suspended solids (TSS), [mg/L]	22	-
Total hardness ($CaCO_3$), [mg/L]	287	260
Chemical oxygen demand [mg/L]	48.5	-
Total organic carbon [mg/L]	4.7	-
Chlorides (Cl^-), [mg/L]	262	26.1
Bicarbonates (HCO_3^-), [mg/L]	278	380
Sulfates (SO_4^{2-}), [mg/L]	62.3	14.7
Phosphates (PO_4^-) [mg/L]	14.9	-
Nitrates (NO_3^-), [mg/L]	2.3	12.4
Bromides (Br^-), [mg/L]	165.6	-
Ca^{2+} [mg/L]	112	92.7
K^+ [mg/L]	15.4	0.6
Na^+ [mg/L]	76.3	5.5

On the other hand, in the presence of BW and WW, the degradation rate of BPS was significantly reduced, achieving only 20% and 15% BPS removal after 30 min in BW and WW, respectively, under the experimental conditions investigated. The reduced performance in real matrices is mainly due to the presence of bicarbonates, natural organic matter, and increased pH, as illustrated in Table 2. This finding is consistent with the results of Wang et al. [36], who studied the degradation of sulfamethoxazole using Cu_3P-doped biochar derived from rice bran, although synthesized through a different process than that employed in the present study, and Alexopoulou et al. [15], who examined the efficiency of the Cu_3P/SPS system for sulfamethoxazole degradation. Both studies demonstrated that the system's performance decreased as the complexity of the water matrix increased.

2.2.4. Examination of Reactive Species Contribution Using Scavengers

To assess the impact of reactive species, quenching experiments were conducted using 9.24 mM methanol (MeOH), 9.24 mM tert-butanol (t-BuOH), and 9.24 mM sodium azide (NaN_3) to scavenge $SO_4^{\bullet-}$, $^{\bullet}OH$ and $^1O_2$, respectively [47–50]. Notably, although NaN_3 is a well-known scavenger for $^1O_2$, it can also react with $HO^{\bullet}$ at a nearly similar kinetic rate [49,51]. In addition, MeOH is an effective scavenger of both $SO_4^{\bullet-}$ and $^{\bullet}OH$ since MeOH reacts nearly 100 times faster with $SO_4^{\bullet-}$ than $^{\bullet}OH$ [50] while t-BuOH can be considered as a potent scavenger for $^{\bullet}OH$ reacting 1000 times faster with $^{\bullet}OH$ than $SO_4^{\bullet-}$ [47,50,51]. As illustrated in Figure 6, the addition of each quenching agent decreased the efficiency of the 2% Cu_3P/BPB/SPS process with NaN_3 showing the greatest inhibition followed by MeOH and t-BuOH. Correspondingly, the k_{app} value was reduced 1.51-fold, 1.73-fold, and 3.91-fold for t-BuOH, MeOH, and NaN_3 compared the k_{app} without a scavenger (Table 3).

These results indicate that both $SO_4^{\bullet-}$ and $^{\bullet}OH$ actively contribute to the degradation of BPS in the 2% Cu_3P/BPB/SPS system. In detail, the possible radical pathway involves the generation of $SO_4^{\bullet-}$ through the activation of SPS via direct electron transfer processes, either on the surface of 2% Cu_3P/BPB catalyst or from BPS molecules adsorbed onto the biochar surface. The $SO_4^{\bullet-}$ may then interact with H_2O and/or HO^- adsorbed molecules to form $^{\bullet}OH$ further promoting the decomposition of BPS. On the other hand, the non-radical mechanism, which appears to be the dominant pathway, likely proceeds through electron transfer to SPS facilitated by surface defects on the 2% Cu_3P/BPB catalyst resulting in the formation of superoxide anions $O_2^{\bullet-}$ which subsequently interact with H^+ to generate singlet oxygen ($^1O_2$) [46,52].

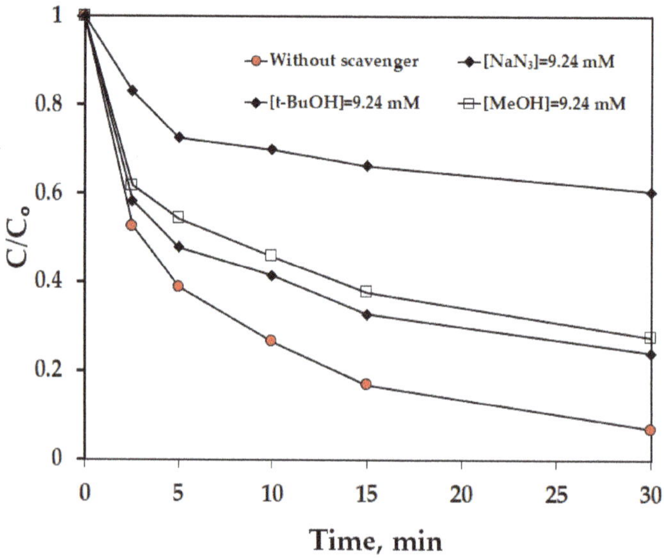

Figure 6. Effect of scavengers on 0.5 mg/L BPS degradation with 250 mg/L 2% Cu_3P/BPB and 250 mg/L SPS in UPW.

Table 3. k_{app} values for 0.5 mg/L BPS degradation for several scavengers at inherent pH. Experimental conditions [2% Cu_3P/BPB] = 250 mg/L and [SPS] = 250 mg/L in UPW, molar ratio $\frac{[scavenger]}{[SPS]} = 8.8$.

Scavenger	k_{app}, min^{-1}	R^2
Without scavenger	0.1295 ± 0.0017	0.95
9.24 mM t-BuOH	0.0857 ± 0.0013	0.92
9.24 mM MeOH	0.0748 ± 0.0011	0.93
9.24 mM NaN_3	0.0331 ± 0.0014	0.91

Similar results were reported by Rong et al. [50], who investigated the reaction mechanism of the γ-Fe_2O_3@banana peel biochar/persulfate system for the degradation of bisphenol A (BPA). In their study, in addition to the mechanism mentioned above, they reported that $O_2^{\bullet-}$ could be produced by molecular oxygen gaining an electron from persistent free radicals (PFRs), such as phenoxyl radicals, generated during the process. Interestingly, the results of this work contrast with those of Wang et al. [36] who studied the reaction mechanism of Cu_3P-doped biochar derived from rice bran. In their study, they found that the dominant pathway involved radicals, while non-radical species ($^1O_2$) played only a minor role in sulfamethoxazole oxidation.

2.3. Proposed Transformation Pathways

To identify the TBPs that are formed during the process as well as to propose the transformation pathways of BPS, an analysis of the various samples at different process stages was performed by UHPLC/TOF-MS system in negative ionization mode. All the data obtained from the analysis are presented in Table 4, providing significant structural information about the generated TBPs. More specifically, four main TBPs were detected during the degradation of BPS, along with their accurate mass values and their corresponding ion molecular formula.

Table 4. High-resolution mass spectrometry (HRMS) data of BPS and its TBPs.

BPS/TBP Code	Ion Molecular Formula	*m/z* [M-H]$^-$	Δ (ppm)	DBE
BPS	$C_{12}H_9O_4S^-$	249.0229	−0.7	8.5
TBP 1	$C_{24}H_{17}O_8S_2{}^-$	497.0357	2.7	16.5
TBP 2	$C_{24}H_{17}O_8S_2{}^-$	497.0364	1.3	16.5
TBP3	$C_8H_9O_5S^-$	217.0176	4.2	4.5
TBP 4	$C_{10}H_7O_3S^-$	207.0119	1.3	7.5

Two isomeric species at *m/z* 497.0357–497.0364 with ion molecular formula $C_{24}H_{17}O_8S_2^-$ were detected, implying the formation of TBPs with higher molecular weight than the parent compound. Transformation of BPS is initiated with the formation of the corresponding radicals that undergo fast, radical-mediated isomerization reactions to produce TBP1 and TBP2. In accordance with our results, similar dimers were identified during the degradation of BPS by heat-activated persulfate [53]. The formation of TBPs with higher molecular weight than the parent compound is well-documented in the literature, especially for molecules bearing the phenolic ring. A representative example also concerns bisphenol-A (BPA), a compound with many structural similarities with BPS, as during its oxidation by different methods, including persulfate-based processes, various intermediates arising mainly from coupling reactions of free radicals were identified [41,54–57].

The other two TBPs involved ring-opening reactions. TBP3 with *m/z* 217.0176 corresponded to ion molecular formula $C_8H_9O_5S^-$. TBP 4 yielded [M-H]$^-$ ions at *m/z* 207.0119 with ion molecular formula of $C_{10}H_7O_3S^-$ and double bond equivalents (DBE) equal to 7.5. Its molecular formula along with the decreased DBE implies the opening of one of the phenolic rings, allowing the suggestion of a structure. TBP3 as well as other ring-opening intermediates were also identified during the photocatalytic oxidation of BPS by Coha et al. [58].

Based on the above, the proposed transformation pathways of BPS include radical coupling and ring opening reactions (Figure 7). As already discussed in Section 2.2.3, both routes are available because of the reactive species that are generated during the process. $SO_4{}^{\bullet -}$ are produced through the activation of SPS via direct electron transfer mechanisms either (i) on the surface of Cu_3P or (ii) from BPS molecules adsorbed by the catalyst. Subsequently, $SO_4{}^{\bullet -}$ can be converted to $HO^{\bullet}$ that also contributes to the transformation of BPS. Thereafter, the H_2O_2 that is produced by the self-quenching of $HO^{\bullet}$ can generate $O_2{}^{\bullet -}$ (with the simultaneous reduction of Cu^{2+}) which is rapidly transformed to $^1O_2$ in a consequent step [36]. The contribution of the reactive species has also been verified by scavenging experiments (Table 3). Although the evolution profiles of the generated TBPs were not possible to follow, due to their fast formation and dissipation, all the TBPs exhibited quite similar relative abundance, thus implying that the cleavage of the aromatic ring as well as the polymerization reactions are very important transformation routes that take place simultaneously, under the adopted experimental conditions.

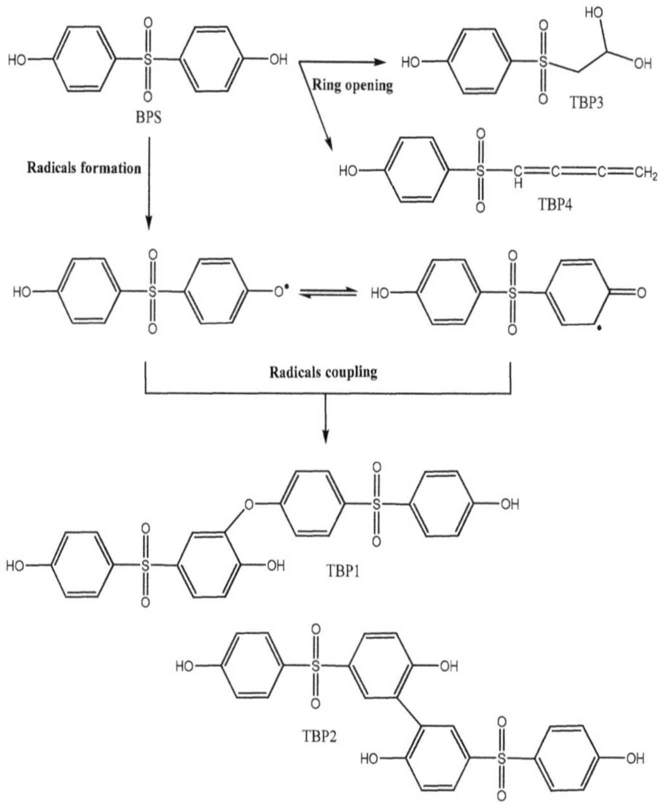

Figure 7. Proposed transformation pathways of BPS.

3. Materials and Methods

3.1. Chemicals and Aqueous Matrices

For the modified catalyst synthesis, copper (II) nitrate trihydrate ($Cu(NO_3)_2 \cdot 3H_2O$ CAS: 13778-31-9) and sodium hypophosphite monohydrate ($NaH_2PO_2 \cdot H_2O$ CAS: 10039-56-2) were purchased from Alfa Aesar, while trisodium citrate dihydrate ($Na_3C_6H_5O_7 \cdot 2H_2O$ CAS: 6132-04-3) and ammonia solution 25% (NH_4OH CAS: 1335-25-3) were provided by Sigma-Aldrich.

For the experimental procedure, Bisphenol-S (BPS) ($C_{12}H_{10}O_4S$ CAS: 80-9-01), sodium persulfate ($Na_2S_2O_8$ CAS: 777-5-27-1), methanol (CH_3OH, CAS: 67-56-1), tert-Butyl alcohol ($C_4H_{10}O$, CAS: 75-65-0), sodium chloride (NaCl, CAS: 7647-14-5), humic acid ($C_{187}H_{186}O_{89}N_9S$, CAS: 1415-93-6) and sodium bicarbonate ($NaHCO_3$, CAS: 144-55-8) were all sourced from Sigma-Aldrich and utilized without any further purification.

3.2. Catalyst Preparation

3.2.1. Banana Peel-Derived Biochar Synthesis

For the preparation of the pristine biochar, banana peels were collected and dried at 60 °C for 2 d in order to reduce moisture content to less than 15%. The dried peels were ground into smaller particles using a mortar, and the resulting powder was subjected to pyrolysis at 850 °C for 3h under inert N_2 atmosphere to prevent combustion. The BC particles were sieved through a No. 270 mesh, corresponding to a particle size of approximately 53 μm and used without any further treatment.

3.2.2. Copper Phosphide-Modified BPB Synthesis

For the growth of Cu_3P crystallites onto biochar's surface, a two-step process was utilized involving the preparation of copper hydroxide $Cu(OH)_2$ followed by low-temperature treatment in the presence of phosphorus precursors. Initially, predetermined amounts of copper nitrate ($Cu(NO_3)_2 \cdot 3H_2O$) and sodium acetate ($Na_3C_6H_5O_7 \cdot 2H_2O$) were added to 100 mL of UPW and vigorously stirred for 15 min. Ammonium hydroxide (NH_4OH) was then introduced drop-wise into the solution followed by the gradual addition of a 2M sodium hydroxide (NaOH) solution until $Cu(OH)_2$ precipitated. The obtained bluish powder was collected by vacuum filtration and dried for 12h.

In the second step, a fixed amount of BPB (2 g) was mixed with a stoichiometric amount of $Cu(OH)_2$ in 30 mL of UPW. The solution was sonicated in an ultrasonic bath for 2 h and the mixture was filtered under vacuum and dried at 60 °C overnight. The resulting powder was subsequently mixed with an appropriate quantity of sodium hypophosphite monohydrate ($Na_2H_2PO_4 \cdot H_2O$) and subjected to thermal treatment at 300 °C for 3 h under N_2 flow.

3.3. *Catalyst Characterization*

The specific surface area (SSA) of the samples was calculated employing the Brunauer–Emmett–Teller method with nitrogen physisorption conducted at liquid nitrogen temperature (77 K). The measurements were performed using a micromeritics Gemini III 2375 instrument (Norcross, GA, USA). X-ray diffraction (XRD) patterns were obtained using a Brucker D8 Advance instrument (Billerica, MA, USA), equipped with Cu Ka radiation source. The patterns were recorded over a 2θ range spanning from 20 to 80°. The identification process was performed according to standards from JCPDS cards. The morphology as well as the elemental composition of the samples was investigated using a scanning electron microscope (SEM) (JEOL 6300, Peabody, MA, USA) equipped with energy dispersive spectrometer (EDS) and the analysis conducted at the Electron Microscopy and Microanalysis Laboratory of the University of Patras.

3.4. *Experimental Procedure*

A stock solution of 10 mg/L Bisphenol-S (BPS) was prepared in UPW and subsequently diluted to create the appropriate working solutions. Degradation experiments were conducted in a Pyrex vessel with a total volume of 100 mL at ambient temperature and pressure. A predetermined amount of catalyst and SPS was simultaneously dispersed in 60 mL of aqueous BPS solution without pH adjustment. Buffer solutions were used only when studying the effect of pH. Samples were withdrawn at fixed time intervals, treated with 0.3 mL of methanol to terminate the reaction, and then filtered using a 0.22 μm PVDF filter.

3.5. *Analytic Techniques*

The concentration of BPS was monitored at 260 nm by means of high-performance liquid chromatography (Waters Alliance 2695, Milford, MA, USA) coupled with UV diode array detector with a C18 reversed-phase column (Kinetex XB-C18, 2.1 mm; 2.6 mm inner diameter × 50 mm length) (Milford, MA, USA). The mobile phase consisting of 70:30 water with 0.1% v/v H_3PO_4: acetonitrile eluted isocratically at 0.2 mL/min and 45 °C, while the injection volume was 100 μL.

3.6. *Analysis of Transformation By-Products (TBPs)*

The identification of TBPs was performed using a Bruker micrOTOF Focus II mass spectrometer coupled to a Dionex (Thermo Scientific, Waltham, MA, USA) Ultimate 3000 UHPLC. The high-resolution mass spectrometer was operated in negative ionization mode and the dry gas flow rate, nebulizer pressure, capillary voltage and dry temperature were set at 8 L min^{-1}, 2.4 bar, 4500 V and 200 °C, respectively. The calibration of the mass spectrometer was performed externally and internally with HCOONa. An AcclaimTMRSLC

120 C18 (2.2 μm, 2.1 × 100 mm) column (Thermo Scientific, MA, USA) protected by an AcclaimTM120 C18 guard cartridge (5 μm, 3.0 × 10 mm) from Thermo Scientific was used as analytical column and the mobile phase was a mixture of UPW with 0.01% formic acid (A) and acetonitrile (B) with a flow rate of 0.15 mL min^{-1}. Gradient elution was performed as follows: % ratio 90/10 (0 min), 10/90 (15 min), 90/10 (17 min) and 90/10 (18 min). The injection volume was 10 μL.

4. Conclusions and Future Work

To summarize, this work demonstrates that even relatively low amounts of copper phosphides (up to 4.00% wt.) on banana peel biochar can significantly accelerate the destruction of the micropollutant bisphenol S in different water matrices in the presence of persulfates. Although the presence of 250 mg/L of chlorides did not affect BPS removal, the presence of 250 mg/L of bicarbonates and 10 mg/L of humic acid showed a low to moderate reduction in the conversion rate. However, the efficiency was significantly hindered in experiments performed in real matrices, such as BW and WW, highlighting the need for more representative investigations of catalytic activity under conditions that reflect real environmental challenges. According to the scavenging experiments, BPS destruction occurs through both radical and non-radical mechanisms. The use of UHPLC/TOF-MS identified four transformation products, two of which had higher molecular weights than the parent compound, suggesting polymerization and indicating the need for further research into the effect of operating parameters on the degradation pathway and the resulting toxicity. This would provide a more comprehensive assessment of the process. Future research should also focus on examining the stability and reusability of the synthesized catalytic materials, as well as comparing the activity of different combinations of biomass sources with metal phosphides.

Supplementary Materials: The following supporting information can be downloaded at: https://www.mdpi.com/article/10.3390/catal14110789/s1, Table S1. Surface and structural analysis of the samples; Figure S1. Characteristic BET isotherms of (A) BPB (B) 0.5% Cu_3P/BPB and (C) 2% Cu_3P/BPB; Figure S2. (A) BPS adsorption at inherent pH = 5.2 by 250 mg/L BPB and 250 mg/L # % wt. Cu_3P/BPB in UPW and inherent pH. (B) Effect of pH on 0.5 mg/L BPS adsorption using 250 mg/L 2% Cu_3P/BPB and 4% Cu_3P/BPB, respectively.

Author Contributions: Conceptualization, A.A.I., Z.F. and D.M.; methodology, A.A.I., A.P., M.A., Z.F. and D.M.; formal analysis, A.A.I., O.L., K.K. and A.P.; investigation, A.A.I., O.L., K.K., M.A. and A.P.; resources, Z.F., M.A. and D.M.; data curation, A.A.I., O.L. and A.P.; writing—original draft preparation, A.A.I., K.K., Z.F., M.A. and A.P.; writing—review and editing, A.A.I., A.P., M.A, Z.F. and D.M.; visualization, A.A.I., O.L. and A.P.; supervision, D.M.; funding acquisition, D.M. All authors have read and agreed to the published version of the manuscript.

Funding: This research received no external funding.

Institutional Review Board Statement: Not applicable.

Informed Consent Statement: Not applicable.

Data Availability Statement: Data are contained within the article and Supplementary Materials.

Conflicts of Interest: The authors declare no conflicts of interest.

References

1. Parida, V.K.; Saidulu, D.; Majumder, A.; Srivastava, A.; Gupta, B.; Gupta, A.K. Emerging Contaminants in Wastewater: A Critical Review on Occurrence, Existing Legislations, Risk Assessment, and Sustainable Treatment Alternatives. *J. Environ. Chem. Eng.* **2021**, *9*, 105966. [CrossRef]
2. Shi, Y.; Xing, Y.; Ma, C.; Sun, Y.; Ding, G. Degradation of Aqueous Organic Pollutants by Dual Oxidant Advanced Oxidation Processes: A Comprehensive Review. *J. Environ. Chem. Eng.* **2024**, *12*, 114174. [CrossRef]
3. Zhou, X.; Guo, Z.; Tang, X.; Wang, W.; Wu, M.; Song, B.; Xiang, Y.; Li, Y.; Xiong, W.; Huang, D.; et al. Sulfate Radical-Based Advanced Oxidation Processes for Simultaneous Removal of Antibiotic-Resistant Bacteria and Antibiotic Resistance Genes and the Affecting Factors. *Chem. Eng. J.* **2024**, *498*, 155149. [CrossRef]

4. Xie, J.; Yang, C.; Li, X.; Wu, S.; Lin, Y. Generation and Engineering Applications of Sulfate Radicals in Environmental Remediation. *Chemosphere* **2023**, *339*, 139659. [CrossRef] [PubMed]
5. Waldemer, R.H.; Tratnyek, P.G.; Johnson, R.L.; Nurmi, J.T. Oxidation of Chlorinated Ethenes by Heat-Activated Persulfate: Kinetics and Products. *Environ. Sci. Technol.* **2007**, *41*, 1010–1015. [CrossRef]
6. Qi, C.; Liu, X.; Ma, J.; Lin, C.; Li, X.; Zhang, H. Activation of Peroxymonosulfate by Base: Implications for the Degradation of Organic Pollutants. *Chemosphere* **2016**, *151*, 280–288. [CrossRef]
7. Gao, Y.; Gao, N.; Deng, Y.; Yang, Y.; Ma, Y. Ultraviolet (UV) Light-Activated Persulfate Oxidation of Sulfamethazine in Water. *Chem. Eng. J.* **2012**, *195–196*, 248–253. [CrossRef]
8. Camargo-Perea, A.L.; Rubio-Clemente, A.; Peñuela, G.A. Use of Ultrasound as an Advanced Oxidation Process for the Degradation of Emerging Pollutants in Water. *Water* **2020**, *12*, 1068. [CrossRef]
9. Moreira, F.C.; Boaventura, R.A.R.; Brillas, E.; Vilar, V.J.P. Electrochemical Advanced Oxidation Processes: A Review on Their Application to Synthetic and Real Wastewaters. *Appl. Catal. B Environ.* **2017**, *202*, 217–261. [CrossRef]
10. Anipsitakis, G.P.; Dionysiou, D.D. Radical Generation by the Interaction of Transition Metals with Common Oxidants. *Environ. Sci. Technol.* **2004**, *38*, 3705–3712. [CrossRef]
11. Chung, H.S.; Jeon, D.; Hwang, I. In Situ Treatment of Contaminated Soil Using Persulfate Activated by Sulfidated Zero-Valent Iron. *Chemosphere* **2024**, *366*, 143440. [CrossRef] [PubMed]
12. Oyekunle, D.T.; Gendy, E.A.; Ifthikar, J.; Chen, Z. Heterogeneous Activation of Persulfate by Metal and Non-Metal Catalyst for the Degradation of Sulfamethoxazole: A Review. *Chem. Eng. J.* **2022**, *437*, 135277. [CrossRef]
13. Liu, M.; Wang, L.; Fan, Q.; Xiao, N.; Chen, Z.; Dong, Y.; Zhao, C.; Sun, Y.; Xu, X.; Wang, W. High performance of N-doped cobalt oxide (N-CoOx) for peroxymonosulfate activation: Unraveling the functional mechanism of Co-Nx sites at molecular level. *Sep. Purif. Technol.* **2025**, *354*, 128881. [CrossRef]
14. Wang, X.; Min, J.; Li, S.; Zhu, X.; Cao, X.; Yuan, S.; Zuo, X.; Deng, X. Sono-Assisted Synthesis of CuO Nanorods–Graphene Oxide as a Synergistic Activator of Persulfate for Bisphenol A Removal. *J. Environ. Chem. Eng.* **2018**, *6*, 4078–4083. [CrossRef]
15. Alexopoulou, C.; Petala, A.; Frontistis, Z.; Drivas, C.; Kennou, S.; Kondarides, D.I.; Mantzavinos, D. Copper Phosphide and Persulfate Salt: A Novel Catalytic System for the Degradation of Aqueous Phase Micro-Contaminants. *Appl. Catal. B Environ.* **2019**, *244*, 178–187. [CrossRef]
16. Zhang, T.; Chen, Y.; Wang, Y.; Le Roux, J.; Yang, Y.; Croué, J.P. Efficient Peroxydisulfate Activation Process Not Relying on Sulfate Radical Generation for Water Pollutant Degradation. *Environ. Sci. Technol.* **2014**, *48*, 5868–5875. [CrossRef]
17. Lei, Y.; Chen, C.; Tu, Y.; Huang, Y.; Zhang, H. Heterogeneous Degradation of Organic Pollutants by Persulfate Activated by CuO-Fe_3O_4: Mechanism, Stability, and Effects of PH and Bicarbonate Ions. *Environ. Sci. Technol.* **2015**, *49*, 6838–6845. [CrossRef]
18. Du, X.; Zhang, Y.; Hussain, I.; Huang, S.; Huang, W. Insight into Reactive Oxygen Species in Persulfate Activation with Copper Oxide: Activated Persulfate and Trace Radicals. *Chem. Eng. J.* **2017**, *313*, 1023–1032. [CrossRef]
19. Phule, A.D.; Zaman, M.W.U.; Elkaee, S.; Kim, S.Y.; Lee, S.G.; Park, G.; Yang, J.H. Carbon-Based Catalysts for Clean Environmental Remediation. *Int. J. Environ. Res.* **2024**, *18*, 3. [CrossRef]
20. Zhang, J.; Shao, X.; Shi, C.; Yang, S. Decolorization of Acid Orange 7 with Peroxymonosulfate Oxidation Catalyzed by Granular Activated Carbon. *Chem. Eng. J.* **2013**, *232*, 259–265. [CrossRef]
21. Chen, J.; Zhang, L.; Huang, T.; Li, W.; Wang, Y.; Wang, Z. Decolorization of Azo Dye by Peroxymonosulfate Activated by Carbon Nanotube: Radical versus Non-Radical Mechanism. *J. Hazard. Mater.* **2016**, *320*, 571–580. [CrossRef] [PubMed]
22. Lee, H.; Lee, H.J.; Jeong, J.; Lee, J.; Park, N.B.; Lee, C. Activation of Persulfates by Carbon Nanotubes: Oxidation of Organic Compounds by Nonradical Mechanism. *Chem. Eng. J.* **2015**, *266*, 28–33. [CrossRef]
23. Bekris, L.; Frontistis, Z.; Trakakis, G.; Sygellou, L.; Galiotis, C.; Mantzavinos, D. Graphene: A New Activator of Sodium Persulfate for the Advanced Oxidation of Parabens in Water. *Water Res.* **2017**, *126*, 111–121. [CrossRef] [PubMed]
24. Zeng, Z.; Umeh, A.; Iyengar, G.A.; Qi, F.; Naidu, R. A Critical Review of Different Types of Biochar-Based Catalysts and Mechanisms in Advanced Oxidation Processes for Organic Contaminants Removal. *J. Environ. Chem. Eng.* **2024**, *12*, 114262. [CrossRef]
25. Nidheesh, P.V.; Gopinath, A.; Ranjith, N.; Praveen Akre, A.; Sreedharan, V.; Suresh Kumar, M. Potential Role of Biochar in Advanced Oxidation Processes: A Sustainable Approach. *Chem. Eng. J.* **2021**, *405*, 126582. [CrossRef]
26. Gallego-Ramírez, C.; Chica, E.; Rubio-Clemente, A. Coupling of Advanced Oxidation Technologies and Biochar for the Removal of Dyes in Water. *Water* **2022**, *14*, 2531. [CrossRef]
27. Ioannidi, A.A.; Frigana, A.; Vakros, J.; Frontistis, Z.; Mantzavinos, D. Persulfate Activation Using Biochar from Pomegranate Peel for the Degradation of Antihypertensive Losartan in Water: The Effects of Pyrolysis Temperature, Operational Parameters, and a Continuous Flow Reactor. *Catalysts* **2024**, *14*, 127. [CrossRef]
28. Giannakopoulos, S.; Vakros, J.; Frontistis, Z.; Manariotis, I.D.; Venieri, D.; Poulopoulos, S.G.; Mantzavinos, D. Biochar from Lemon Stalks: A Highly Active and Selective Carbocatalyst for the Oxidation of Sulfamethoxazole with Persulfate. *Catalysts* **2023**, *13*, 233. [CrossRef]
29. Grilla, E.; Vakros, J.; Konstantinou, I.; Manariotis, I.D.; Mantzavinos, D. Activation of Persulfate by Biochar from Spent Malt Rootlets for the Degradation of Trimethoprim in the Presence of Inorganic Ions. *J. Chem. Technol. Biotechnol.* **2020**, *95*, 2348–2358. [CrossRef]

30. Zhu, S.; Huang, X.; Ma, F.; Wang, L.; Duan, X.; Wang, S. Catalytic Removal of Aqueous Contaminants on N-Doped Graphitic Biochars: Inherent Roles of Adsorption and Nonradical Mechanisms. *Environ. Sci. Technol.* **2018**, *52*, 8649–8658. [CrossRef]
31. Hussain, I.; Zhang, Y.; Huang, S.; Du, X. Degradation of P-Chloroaniline by Persulfate Activated with Zero-Valent Iron. *Chem. Eng. J.* **2012**, *203*, 269–276. [CrossRef]
32. Luo, R.; Liu, C.; Li, J.; Wang, J.; Hu, X.; Sun, X.; Shen, J.; Han, W.; Wang, L. Nanostructured CoP: An Efficient Catalyst for Degradation of Organic Pollutants by Activating Peroxymonosulfate. *J. Hazard. Mater.* **2017**, *329*, 92–101. [CrossRef] [PubMed]
33. Yang, Z.; Liu, L.; Wang, X.; Yang, S.; Su, X. Stability and Electronic Structure of the Co–P Compounds from First-Principle Calculations. *J. Alloys Compd.* **2011**, *509*, 165–171. [CrossRef]
34. Callejas, J.F.; Read, C.G.; Roske, C.W.; Lewis, N.S.; Schaak, R.E. Synthesis, Characterization, and Properties of Metal Phosphide Catalysts for the Hydrogen-Evolution Reaction. *Chem. Mater.* **2016**, *28*, 6017–6044. [CrossRef]
35. Xu, Y.-F.; Duchesne, P.N.; Wang, L.; Tavasoli, A.; Ali, F.M.; Xia, M.; Liao, J.-F.; Kuang, D.-B.; Ozin, G.A. High-Performance Light-Driven Heterogeneous CO_2 Catalysis with near-Unity Selectivity on Metal Phosphides. *Nat. Commun.* **2020**, *11*, 5149. [CrossRef]
36. Wang, X.; Li, W.; Wang, Y.; Yang, L.; Zhang, J.; Zhao, G.; Zhao, Q. Efficient removal of sulfamethoxazole by biomass phosphorus-derived Cu_3P/biochar catalyst combined with persulfate. *Sep. Purif. Technol.* **2023**, *326*, 124768. [CrossRef]
37. Petala, A.; Bampos, G.; Frontistis, Z. Using Sawdust Derived Biochar as a Novel 3D Particle Electrode for Micropollutants Degradation. *Water* **2022**, *14*, 357. [CrossRef]
38. Ismail, S.N.; Ali, E.M.; Alwan, B.J.; Abd, A.N. Potassium Chloride Nanoparticles: Synthesis, Characterization, and Study the Antimicrobial Applications. *Macromol. Symp.* **2022**, *401*, 2100312. [CrossRef]
39. Soares Dias, A.P.; Pedra, I.; Salvador, É.; Rijo, B.; Costa Pereira, M.F.; Serralha, F.; Nogueira, I. Biodiesel Production over Banana Peel Biochar as a Sustainable Catalyst. *Catalysts* **2024**, *14*, 266. [CrossRef]
40. Xie, Y.; Wang, X.; Tong, W.; Hu, W.; Li, P.; Dai, L.; Wang, Y.; Zhang, Y. FexP/biochar composites induced oxygen-driven Fenton-like reaction for sulfamethoxazole removal: Performance and reaction mechanism. *J. Chem. Eng.* **2020**, *396*, 125321. [CrossRef]
41. Ioannidi, A.A.; Bampos, G.; Antonopoulou, M.; Oulego, P.; Mantzavinos, D.; Frontistis, Z. Peroxydisulfate activation by cerium (IV) oxide-supported palladium (Pd/CeO_2) for bisphenol A oxidation and E. coli inactivation from aquatic matrices. *J. Environ. Chem. Eng.* **2024**, *12*, 111851. [CrossRef]
42. Ioannidi, A.; Frontistis, Z.; Mantzavinos, D. Destruction of propyl paraben by persulfate activated with UV-A light emitting diodes. *Environ. Chem. Eng.* **2018**, *6*, 2992–2997. [CrossRef]
43. Wu, L.-H.; Zhang, X.-M.; Wang, F.; Gao, C.-J.; Chen, D.; Palumbo, J.R.; Guo, Y.; Zeng, E.Y. Occurrence of bisphenol S in the environment and implications for human exposure: A short review. *Sci. Total Environ.* **2018**, *615*, 87–98. [CrossRef] [PubMed]
44. Kanigaridou, Y.; Petala, A.; Frontistis, Z.; Antonopoulou, M.; Solakidou, M.; Konstantinou, I.; Deligiannakis, Y.; Mantzavinos, D.; Kondarides, D.I. Solar photocatalytic degradation of bisphenol A with CuOx/$BiVO_4$: Insights into the unexpectedly favorable effect of bicarbonates. *Chem. Eng. J.* **2017**, *318*, 39–49. [CrossRef]
45. Pueyo, N.; Ormad, M.P.; Miguel, N.; Kokkinos, P.; Ioannidi, A.; Mantzavinos, D.; Frontistis, Z. Electrochemical oxidation of butyl paraben on boron doped diamond in environmental matrices and comparison with sulfate radical-AOP. *J. Environ. Manag.* **2020**, *269*, 110783. [CrossRef]
46. Zhong, Q.; Lin, Q.; Huang, R.; Fu, H.; Zhang, X.; Luo, H.; Xiao, R. Oxidative degradation of tetracycline using persulfate activated by N and Cu codoped biochar. *Chem. Eng. J.* **2020**, *380*, 122608. [CrossRef]
47. Ghanbari, F.; Moradi, M. Application of peroxymonosulfate and its activation methods for degradation of environmental organic pollutants: Review. *Chem. Eng. J.* **2017**, *310*, 41–62. [CrossRef]
48. Kemmou, L.; Frontistis, Z.; Vakros, J.; Manariotis, I.D.; Mantzavinos, D. Degradation of antibiotic sulfamethoxazole by biocharactivated persulfate: Factors affecting the activation and degradation processes. *Catal. Today* **2018**, *313*, 128–133. [CrossRef]
49. Wang, J.; Wang, C.; Guo, H.; Ye, T.; Liu, Y.; Cheng, X.; Li, W.; Yang, B.; Du, E. Crucial roles of oxygen and superoxide radical in bisulfite-activated persulfate oxidation of bisphenol AF: Mechanisms, kinetics and DFT studies. *J. Hazard. Mater.* **2020**, *391*, 122228. [CrossRef]
50. Rong, X.; Xie, M.; Kong, L.; Natarajan, V.; Ma, L.; Zhan, J. The magnetic biochar derived from banana peels as a persulfate activator for organic contaminants degradation. *Chem. Eng. J.* **2019**, *372*, 294–303. [CrossRef]
51. Zhu, S.; Li, X.; Kang, J.; Duan, X.; Wang, S. Persulfate Activation on Crystallographic Manganese Oxides: Mechanism of Singlet Oxygen Evolution for Nonradical Selective Degradation of Aqueous Contaminants. *Environ. Sci. Technol.* **2019**, *53*, 307–315. [CrossRef] [PubMed]
52. Liu, Z.; Tan, C.; Zhao, Y.; Song, C.; Lai, J.; Song, M. Singlet oxygen in biochar-based catalysts-activated persulfate process: From generation to detection and selectivity removing emerging contaminants. *Chem. Eng. J.* **2024**, *485*, 149724. [CrossRef]
53. Wang, Q.; Lu, X.; Cao, Y.; Ma, J.; Jiang, J.; Bai, X.; Hu, T. Degradation of Bisphenol S by Heat Activated Persulfate: Kinetics Study, Transformation Pathways and Influences of Co-Existing Chemicals. *Chem. Eng. J.* **2017**, *328*, 236–245. [CrossRef]
54. Potakis, N.; Frontistis, Z.; Antonopoulou, M.; Konstantinou, I.; Mantzavinos, D. Oxidation of Bisphenol A in Water by Heat-Activated Persulfate. *J. Environ. Manag.* **2017**, *195*, 125–132. [CrossRef]
55. Outsiou, A.; Frontistis, Z.; Ribeiro, R.S.; Antonopoulou, M.; Konstantinou, I.K.; Silva, A.M.T.; Faria, J.L.; Gomes, H.T.; Mantzavinos, D. Activation of Sodium Persulfate by Magnetic Carbon Xerogels (CX/CoFe) for the Oxidation of Bisphenol A: Process Variables Effects, Matrix Effects and Reaction Pathways. *Water Res.* **2017**, *124*, 97–107. [CrossRef]

56. Poerschmann, J.; Trommler, U.; Górecki, T. Aromatic Intermediate Formation during Oxidative Degradation of Bisphenol A by Homogeneous Sub-Stoichiometric Fenton Reaction. *Chemosphere* **2010**, *79*, 975–986. [CrossRef]
57. Deborde, M.; Rabouan, S.; Mazellier, P.; Duguet, J.-P.; Legube, B. Oxidation of Bisphenol A by Ozone in Aqueous Solution. *Water Res.* **2008**, *42*, 4299–4308. [CrossRef]
58. Coha, M.; Dal Bello, F.; Fabbri, D.; Calza, P.; Medana, C. Structural Elucidation of Bisphenol E and Bisphenol S Photoinduced By-Products by High-Resolution Electrospray Ionisation Mass Spectrometry and Tandem Mass Spectrometry. *Rapid Commun. Mass Spectrom.* **2021**, *35*, e9039. [CrossRef]

catalysts MDPI

Article

Kinetic Modelling of Aromaticity and Colour Changes during the Degradation of Sulfamethoxazole Using Photo-Fenton Technology

Natalia Villota [1,*], Unai Duoandicoechea [1], Jose Ignacio Lombraña [2] and Ana María De Luis [3]

1 Department of Environmental and Chemical Engineering, Faculty of Engineering Vitoria-Gasteiz, University of the Basque Country UPV/EHU, Nieves Cano 12, 01006 Vitoria-Gasteiz, Spain; unai.duoandicoechea@ehu.eus
2 Department of Chemical Engineering, Faculty of Science and Technology, University of the Basque Country UPV/EHU, Barrio Sarriena, s/n, 48940 Leioa, Spain; ji.lombrana@ehu.eus
3 Department of Environmental and Chemical Engineering, Bilbao School of Engineering, University of the Basque Country UPV/EHU, Rafael Moreno "Pitxitxi" 2, 48013 Bilbao, Spain; ana.deluis@ehu.eus
* Correspondence: natalia.villota@ehu.eus; Tel.: +34-945013248

Abstract: Sulfamethoxazole (SMX) is an antibiotic that is extensively used in veterinary medicine, and its occurrence in wastewater and surface water can reach up to 20 μg/L. SMX is categorized as a pollutant of emerging concern by the US EPA due to its persistence and effects on humans and the environment. In this study, photo-Fenton technology is proposed for the removal of SMX. Aqueous solutions of SMX (50.0 mg/L) are treated in a 150 W UV photoreactor, using $[Fe^{2+}]_0$ = 0.5 mg/L and varying $[H_2O_2]_0$ = 0–3.0 mM. During the reaction, colour (AU) was assessed along with SMX (mg/L), turbidity (NTU), and TC (mg/L). SMX degrades to aromatic intermediates with chromophoric groups, exhibiting colour (yellow to brown) and turbidity. As these intermediates are mineralized into CO_2 and H_2O, the colour and turbidity of the water lose intensity. Using a molar ratio of 1 mol SMX:10 mol H_2O_2, the maximum degradation of aromatic species takes place (71% elimination), and colourless water with turbidity < 1 NTU is obtained. A kinetic modelling for aromaticity loss and colour formation as a function of the oxidant concentration has been proposed. The application of this model allows the estimation of oxidant amounts for an efficient removal of SMX under environmentally friendly conditions.

Keywords: aromaticity; colour; photo-Fenton; sulfamethoxazole

1. Introduction

Antibiotic residues have been a source of concern in recent years due to the environmental pollution they generate. Antibiotics, even at low concentrations, form a part of the group of emerging pollutants that cause detrimental effects on aquatic organisms [1,2] due to their toxic, carcinogenic, and mutagenic nature [3]. The overuse of antibiotics has caused increasing concern for public health and the balance of natural aquatic ecosystems [4,5]. This is because antibiotics are only partially absorbed by the human body and animals and are excreted into water and the environment through urine and faeces [6].

Recent research has found a global antibiotic consumption rate of 14.3 defined daily doses (DDDs) per 1000 people per day in 2018. Considering that this rate was 9.8 DDDs per 1000 people per day in 2000, this represents an increase of 46% [7,8]. In addition, it should be noted that, during the COVID-19 pandemic, global antibiotic use has seen a sharp spike [9]. Therefore, global antibiotic consumption is on the rise, with a 67% increase in antibiotic consumption predicted by 2030 [10]. Among the most commonly used antibiotics are sulfamethoxazole, tetracycline, ciprofloxacin, and trimethoprim, which are found in surface water and groundwater in varying concentrations. Some of their most notable characteristics are that they are toxic, poorly or non-biodegradable, and, consequently, persistent [11–13].

Specifically, sulfamethoxazole (SMX) is a common sulphonamide antibiotic that has been widely used for the treatment of bronchitis and other infections [14,15]. About 90% of this drug is excreted in the urine within 24 h of oral administration, which increases its toxicity in environmental matrices [16]. It is important to note that, although SMX concentrations in the environment are below those that can cause a toxic effect on humans [17], its accumulation in fish and birds can lead, even at trace levels, to the development of antibiotic resistance [16]. This would have negative consequences for humans and represent a serious threat to the ecosystems [18–21]. The use of SMX is so widespread that it has been detected in the outflow of wastewater treatment plants (WWTPs) [12]. The biological processes that are commonly used in wastewater treatment plants do not achieve an efficient removal of this type of contaminant [18,22], which is related to its recalcitrant characteristics. Therefore, there is a need to develop new technologies for the degradation of antibiotic residues in the aquatic environment that guarantee safe water [23]. In this context, advanced oxidation processes (AOPs) represent an efficient and environmentally compatible alternative to eliminate this persistent pollutant.

Advanced oxidation processes (AOPs) are based on the oxidation of organic pollutants by hydroxyl radicals ($HO^{\bullet}$), which react rapidly and indiscriminately with various organic compounds. In this case, their main objective is to mineralize SMX into carbon dioxide and water or to transform it into by-products that are less harmful to human health and the aquatic environment. The oxidation capacity of these processes usually increases in the presence of catalysts, leading to higher levels of mineralization [24]. In this context, a typical AOP with the presence of a catalyst for SMX removal is the photo-Fenton process. This process is a variant of the Fenton process, which combines the reaction of hydrogen peroxide and Fe^{2+} under UV irradiation, achieving a more complete degradation of SMX compared to conventional treatments [25–28].

Various AOPs have been developed and applied for wastewater treatment, but the Fenton and photo-Fenton processes stand out as the most powerful, efficient, and cost-effective options. These methods are particularly effective in treating persistent pollutants, whether used alone or in combination with conventional or biological treatments [29]. The photo-Fenton process offers several key advantages over the conventional Fenton method. By utilizing light energy, it generates more hydroxyl radicals, leading to faster and more efficient pollutant degradation. It also enhances the regeneration of ferrous iron, reducing the need for additional chemicals like hydrogen peroxide and iron. Unlike the traditional Fenton process, which works best in acidic conditions, the photo-Fenton process can operate effectively at a wider pH range, including near-neutral conditions, making it more versatile and environmentally friendly. Based on this, studies reported in the literature show that the photo-Fenton process is an effective treatment both in the degradation of micropollutants and in bacterial inactivation [30], verifying that in the case of the elimination of SMX, both the degradation performance and the speed at which it is eliminated are much greater using the photo-Fenton technology than using the conventional Fenton process [31].

Several authors have proposed kinetic models to describe the degradation of antibiotics by the photo-Fenton process, with the aim of optimising the design and operation of this technology. These models include studies on kinetic constants and the influence of key operational parameters such as pH, hydrogen peroxide concentration, iron concentration, and irradiation time on contaminant degradation [32,33]. Such models are essential for predicting process behaviour under different conditions and maximising pollutant removal efficiency. However, in addition to the removal of SMX, it is important to study the different mechanisms of the reaction that takes place, taking into account the intermediates that are generated throughout the process before complete mineralisation [34–36]. These intermediates can contribute to the colour and turbidity of the treated water, as it has been detected in the treatment of other types of medicines [37–40].

This study aims to complement previous studies on the degradation of SMX with UV light combined with hydrogen peroxide and iron salts. Specifically, the effect of this technology on the organoleptic parameters of water quality, such as colour and turbidity, will be analysed to ensure that the treated water complies with environmental regulations. In addition, based on the results obtained and the SMX degradation mechanisms published in the literature, the theoretical pathways of SMX degradation leading to the formation of intermediate products responsible for the high turbidity and colouration that arise during the process are studied.

2. Results and Discussion

2.1. Parameters Indicating the Water Quality during the SMX Oxidation

Figure 1 shows the kinetics of several of the parameters that indicate water quality when aqueous solutions of SMX are oxidized using the photo-Fenton technology. Figure 1a shows that the photo-Fenton treatment is capable of completely degrading the SMX load present in the water in less than 15 min. As can be observed in the graph, the water's aromaticity experiences a slight initial increase during the first 5 min of the reaction, followed by a decrease over time until reaching a stable value that persists in the treated water. This is because, in the early stages of the photo-Fenton reaction, hydroxyl radicals attack the aromatic ring of SMX, generating partially oxidized intermediates. In these early stages, the hydroxylation of SMX occurs mainly at the benzene ring and the amino group, producing several oxidized mono- and dihydroxylated intermediates, such as 3-hydroxysulfamethoxazole, 5-hydroxysulfamethoxazole, and the oxidized derivatives of isoxazole [35,41]. These intermediate by-products may have greater aromaticity or even temporarily form more aromatic structures. This increase occurs because, by breaking certain non-aromatic bonds of the compound, intermediate structures with more pronounced or additional aromatic rings are formed.

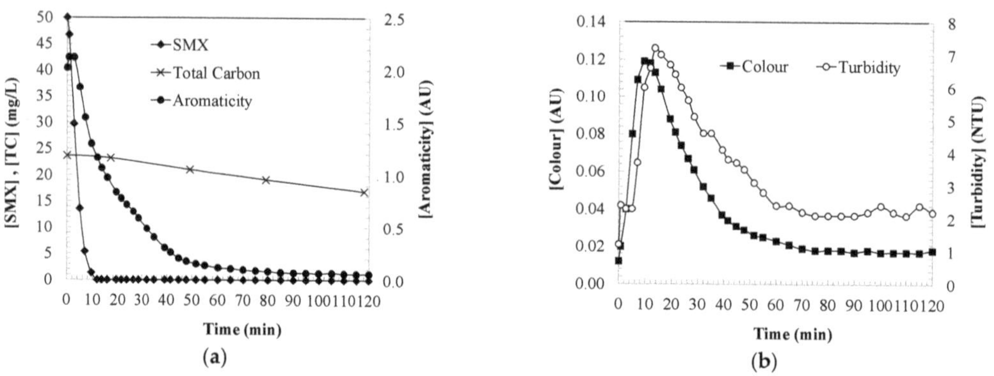

Figure 1. Parameters indicating the water quality during the SMX oxidation by the photo-Fenton technology: (**a**) SMX concentration (mg/L), aromaticity (AU), and total carbon (mg/L) and (**b**) colour (AU) and turbidity (NTU). Experimental conditions: $[SMX]_0$ = 50.0 mg/L; [UV] = 150 W; $[H_2O_2]_0$ = 3.0 mM; $[Fe^{2+}]_0$ = 0.5 mg/L; [pH] = 3.0; and [T] = 30 °C.

As the photo-Fenton reaction progresses, hydroxyl radicals continue to attack these aromatic intermediates, leading to the opening of the benzene rings or the complete oxidation of aromatic compounds into simpler products such as organic acids, CO_2, and water. This explains the subsequent decrease in aromaticity, as the aromatic structures are being destroyed. However, in some cases, certain resistant aromatic by-products may form during the process and are not easily degraded by hydroxyl radicals. The presence of certain substituent groups on aromatic molecules such as nitro, sulfonate, or carboxylate groups can affect the reactivity of hydroxyl radicals because they may induce inductive or resonance effects that further stabilize the aromatic structure, making radical attack less likely. These groups can also increase the electron density at certain positions on the ring, protecting the compound from oxidative attack [42]. Furthermore, during the oxidation process, oxidized intermediates are generated that have a very stable structure, such as quinones or aromatic acids that can be especially resistant [38]. These refractory residues are more complex, oxidized, and stable molecules that do not easily react under the conditions of the photo-Fenton treatment. This may be due to the inability of hydroxyl radicals to break certain highly condensed or resistant aromatic structures.

Figure 2a shows the residual aromaticity value of water samples oxidized using different concentrations of hydrogen peroxide. As the concentration of the oxidant used in the treatment increases, the degradation level of aromatic species increases, as indicated by the relationship shown in Equation (1). As confirmed, the maximum degradation of aromatic species (71% removal) is achieved by dosing an oxidant concentration of 2.0 mM, corresponding to a molar ratio of 1 mol SMX:10 mol H_2O_2.

$$[\text{Aromaticity}]_{\text{final}} = -0.0114\,[H_2O_2]_0^3 + 0.169\,[H_2O_2]_0^2 - 0.6344\,[H_2O_2]_0 + 0.7407 \quad (1)$$

$$[\text{TC}]_{\text{final}} = 1.23\,[H_2O_2]_0^2 - 8.9597\,[H_2O_2]_0 + 24.586 \quad (2)$$

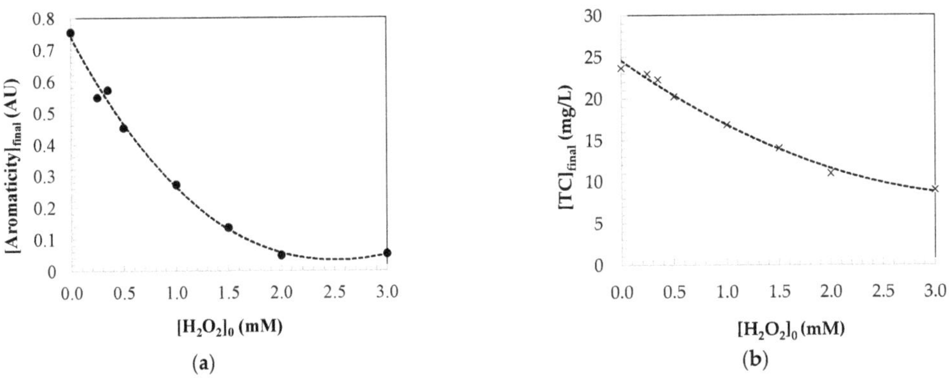

Figure 2. Water quality parameters analysed in SMX aqueous solutions oxidized by the photo-Fenton technology at the steady state: (**a**) aromaticity (AU) and (**b**) total carbon (mg/L). Experimental conditions: $[SMX]_0 = 50.0$ mg/L; [UV] = 150 W; $[Fe^{2+}]_0 = 0.5$ mg/L; [pH] = 3.0; and [T] = 30 °C.

On the other hand, when analysing the mineralization of the treated water, Figure 1a shows that the concentration of total carbon (TC, mg/L) decreases as SMX degradation progresses. Figure 2b illustrates the dependence of total carbon in the water on the concentration of the oxidant used in the treatment (Equation (2)). This occurs because, when SMX is attacked by hydroxyl radicals, its C-C and C-H bonds are broken, fragmenting the molecule into simpler compounds such as organic acids, aldehydes, alcohols, and other intermediate species. During the advanced oxidation process, the intermediate products generated (fragments of the initial molecule) continue to be attacked by hydroxyl radicals. If the process is sufficiently efficient, these intermediates are further degraded, eventually transforming into CO_2, H_2O, and inorganic salts (mineralization). The conversion of organic carbon into carbon dioxide contributes to the reduction in total carbon in the solution.

As the photo-Fenton process progresses, more carbon from the initial organic molecules and their intermediate products is oxidized into CO_2, which is released into the atmosphere as a gas. This phenomenon leads to a continuous decrease in total carbon concentration in the aqueous phase. UV light regenerates the Fe^{2+} ion from Fe^{3+} and promotes the continuous production of hydroxyl radicals, which sustains the degradation of organic compounds, thereby accelerating the reduction in total carbon.

A notable phenomenon within this reaction system is that, during the oxidation of SMX, brown colour and high turbidity are generated in the water (see Figure 1b). As observed, both the colour and turbidity in the water increase immediately after the reaction begins, reaching a maximum intensity around 15–20 min into the reaction. Subsequently, they decrease over time until reaching a stable value that remains once the steady state is achieved (see Figure 3a,b). It is confirmed that both parameters are dependent on the concentration of the oxidant used in the treatment, as shown in Equations (3) and (4). Operating with an oxidant concentration of 2.0 mM, which corresponds to a molar ratio of 1 mol SMX:10 mol H_2O_2, results in colourless water with turbidity < 1 NTU, making it suitable for discharge into natural watercourses.

$$[\text{Turbidity}]_{\text{final}} = -0.2027\,[H_2O_2]_0^3 + 2.2859\,[H_2O_2]_0^2 - 7.5725\,[H_2O_2]_0 + 8.0725 \quad (3)$$

$$[\text{Colour}]_{\text{final}} = 0.90 \times \exp(-1.557\,[H_2O_2]_0) \quad (4)$$

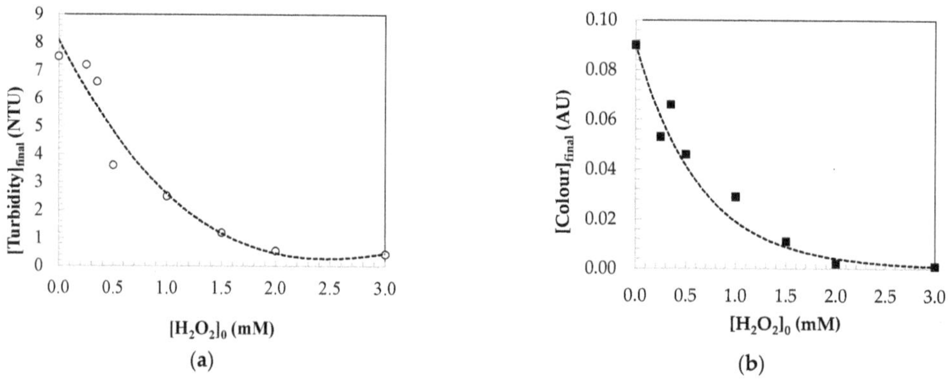

Figure 3. Water quality parameters analysed in SMX aqueous solutions oxidized by the photo-Fenton technology at the steady state: (**a**) turbidity (NTU) and (**b**) colour (AU). Experimental conditions: $[SMX]_0$ = 50.0 mg/L; [UV] = 150 W; $[Fe^{2+}]_0$ = 0.5 mg/L; [pH] = 3.0; and [T] = 30 °C.

During the oxidation process, hydroxyl radicals attack the benzene ring (aromatic structure) of SMX, leading to the formation of oxidized aromatic derivatives, such as quinones or polycyclic compounds, which have chromophoric properties. These compounds absorb visible light, resulting in the appearance of colour in the solution. Sometimes, instead of fully decomposing, fragments of organic molecules can react with each other to form larger and more complex compounds, such as condensation products, which may also contain chromophoric groups. These larger intermediate products tend to be more coloured and can temporarily accumulate during SMX degradation. Another group of intermediates that may appear during degradation includes aldehydes and unsaturated organic acids. Some of these compounds may contain conjugated double bonds, which can also contribute to the appearance of colour in the solution. Additionally, SMX contains a heterocyclic ring (a structure that includes sulphur and nitrogen). Oxidation of this type of ring can generate intermediate products with complex structures that also possess chromophoric properties. As these intermediates are mineralized into CO_2 and H_2O, the colour of the water gradually decreases.

The turbidity generated during SMX oxidation is primarily due to the precipitation of iron hydroxides and the formation of colloidal particles and insoluble organic intermediate products that remain suspended in the water. Turbidity decreases at the end of the process, when the intermediates are fully degraded and the iron precipitates are either removed or dissolved. During the process, Fe^{2+} is oxidized to Fe^{3+}, which can form precipitates of iron hydroxides ($Fe(OH)_3$). These hydroxides are solid particles that contribute to turbidity by remaining suspended as colloidal particles in the water. Furthermore, as SMX degrades, intermediates with low water solubility, such as organic acids, aldehydes, or polymeric compounds, are formed. These products can aggregate and form colloidal particles, increasing the turbidity of the solution. In some cases, intermediates can react with each other or with inorganic species present in the water, forming larger aggregates or structures that contribute to the formation of suspended particles. Additionally, products of incomplete oxidation, which have not yet been fully mineralized, can form larger or more complex substances that are less soluble in water, increasing turbidity until they are fully degraded [39].

2.2. *Kinetic Modelling of the Parameters Indicating the Water Quality*

2.2.1. Kinetic Modelling of Pseudo-First Order for SMX Oxidation

A pseudo-first-order kinetic model has been proposed to predict the degradation kinetics of SMX, where SMX degrades to reaction intermediates according to a kinetic constant k_{SMX} (1/min) (see Equation (5)).

$$\text{SMX} \overset{k_{SMX}}{\rightarrow} \text{Reaction intermediates} \quad (5)$$

By setting up the mass balance (Equation (6)) and integrating, the first-order kinetic equation for SMX degradation is obtained (Equation (7)), where $[SMX]_0$ is the initial concentration of SMX in the water (=50.0 mg/L), and t is the reaction time (min).

$$d[SMX]\ /\ dt = -\ k_{SMX}\ [SMX] \quad (6)$$

$$[SMX] = [SMX]_0 \times \exp\left(-\ k_{SMX}\ t\right) \quad (7)$$

Figure 4 shows the model predictions, which allow for determining the dependence of the SMX degradation rate constant on the concentration of the oxidant used in the treatment $[H_2O_2]_0$ in mM (Equation (8)). Table 1 presents the estimated values of the rate constants.

$$k_{SMX} = -\ 0.0627\ [H_2O_2]_0^2 + 0.2325\ [H_2O_2] + 0.1553\ (r^2 = 0.9777) \quad (8)$$

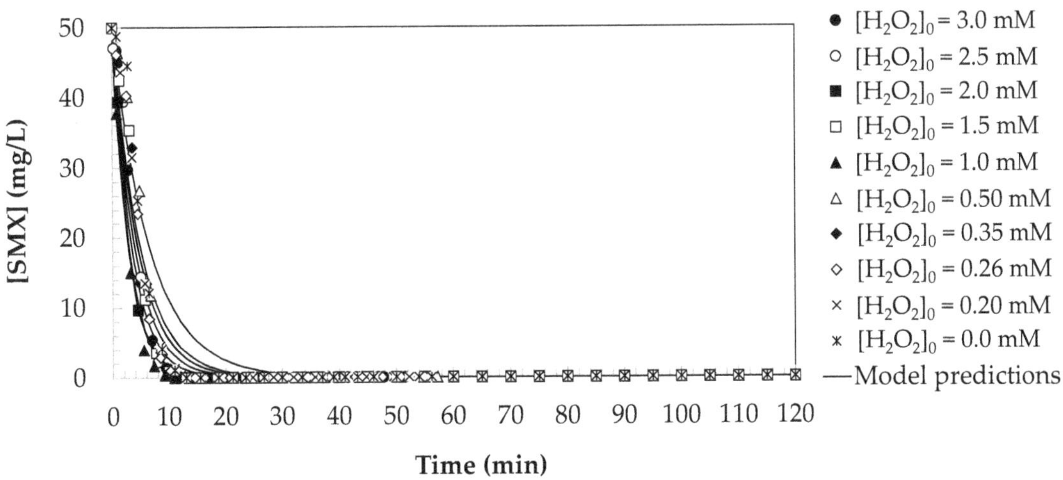

Figure 4. Predictions of the proposed first-order kinetic model for the oxidation of SMX by a photo-Fenton treatment. Experimental conditions: $[SMX]_0$ = 50.0 mg/L; [UV] = 150 W; $[Fe^{2+}]_0$ = 0.5 mg/L; [pH] = 3.0; and [T] = 30 °C.

Table 1. Kinetic parameters estimated for the SMX oxidation by the photo-Fenton technology. Experimental conditions: $[SMX]_0$ = 50.0 mg/L; [UV] = 150 W; $[Fe^{2+}]_0$ = 0.5 mg/L; [pH] = 3.0; and [T] = 30 °C.

$[H_2O_2]_0$ (mM)	k_{SMX} (1/min)	k_{arom} (1/min)	$k_{colour,form}$ (1/min)	$k_{colour,deg}$ (1/min)	α_{colour} (-)
0.00	0.15	7.0×10^{-2}	1.9×10^{-4}	2.9×10^{-2}	2.00
0.20	0.20	6.3×10^{-2}	3.7×10^{-4}	3.0×10^{-2}	1.83
0.25	0.22	6.5×10^{-2}	3.4×10^{-4}	3.0×10^{-2}	1.80
0.35	0.22	6.0×10^{-2}	3.6×10^{-4}	3.5×10^{-2}	1.77
0.50	0.25	5.5×10^{-2}	4.8×10^{-4}	3.5×10^{-2}	1.62
1.00	0.35	4.7×10^{-2}	8.0×10^{-4}	4.0×10^{-2}	1.25
1.50	0.35	4.0×10^{-2}	9.9×10^{-4}	4.6×10^{-2}	1.11
2.00	0.36	4.5×10^{-2}	1.0×10^{-3}	5.5×10^{-2}	1.00
2.50	0.35	4.4×10^{-2}	9.6×10^{-4}	5.8×10^{-2}	1.00
3.00	0.29	5.0×10^{-2}	9.0×10^{-4}	5.8×10^{-2}	1.03

2.2.2. Kinetic Modelling for Aromaticity Loss

A pseudo-first-order model has been proposed to predict the kinetics of aromaticity loss in aqueous SMX solutions oxidized by a photo-Fenton treatment. It is important to note that, to simplify the model, the initial increase in aromaticity observed in the water during the first 5 min of the reaction has been neglected, considering only the phase corresponding to the loss of aromaticity. Based on this approach, a mass balance has been proposed, where the initial aromaticity of the water containing 50.0 mg/L of SMX decreases as the aromatic functional groups are broken down by the action of hydroxyl radicals, according to a pseudo-first-order rate constant k_{arom} (1/min) (see Equation (9)).

$$\text{SMX} \overset{k_{arom}}{\rightarrow} \text{non-aromatic intermediates} \quad (9)$$

The mass balance proposed in Equation (10), corresponding to the decrease in aromaticity following first-order kinetics, has been corrected with the term $[\text{Aromaticity}]_{final}$ (AU) (estimated in Equation (1)). This adjustment is necessary because, given the characteristics of the oxidation treatment, the dosed hydrogen peroxide is consumed without fully eliminating the aromatic load. As a result, a refractory aromatic residue remains in the treated water, and its value must be accounted for in the mass balance to improve the accuracy of the model prediction.

$$d[\text{Aromaticity}] / dt = -\ k_{arom}\ ([\text{Aromaticity}] - [\text{Aromaticity}]_{final}) \quad (10)$$

By integrating the mass balance, the kinetic equation for the loss of aromaticity in aqueous SMX solutions oxidized by a photo-Fenton treatment is obtained (Equation (11)). A mean value for the initial aromaticity of water containing 50.0 mg/L of SMX can be considered as $[\text{Aromaticity}]_o$ = 1.958 AU.

$$[\text{Aromaticity}] = ([\text{Aromaticity}]_0 - [\text{Aromaticity}]_0) \times \exp(-k_{arom}\ t) + [\text{Aromaticity}]_{final} \quad (11)$$

Figure 5 shows the predictions of the proposed model. Based on the experiments conducted, the rate constant for the loss of aromaticity in the water has been estimated as a function of the initial oxidant concentration dosed in the treatment (Equation (12)). Table 1 presents the estimated values of the rate constants.

$$k_{arom} = -0.0079\ [H_2O_2]_0^2 - 0.0301\ [H_2O_2]_0 + 0.0696\ (r^2 = 0.9679) \quad (12)$$

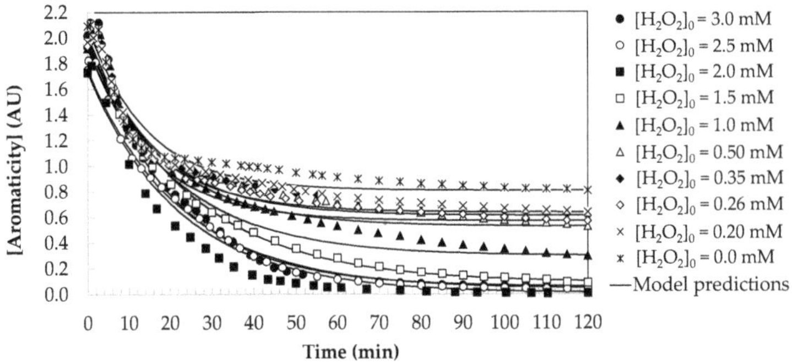

Figure 5. Predictions of the proposed kinetic model for the aromaticity changes during the SMX oxidation by a photo-Fenton treatment. Experimental conditions: $[SMX]_0$ = 50.0 mg/L; [UV] = 150 W; $[Fe^{2+}]_0$ = 0.5 mg/L; [pH] = 3.0; and [T] = 30 °C.

2.2.3. Kinetic Modelling for Colour Changes

A series reaction kinetic model has been proposed for the colour changes observed in oxidized water, where SMX is oxidized to coloured intermediates according to a pseudo-first-order kinetic constant $k_{colour,form}$ (1/min). Next, the intermediates of coloured nature degrade to colourless species according to a pseudo-first-order kinetic constant $k_{colour,deg}$ (1/min) (Equation (13)).

$$\text{SMX} \overset{k\ colour,\ form}{\rightarrow} \text{Coloured intermediates} \overset{k\ colour,deg}{\rightarrow} \text{Colourless species} \quad (13)$$

Equation (14) shows the proposed mass balance, considering that the colour of the water would maintain a behaviour similar to a reaction intermediate. Since the mass balance uses concentration units for SMX concentration and absorbance units for colour, the term α_{colour} (AU L/mg) has been introduced to homogenize units. Furthermore, because in several tests the oxidant used in the treatment is consumed without completely degrading into coloured species, leaving a refractory-coloured residue in the treated water, a correction has been made in the mass balance, introducing the term $[Colour]_{final}$ (AU) estimated in Equation (4).

$$d[Colour] / dt = k_{colour,form} \times \alpha_{colour} \times [SMX] - k_{colour,deg} \times ([Colour] - [Colour]_{final}) \quad (14)$$

Integrating the mass balance, the kinetic equation for the colour changes in the aqueous solutions containing SMX oxidized with the photo-to-Fenton technology is obtained (Equation (15)).

$$[Colour] = [Colour]_{final} + \left[\frac{k_{colour,form}[SMX]_0}{k_{colour,deg} - k_{SMX}}\right] \times \left[\alpha_{colour} \times \exp(-k_{SMX}\, t) - \exp(-k_{colour,deg}\, t)\right] \quad (15)$$

$$k_{colour,form} = -0.0002\, [H_2O_2]_0^2 + 0.0008\, [H_2O_2]_0 + 0.0001\ (r^2 = 0.9889) \quad (16)$$

$$k_{colour,deg} = -0.0017\, [H_2O_2]_0^2 + 0.0154\, [H_2O_2]_0 + 0.028\ (r^2 = 0.9744) \quad (17)$$

$$\alpha_{colour} = \lambda_{455}/\lambda_{260} = 0.2098\, [H_2O_2]_0^2 - 0.9689\, [H_2O_2]_0 + 2.0639\ \left(r^2 = 0.9932\right) \quad (18)$$

The predictions of the proposed model are shown in Figure 6. From the tests carried out, the kinetic constants for the formation and degradation of the colour of the water have been estimated, as well as the correlation parameter of units, depending on the concentration of oxidant used in the treatment (see Equations (16)–(18)). The estimated parameter values are shown in Table 1.

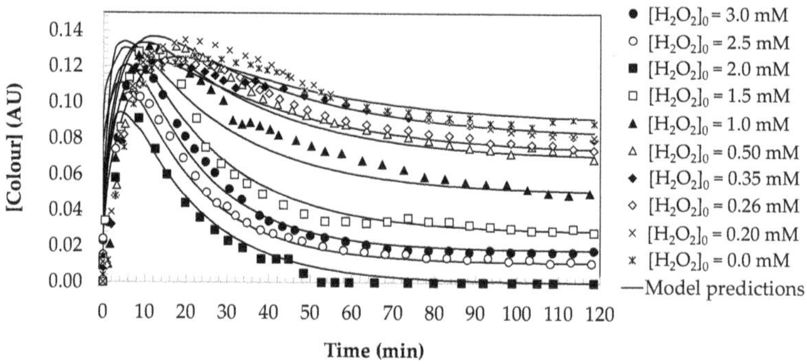

Figure 6. Predictions of the proposed kinetic model for the colour changes during the SMX oxidation by a photo-Fenton treatment. Experimental conditions: $[SMX]_0$ = 50.0 mg/L; [UV] = 150 W; $[Fe^{2+}]_0$ = 0.5 mg/L; [pH] = 3.0; and [T] = 30 °C.

3. Materials and Methods

The oxidation assays by a photo-Fenton treatment were conducted in a photo-catalytic reactor of 1.0 L oxidizing aqueous solutions of sulphamethoxazole $[SMX]_0 = 50.0\ mg{\cdot}L^{-1}$ (Fragon, Rotterdam, The Netherlands, $C_{10}H_{11}N_3O_3S$ 100.6%) by a 150 W medium-pressure mercury lamp (Heraeus, Hanau, Germany, 85.8 V, 148.8 W, 1.79 A, 95% transmission between 300 and 570 nm) combined with hydrogen peroxide $[H_2O_2]_0$ = 0–3.0 mM (Panreac, Castellar del Vallès, Spain, H_2O_2 30% *w*/*v*) and ferrous ion dosage $[Fe^{2+}]_0 = 5.0\ mg\ L^{-1}$ (Panreac, $FeSO_4{\cdot}7H_2O$ 99.0%). The reactor, operating in batch mode, was stirred using a magnetic stirrer. All reactions were performed at pH = 3.0 (using a pH-meter Kent EIL9142, Cambridge, UK) and the temperature was kept at around 30 °C using a heating bath (Frigiterm-P Selecta, Barcelona, Spain).

During 120 min of reaction, the parameters that were measured are as follows: the aromaticity of the water at λ = 254 nm ([Aromaticity], AU) and colour at λ = 455 nm ([Colour], AU) using a UV/Vis Spectrophotometer (Model V-630, Jasco, Madrid, Spain) [39]; the water turbidity ([Turbidity], NTU) with a nephelometric turbidimeter (Model HI88703, Hanna Instruments S.L., Eibar, Spain); the dissolved oxygen ([DO], mg/L) with a dissolved oxygen-meter (Model HI 9142, Hanna Instruments S.L., Eibar, Spain), and the total carbon ([TC], mg/L) using a TOC meter (Shimadzu TOC-V, Shimadzu Corporation, Kyoto, Japan).

The SMX concentration was determined by High-Performance Liquid Chromatography (Model 2695, Waters Cromatografía S.A., Cerdanyola del Vallès, Spain) with a Dual λ Absorbance Detector (Model 2487, Waters Cromatografía S.A., Cerdanyola del Vallès, Spain). A ZORBAX Eclipse PAH column (150 mm, 4.6 mm, particle size 5 µm) and a ZORBAX Eclipse PAH guard column (4.6 mm, 12.5 mm) supplied by Agilent (Santa Clara, CA, USA) were used. The mobile phase consisted of water and acetonitrile (ACN), with a flow rate of 0.8 mL/min. Initial gradient conditions were set at 20% ACN, maintained for 1 min, then increased to 50% *v*/*v* ACN over 7 min, and finally returned to 20% *v*/*v* ACN in 1 min. Total run time was 9 min. The injection volume was 50 µL, and all separations were performed at room temperature. DCF identification was performed by comparison with a standard. Detection was carried out at 275 nm.

4. Conclusions

The photo-Fenton treatment is capable of completely degrading the SMX load contained in the water. During oxidation, the aromaticity of water experiences a slight initial increase because partially oxidized intermediates are generated that contain greater aromaticity or that can even generate more aromatic structures. In addition, it is observed that a brown colour is being generated in the water. The colour is caused by some of the aromatic intermediates, such as quinones or polycyclic compounds, which are chromophoric in nature. It can also be caused by fragments of organic molecules reacting with each other forming condensation products. On the other hand, it must be considered that SMX contains a heterocyclic ring (containing sulphur and nitrogen) that can generate intermediate products with chromophoric properties. Simultaneously with the colour, turbidity is also generated in the water, which can be caused by the precipitation of iron hydroxides and the formation of colloidal particles and insoluble organic intermediates, which remain suspended in the water. As the photo-Fenton reaction proceeds, hydroxyl radicals continue to attack the aromatic intermediates, causing the opening of the benzene rings or the complete oxidation of the aromatic compounds to simpler products. As these intermediates are mineralized into CO_2 and H_2O, the colour and turbidity of the water lose intensity. It is verified that, using a molar ratio of 1 mol SMX:10 mol H_2O_2, the maximum degradation of aromatic species takes place (71% elimination), and colourless water with turbidity < 1 NTU is obtained, which enables them to be discharged into natural channels. A kinetic modelling of colour formation as a function of the oxidant concentration has been proposed, following a series reaction model. The application of this model allows the estimation of oxidant amounts for an efficient removal of SMX under environmentally friendly conditions.

Author Contributions: Conceptualization, N.V.; methodology, U.D.; software, U.D. and N.V.; validation, N.V.; formal analysis, U.D. and N.V.; investigation, N.V. and U.D.; resources, U.D., N.V. and A.M.D.L.; data curation, U.D. and N.V.; writing—original draft preparation, N.V., A.M.D.L., U.D. and J.I.L.; writing—review and editing, N.V.; visualization, N.V., A.M.D.L. and U.D.; supervision, A.M.D.L. and J.I.L.; project administration, N.V., A.M.D.L. and J.I.L.; funding acquisition, J.I.L. All authors have read and agreed to the published version of the manuscript.

Funding: This research was funded by PIBA_2023_01_0032—Basic and/or Applied Research Project. Financing entity: Basque Government 2023.

Data Availability Statement: The data supporting the findings of this study can be found within the article.

Acknowledgments: The authors thank the Department of Chemical and Environmental Engineering and the Department of Chemical Engineering of the University of the Basque Country UPV/EHU for their support.

Conflicts of Interest: The authors declare no conflicts of interest.

References

1. Lauzurique, Y.; Miralles-Cuevas, S.; Godoy, M.; Sepúlveda, P.; Bollo, S.; Cabrera-Reina, A.; Huiliñir, C.; Malato, S.; Oller, I.; Salazar-González, R. Elimination of sulfamethoxazole by anodic oxidation using mixed metal oxide anodes. *J. Water Process Eng.* **2023**, *54*, 103922. [CrossRef]
2. Cabello, F.C.; Godfrey, H.P.; Buschmann, A.H.; Dölz, H.J. Aquaculture as yet another environmental gateway to the development and globalisation of antimicrobial resistance. *Lancet Infect. Dis.* **2016**, *16*, 127–133. [CrossRef]
3. Isidori, M.; Lavorgna, M.; Nardelli, A.; Pascarella, L.; Parrella, A. Toxic and genotoxic evaluation of six antibiotics on non-target organisms. *Sci. Total Environ.* **2005**, *346*, 87–98. [CrossRef]
4. Kohantorabi, M.; Moussavi, G.; Oulego, P.; Giannakis, S. Radical-based degradation of sulfamethoxazole via UVA/PMS-assisted photocatalysis, driven by magnetically separable Fe_3O_4@CeO_2@BiOI nanospheres. *Sep. Purif. Technol.* **2021**, *267*, 118665. [CrossRef]
5. Huang, Z.; Dai, X.; Huang, Z.; Wang, T.; Cui, L.; Ye, J.; Wu, P. Simultaneous and efficient photocatalytic reduction of Cr(VI) and oxidation of trace sulfamethoxazole under LED light by rGO@Cu_2O/$BiVO_4$ p-n heterojunction composite. *Chemosphere* **2019**, *221*, 824–833. [CrossRef] [PubMed]
6. Dong, H.; Guo, X.; Yang, C.; Ouyang, Z. Synthesis of g-C_3N_4 by different precursors under burning explosion effect and its photocatalytic degradation for tylosin. *Appl. Catal. B Environ.* **2018**, *230*, 65–76. [CrossRef]
7. Tang, K.W.K.; Millar, B.C.; Moore, J.E. Antimicrobial resistance (AMR). *Br. J. Biomed. Sci.* **2023**, *80*, 11387. [CrossRef]
8. Gahrouei, A.E.; Vakili, S.; Zandifar, A.; Pourebrahimi, S. From wastewater to clean water: Recent advances on the removal of metronidazole, ciprofloxacin, and sulfamethoxazole antibiotics from water through adsorption and advanced oxidation processes (AOPs). *Environ. Res.* **2024**, *252*, 119029. [CrossRef]
9. Zyoud, S.H. The state of current research on COVID-19 and antibiotic use: Global implications for antimicrobial resistance. *J. Health Popul. Nutr.* **2023**, *42*, 42. [CrossRef]

10. Oharisi, O.L.; Ncube, S.; Nyoni, H.; Madikizela, M.L.; Olowoyo, O.J.; Maseko, B.R. Occurrence and Prevalence of Antibiotics in Wastewater Treatment Plants and Effluent Receiving Rivers in South Africa Using UHPLC-MS Determination. *J. Environ. Manag.* **2023**, *345*, 118621. [CrossRef]
11. Mertah, O.; Gómez-Avilés, A.; Kherbeche, A.; Belver, C.; Bedia, J. Peroxymonosulfate enhanced photodegradation of sulfamethoxazole with TiO_2@$CuCo_2O_4$ catalysts under simulated solar light. *J. Environ. Chem. Eng.* **2022**, *10*, 108438. [CrossRef]
12. de Matos, M.H.; Afonso, P.; Borges, K.C.M.; de Melo, L.; de Fátima, R.; Daldin, M.; Vilella, F.; Maribondo, R.; Godinho, M. Enhanced degradation of the antibiotic sulfamethoxazole by heterogeneous photocatalysis using $C_{e0,8}$ $Gd_{0,2}$ $O_{2\text{-}d}$/TiO_2 particles. *J. Alloys Compd.* **2019**, *808*, 151711. [CrossRef]
13. Bao, Y.; Lim, T.T.; Goei, R.; Zhong, Z.; Wang, R.; Hu, X. One-step construction of heterostructured metal-organics@Bi_2O_3 with improved photoinduced charge transfer and enhanced activity in photocatalytic degradation of sulfamethoxazole under solar light irradiation. *Chemosphere* **2018**, *205*, 396–403. [CrossRef] [PubMed]
14. Yazdanbakhsh, A.R.; Eslami, A.; Mohamadreza Massoudinejad, M.; Avazpour, M. Enhanced degradation of sulfamethoxazole antibiotic from aqueous solution using Mn-WO_3/LED photocatalytic process: Kinetic, mechanism, degradation pathway and toxicity reduction. *Chem. Eng. J.* **2020**, *380*, 122497. [CrossRef]
15. Murillo-Sierra, J.C.; Ruiz-Ruiz, E.; Hinojosa-Reyes, L.; Guzmán-Mar, J.L.; Machuca-Martínez, F.; Hernández-Ramírez, A. Sulfamethoxazole mineralization by solar photo electro-Fenton process in a pilot plant. *Catalysis Today.* **2018**, *313*, 175–181. [CrossRef]
16. Keerthanan, S.; Jayasinghe, C.; Bolan, N.; Rinklebe, J.; Vithanage, M. Retention of sulfamethoxazole by cinnamon wood biochar and its efficacy of reducing bioavailability and plant uptake in soil. *Chemosphere* **2022**, *297*, 134073. [CrossRef]
17. Dirany, A.; Sirés, I.; Oturan, N.; Oturan, M.A. Electrochemical abatement of the antibiotic sulfamethoxazole from water. *Chemosphere* **2010**, *81*, 594–602. [CrossRef]
18. Rizzo, L.; Manaia, C.; Merlin, C.; Schwartz, T.; Dagot, C.; Ploy, M.C.; Michael, I.; Fatta-Kassinos, D. Urban wastewater treatment plants as hotspots for antibiotic resistant bacteria and genes spread into the environment: A review. *Sci. Total Environ.* **2013**, *447*, 345–360. [CrossRef]
19. Liu, F.; Zhou, H.; Pan, Z.; Liu, Y.; Yao, G.; Guo, Y.; Lai, B. Degradation of sulfamethoxazole by cobalt-nickel powder composite catalyst coupled with peroxymonosulfate: Performance, degradation pathways and mechanistic consideration. *J. Hazard. Mater.* **2020**, *400*, 123322. [CrossRef] [PubMed]
20. Mirzaei, A.; Yerushalmi, L.; Chen, Z.; Haghighat, F.; Guo, J. Enhanced photocatalytic degradation of sulfamethoxazole by zinc oxide photocatalyst in the presence of fluoride ions: Optimization of parameters and toxicological evaluation. *Water Res.* **2018**, *132*, 241–251. [CrossRef]
21. Gong, H.; Chu, W. Permanganate with a double-edge role in photodegradation of sulfamethoxazole: Kinetic, reaction mechanism and toxicity. *Chemosphere* **2018**, *191*, 494–502. [CrossRef] [PubMed]
22. Poza-Nogueiras, V.; Gomis-Berenguer, A.; Pazos, M.; Sanroman, A.; Ania, C.O. Exploring the use of carbon materials as cathodes in electrochemical advanced oxidation processes for the degradation of antibiotics. *J. Environ. Chem. Eng.* **2022**, *10*, 107506. [CrossRef]
23. Wei, R.; Ge, F.; Huang, S.; Chen, M.; Wang, R. Occurrence of veterinary antibiotics in animal wastewater and surface water around farms in Jiangsu Province, China. *Chemosphere* **2011**, *82*, 1408–1414. [CrossRef] [PubMed]
24. Martini, J.; Orge, C.A.; Faria, J.L.; Pereira, M.F.R.; Soares, O.S.G.P. Catalytic Advanced Oxidation Processes for Sulfamethoxazole Degradation. *Appl. Sci.* **2019**, *9*, 2652. [CrossRef]
25. González, O.; Sans, C.; Esplugas, S. Sulfamethoxazole abatement by photo-Fenton Toxicity, inhibition and biodegradability assessment of intermediates. *J. Hazard. Mater.* **2007**, *146*, 459–464. [CrossRef]
26. Wu, C.H.; Wu, J.T.; Lin, Y.H. Mineralization of sulfamethizole in photo-Fenton and photo-Fenton-like systems. *Water Sci. Technol.* **2016**, *73*, 746–750. [CrossRef]
27. Villegas-Guzman, P.; Oppenheimer-Barrot, S.; Silva-Agredo, J.; Torres-Palma, R.A. Comparative Evaluation of Photo-Chemical AOPs for Ciprofoxacin Degradation: Elimination in Natural Waters and Analysis of pH Effect, Primary Degradation By-Products, and the Relationship with the Antibiotic Activity. *Water Air. Soil. Pollut.* **2017**, *228*, 209. [CrossRef]
28. Sun, S.; Yao, H.; Li, X.; Deng, S.; Zhao, S.; Zhang, W. Enhanced Degradation of Sulfamethoxazole (SMX) in Toilet Wastewater by Photo-Fenton Reactive Membrane Filtration. *Nanomaterials* **2020**, *10*, 180. [CrossRef]
29. Ameta, R.; Chohadia, A.K.; Jain, A.; Punjabi, P.B. Chapter 3—Fenton and Photo-Fenton Processes. In *Advanced Oxidation Processes for Waste Water Treatment*; Ameta, S.C., Ameta, R., Eds.; Academic Press: Cambridge, MA, USA, 2018; pp. 49–87. [CrossRef]
30. Soriano-Molina, P.; De la Obra, I.; Miralles-Cuevas, S.; Gualda-Alonso, E.; Casas López, J.L.; Sánchez Pérez, J.A. Assessment of Different Iron Sources for Continuous Flow Solar Photo-Fenton at Neutral pH for Sulfamethoxazole Removal in Actual MWWTP Effluents. *J. Water Process Eng.* **2021**, *42*, 102109. [CrossRef]
31. Martínez-Costa, J.I.; Rivera-Utrilla, J.; Leyva-Ramos, R.; Sánchez-Polo, M.; Velo-Gala, I.; Mota, A.J. Individual and Simultaneous Degradation of the Antibiotics Sulfamethoxazole and Trimethoprim in Aqueous Solutions by Fenton, Fenton-like and Photo-Fenton Processes Using Solar and UV Radiations. *J. Photochem. Photobiol. A Chem.* **2018**, *360*, 95–108. [CrossRef]
32. Elmolla, E.S. Effect of Photo-Fenton Operating Conditions on the Performance of Photo-Fenton-SBR Process for Recalcitrant Wastewater Treatment. *J. Appl. Sci.* **2010**, *10*, 3236–3242. [CrossRef]
33. Tian, L.; Wang, L.; Wei, S.; Zhang, L.; Dong, D.; Guo, Z. Enhanced degradation of enoxacin using ferrihydrite-catalyzed heterogeneous photo-Fenton process. *Environ. Res.* **2024**, *251*, 118650. [CrossRef] [PubMed]

34. Ji, Y.; Fan, Y.; Liu, K.; Kong, D.; Lu, J. Thermo activated persulfate oxidation of antibiotic sulfamethoxazole and structurally related compounds. *Water Res.* **2015**, *87*, 1–9. [CrossRef] [PubMed]
35. Moradi, M.; Moussavi, G. Investigation of chemical-less UVC/VUV process for advanced oxidation of sulfamethoxazole in aqueous solutions: Evaluation of operational variables and degradation mechanism. *Sep. Purif. Technol.* **2018**, *190*, 90–99. [CrossRef]
36. Trovó, A.G.; Nogueira, R.F.P.; Agüera, A.; Fernandez-Alba, A.R.; Sirtori, C.; Malato, S. Degradation of sulfamethoxazole in water by solar photo-Fenton. Chemical and toxicological evaluation. *Water Res.* **2009**, *43*, 3922–3931. [CrossRef]
37. Villota, N.; Echevarria, B.; Duoandicoechea, U.; Lombrana, J.I.; De Luis, A.M. Kinetic Study of the Water Quality Parameters during the Oxidation of Diclofenac by UV Photocatalytic Variants. *Catalysts* **2024**, *14*, 580. [CrossRef]
38. Villota, N.; Cruz-Alcalde, A.; Ferreiro, C.; Lombrana, J.I.; Esplugas, S. Changes in solution turbidity and color during paracetamol removal in laboratory and pilot-scale semicontinuous ozonation reactors. *Sci. Total. Environ.* **2023**, *854*, 158682. [CrossRef]
39. Villota, N.; Jankelevitch, S.; Lomas, J.M. Kinetic modelling of colour and turbidity formation in aqueous solutions of sulphamethoxazole degraded by UV/H_2O_2. *Environ. Technol.* **2024**, *45*, 349–359. [CrossRef] [PubMed]
40. Villota, N.; Ferreiro, C.; Qulatein, H.A.; Lomas, J.M.; Camarero, L.M.; Lombrana, J.I. Colour Changes during the Carbamazepine Oxidation by Photo-Fenton. *Catalysts* **2021**, *11*, 386. [CrossRef]
41. Peleyeju, M.G.; Umukoro, E.H.; Tshwenya, L.; Moutloali, R.; Babalola, J.O.; Arotiba, O.A. Photoelectrocatalytic Water Treatment Systems: Degradation, Kinetics and Intermediate Products Studies of Sulfamethoxazole on a TiO_2–Exfoliated Graphite Electrode. *RSC Adv.* **2017**, *7*, 40571–40583. [CrossRef]
42. Ren, M.; Sun, S.; Wu, Y.; Shi, Y.; Wang, Z.; Cao, H.; Xie, Y. The structure-activity relationship of aromatic compounds in advanced oxidation processes: A review. *Chemosphere* **2022**, *296*, 134071. [CrossRef] [PubMed]

catalysts

MDPI

Article

Modified Gasification-Slag-Driven Persulfate Activation for Highly Efficient Degradation of Acetaminophen: N/O Active Site Regulation and Nonradical Oxidation

Wenhao Si [1,2,†], Fei Qi [2,†], Kangjun Wang [1], Qiang Wang [2], Zequan Zeng [2,*], Yuting Niu [2] and Zhanggen Huang [2,3,*]

1 College of Chemical Engineering, Shenyang University of Chemical Technology, Shenyang 110142, China; 13555855295@163.com (W.S.)
2 State Key Laboratory of Coal Conversion, Institute of Coal Chemistry, Chinese Academy of Sciences, Taiyuan 030001, China; qifei@sxicc.ac.cn (F.Q.)
3 University of Chinese Academy of Sciences, Beijing 100049, China
* Correspondence: zengzequan@sxicc.ac.cn (Z.Z.); zghuang@sxicc.ac.cn (Z.H.)
† These authors contributed equally to this work.

Abstract: With the development of coal chemical technology, a large amount of gasification slag and wastewater are produced through coal gasification. Efficient gasification slag utilization and wastewater treatment have attracted much attention. In this study, gasification slag was modified and used as a low-cost and efficient catalyst to activate persulfate for acetaminophen degradation. Via the analysis of high-resolution X-ray photoelectron spectroscopy, the surfaces of nitric acid and calcined modified gasification slag retained a considerable number of carbonyl and graphite N functional groups. These proved to be effective active sites for the activation of persulfate. X-ray diffraction analysis revealed that the gasification slag was composed of carbon and SiO_2. The evaluation of catalytic activity and application of density functional theory proved that the interaction between carbonyl and graphitic nitrogen significantly affected the catalyst activity. When the ratio of graphitic nitrogen to carbonyl was 1:3, the adsorption and activation of persulfate were significantly enhanced. The results of the quenching experiments also confirmed that the non-free radical pathway is the main pathway to activate persulfate using the gasification slag. This study provides a new approach to industrial waste utilization in wastewater treatment.

Keywords: persulfate; gasification slag; nonradical oxidation; acetaminophen

1. Introduction

It is well known that acetaminophen (APAP) is a pharmaceutical compound used extensively as an analgesic and antipyretic drug and is one of the most commonly detected pharmaceutical compounds in aquatic environments [1,2]. APAP is also an emerging environmental endocrine disruptor that is likely to pose a threat to the environment and human health [3,4]. APAP is difficult to remove using traditional bio-treatment processes due to its stable chemical structure [5,6].

Persulfate-based advanced oxidation technology (PS-AOP) has been widely considered an effective method for pharmaceutical wastewater treatment because of its high oxidation potential, wide pH range, and long half-life of sulfate radicals [7,8]. Several approaches, such as catalysts (transition metal ions, metal oxides, and carbon), ultraviolet (UV), and ultrasonic (US) methods, have been proposed to activate persulfate for the formation of oxidative active species [9]. Carbon catalysts are effective for persulfate activation with no secondary metal leaching and have been widely investigated [10].

Carbonyl and graphitic N functional groups are the key sites to activate persulfate for the treatment of tetracycline (TC), and 90% degradation is achieved in 20 min at 25 °C [11,12]. However, the high cost and radical interference caused by water impurities have hindered the application of PS-AOPs with carbon catalysts. To conquer these limitations, a cheap carbon catalyst was prepared with urea as the raw material and used to activate persulfate for cost reduction [13]. Nitrogen doping can also regulate the persulfate activation path to obtain nonradical active species with an anti-interference property [14]. Xu et al. [15] proposed a 3D N-doped porous carbon to activate peroxymonosulfate via a nonradical pathway to oxidize chlorophenols from wastewater. However, carbon catalysts with large specific surface areas and high activities are still expensive [16], and the regulation mechanism of the PS activation path is still unclear.

Coal gasification slag (CGS) is a mixture of residual carbon produced by the coal gasification process [17], with a high carbon and nitrogen content, well-developed porous structure, low price, and wide application [18–20]. It has been proven that modified CGS is an effective and inexpensive catalyst for pollutant removal [21,22]. Liu et al. [23] prepared a microsphere material for the adsorption of Pb (II) and Congo red by combining fine CGS with silica. Fe-doped CGS has been successfully used to activate PMS for the removal of sulfamethoxazole [22]. As a metal-free and high-efficiency carbon catalyst, the performance of persulfate activation via CGS for pharmaceutical wastewater degradation has not been evaluated.

As CGS has a variety of nitrogen-containing functional groups and the proportion of nitrogen–oxygen functional groups in CGS is not low, the mechanism of the nitrogen–oxygen functional groups in CGS is still unclear. Thus, the effect of nitrogen and oxygen functional groups on the activation mechanism needs to be further studied to treat complex wastewater.

In this study, nitric acid treatment combined with calcination in nitrogen was used to modify the nitrogen and oxygen functional groups of the CGS. The surface of the active sites was regulated by calcination, and the mechanism of the catalytic activation of persulfate was further regulated to clarify the internal relationship between heteroatom doping and the activation mechanism. The regulatory mechanism of radical oxidation and the nonradical oxidation pathway via heteroatom doping modification was further clarified.

2. Results and Discussion

2.1. Materials Characterization

Through the XRD patterns of the CGS, CGSO, and CGSO300 (Figure 1), it was found that the main peaks of the gasification slag were at 26.603°, 42.464°, and 47.305°. These peaks are ascribed to carbon, indicating that carbon is the main component of gasification slag. The peaks at 20.859°, 36.543°, 38.852°, and 66.546° are ascribed to SiO_2; this shows that the gasification slag comprises some ash, such as SiO_2.

The CGS and modified gasification slag (CGSO and CGSO300) were characterized through BET. It can be seen from Table 1 that the specific surface area and pore volume of the CGS increased after treatment with nitric acid, which is most probably owing to the removal of ash to form new pore channels. The increase in the specific surface areas facilitates the adsorption of pollutants on the CGS. After calcination, some volatile surface oxygen-containing functional groups were decomposed or reacted with charcoal, resulting in the collapse of the pore channels, leading to a decline in the specific surface area and pore volume of CGSO300.

Table 1. BET data of modified CGS.

Samples	S_{BET} (m^2g^{-1})	V_p (cm^3g^{-1})	Pore Size (nm)
CGS	200	0.22	4.3
CGSO	360	0.32	3.6
CGSO300	208	0.21	3.9

As shown in Figure 2a, absorption and desorption experiments were conducted on CGS, CGSO, and CGSO300. The figure shows that their adsorption type is type IV isothermal adsorption, and the pore structure of the catalysts mainly comprises mesoporous and microporous structures. CGSO300 has a larger micropore proportion, more micropores, and contributes a more specific surface area, which is more conducive to the reaction. This is also confirmed by the aperture size distribution curve (Figure 2b). The mesoporous and microporous structure of the catalyst is conducive to the transport and activation of persulfate for the degradation of APAP.

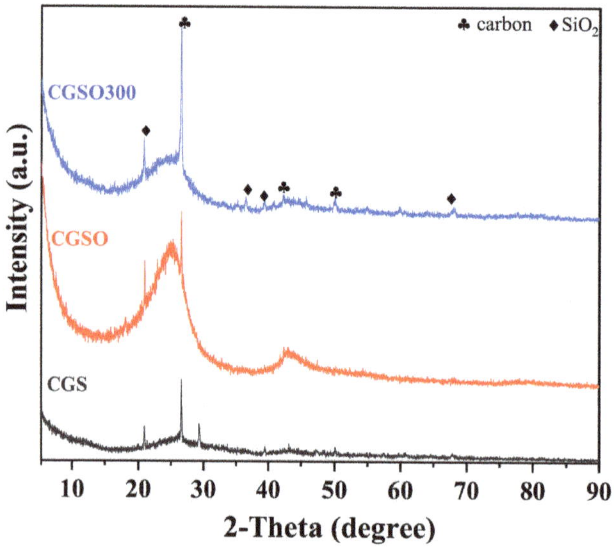

Figure 1. XRD patterns of modified gasification slag.

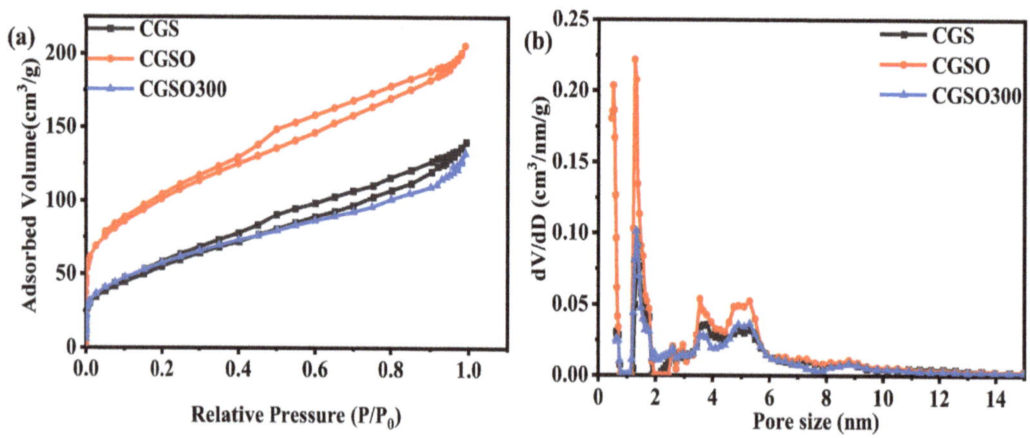

Figure 2. (**a**) N_2 sorption isotherms and (**b**) pore size distribution of modified CGS.

Raman spectra can show the degree of crystallization of a material, reflecting the defects as well as the disorder of the material [24]. As seen in Figure 3, the characteristic D and G bands appear at 1350 cm^{-1} and 1580 cm^{-1}, respectively. The G band represents the in-plane tensile vibration of sp^2 hybrid carbon, which is ascribed to the graphitic properties and reveals the structure of sp^2-bonded carbon. The D band is a disorder-induced vibrational mode ascribed to the fundamental structural defects in the amorphous carbon material. Furthermore, the degree of graphitization and disorder in the carbon structure is assessed based on the relative intensity ratio of the D spectral band to the G spectral band (I_D/I_G) [25]. It can be seen in Figure 3 that the I_D/I_G values of CGS, CGSO, and CGSO300 are 1.13, 1.20, and 1.29, respectively. Among them, CGS has the highest degree of graphitization, while CGSO300 has the lowest degree of graphitization, which indicates that the crystal structure of CGSO300 is more disordered. A considerable number of oxygen-containing functional groups were formed through nitric acid treatment on the surface of the gasification slag. Further calcination resulted in the decomposition of oxygen-containing functional groups to form a considerable number of defect sites, increasing the disordering of the carbon materials. Demiral et al. [26] found that with the nitric acid modification on activated carbon, the number of carboxyl groups, lactones, and phenyl groups on the surface of activated carbon significantly increased, which increased the disorder of the activated carbon. He et al. [27] found that the modification of graphite with HNO_3 can introduce more defect sites and oxygen/nitrogen-containing groups on graphite. The reason for this is probably the decomposition of the oxygen-containing functional group providing more active sites. More defect sites are conducive to the adsorption and activation of PS.

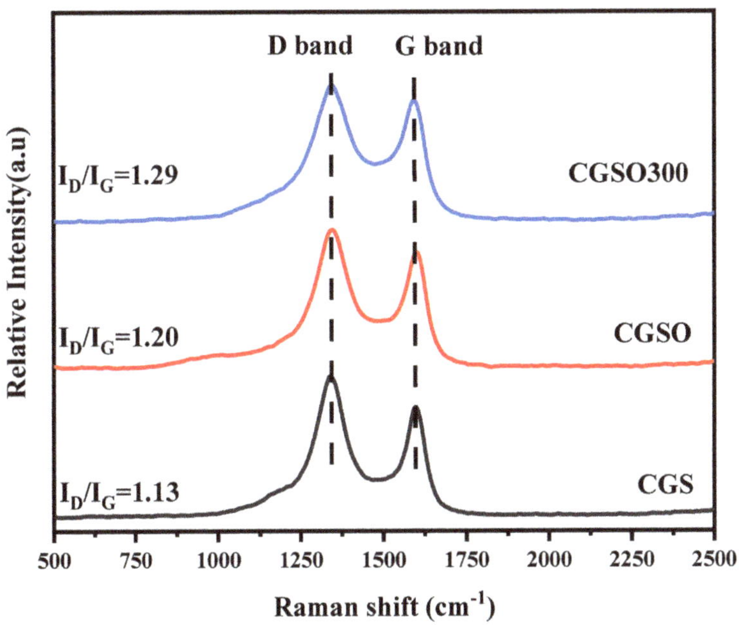

Figure 3. Raman spectrogram of the modified gasification slag.

The modified gasification slags were also characterized using XPS to determine the surface chemical composition and functional group species. Figure 4 shows the XPS broad-sweep curve profile of the modified gasification slags, in which the peaks with binding energies of 284.5 eV, 400.9 eV, and 532.5 eV were ascribed to the elements of carbon, nitrogen, and oxygen, respectively. The atomic concentration of the elements in Table 2 is given through XPS analysis. Since the atomic concentration of elements is obtained using the normalization method, the relative atomic ratio can better reflect the change in element distribution than atomic concentration. It can be observed from Table 2 that, after calcination, both the C/N ratio and O/N ratio significantly decreased. This is because the oxygen-containing surface functional groups formed by nitric acid treatment (CGSO) contain a considerable number of unstable functional groups such as carboxyl groups, which are released in the form of CO_2 and other products during calcination, thus reducing the C and O ratio on the surface. However, the nitrogen-containing functional groups in CGSO are formed during the formation of gasification slag at high temperatures (800–1400 °C), which has high stability during calcination at 300 °C. So, the relative concentration of the N element increased and the relative concentration of C and O decreased. The other reason is presumably the oxidation of CGS by nitric acid. Compared with the CGS, the oxygen content of CGSO and CGSO300 was substantially enhanced, indicating that a considerable number of oxygen-containing functional groups were successfully formed on the surface of the carbon material [28–30]. The reduction in elemental C and O in CGSO300 compared to CGSO is probably due to the decomposition of the oxygen-containing functional group to CO_2, CO, and H_2O during the calcination process, which indirectly leads to the increase in elemental N content in the modified gasification slag.

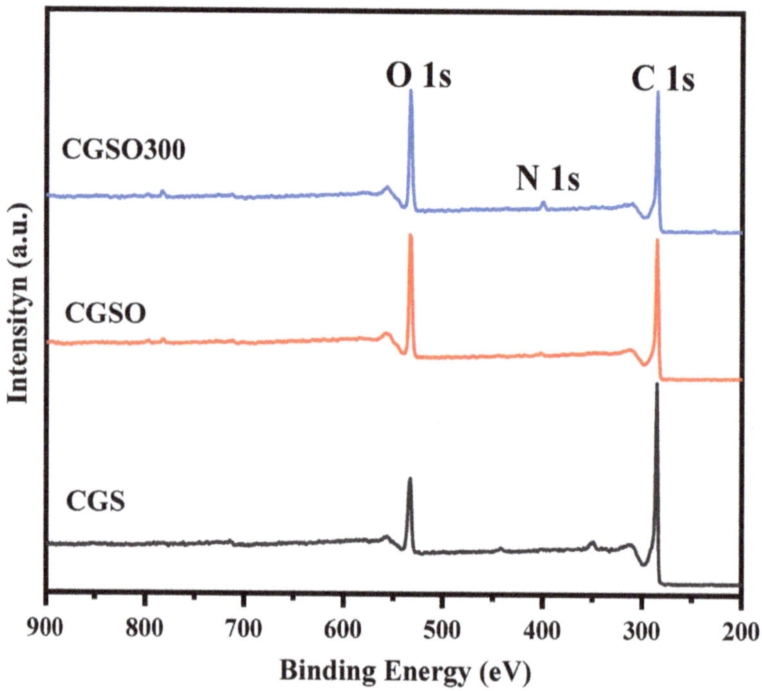

Figure 4. XPS wide-scan profile of modified gasification slag.

Table 2. Surface element composition of modified gasification slag.

Samples	C (at.%)	O (at.%)	N (at.%)	C/N Ratio	O/N Ratio
CGS	82.59	16.27	1.14	72	14
CGSO	71.84	26.71	1.45	49	18
CGSO300	70.15	26.50	3.35	20	7

Figure 5a and Table 3 show the N 1s spectra and the corresponding peak-splitting results. The N 1s spectra can be fitted by four peaks: the peak at 398.5 eV, 400 eV, 401.5 eV, and 404.5 eV are ascribed to pyridine N, pyrrole N, graphitic N, and nitrogen oxides, respectively [31]. On the one hand, the results showed that the proportions of pyridine N, pyrrole N, and nitrogen oxide increased after the modification of CGS with nitric acid due to its strong oxidizing properties. On the other hand, the calcination of CGSO leads to increasing graphitic N owing to the decomposition of pyrrole N and nitrogen oxide.

As shown in Figure 5b and Table 4, it can be seen that the O 1s spectrum can be fitted by five peaks: the peak of 531.1 eV is ascribed to C=O in carbonyl and quinone groups; the peak of 532.2 eV is ascribed to C=O in anhydride and ester groups; the peak of 533.3 eV is ascribed to C–O in phenol, anhydride, and ester groups; the peak of 535.1 eV is ascribed to oxygen in the form of –COOH; and the peak of 537 eV is ascribed to water in the adsorbed state and CO_2 [32–35]. It is found that the C–O functional group content increases significantly after modification with nitric acid. The carbonyl group content decreased after the CGS was modified by nitric acid to form CGSO and increased after heat treatment to form CGSO300. Carbonyl is a potential active site for PS activation. The loss of carbonyl may lead to a decrease in the catalytic activity of CGSO.

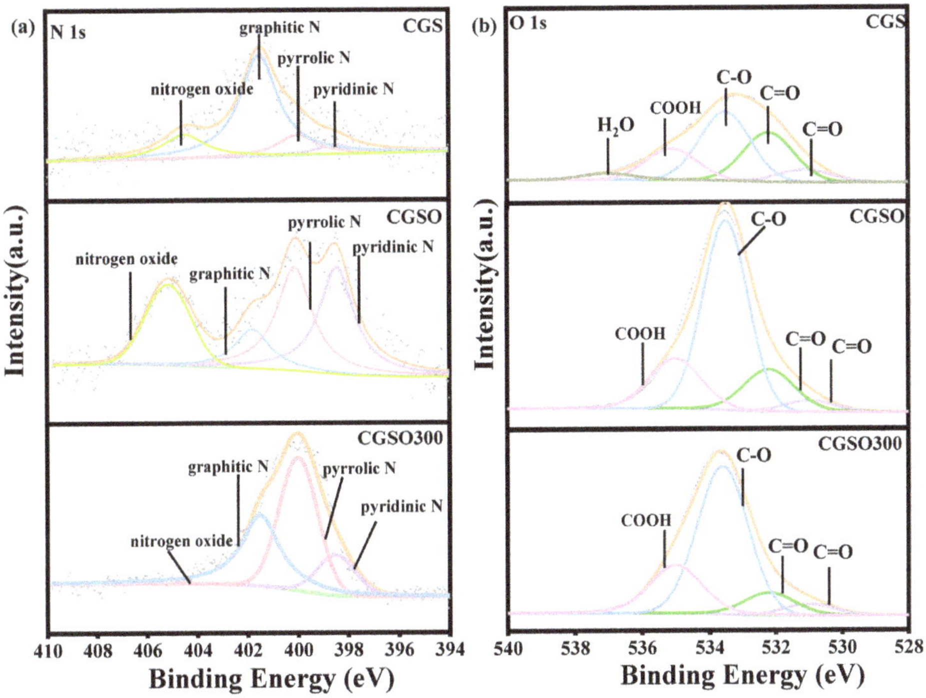

Figure 5. (**a**) N 1s and (**b**) O 1s spectra of modified gasification slag.

Table 3. Nitrogen-containing functional group proportion of modified gasification slag.

Samples	Pyridine N (at.%)	Pyrrole N (at.%)	Graphitic N (at.%)	Nitrogen Oxide (at.%)
CGS	0.07	0.01	0.73	0.19
CGSO	0.48	0.47	0.17	0.32
CGSO300	0.49	0.24	1.26	0.08

Table 4. Oxygen functional group content of modified gasification slag.

Samples	C=O (Carbonyl) (at.%)	C=O (Ester, Anhydride) (at.%)	C–O (Ester, Anhydride, Phenol) (at.%)	COOH (at.%)
CGS	1.3	4.8	6.6	3.0
CGSO	1.1	4.3	16.2	5.1
CGSO300	1.3	2.7	16.4	6.1

2.2. Catalytic Activity for PS Activation

The catalytic activity was evaluated by examining the efficiency of PS activation for the degradation of acetaminophen (APAP), as shown in Figure 6. The insert in Figure 6a shows the adsorption curve of APAP on the modified gasification slag and in the CGS/PS system. APAP could be completely degraded in 40 min with a kinetic rate constant of 0.086 min^{-1}, indicating that the pristine CGS could also activate PS. After the CGSO, the APAP was not completely removed within 50 min with the kinetic rate constant of 0.033 min^{-1}. The reason for this may be the loss of carbonyl, which is an effective site for PS activation. This also may be because the introduction of a considerable number of acidic oxygen-containing functional groups is detrimental to the transfer of electrons from the catalyst to the peroxy-bonds in PS, which greatly inhibits the activation of PS. After calcination at 300 °C, the activation activity of CGSO300 greatly increased with the highest kinetic rate constant of 0.148 min^{-1}. As the nitrogen content of CGSO300 was increased to 6.1% with graphitic N as the dominant nitrogen-containing functional group, the positive charge density of the neighboring oxygen atoms was increased by graphitic N and more readily interacted with PS to form reactive complexes, which enhanced the oxidation of pollutants via the electron transfer pathway.

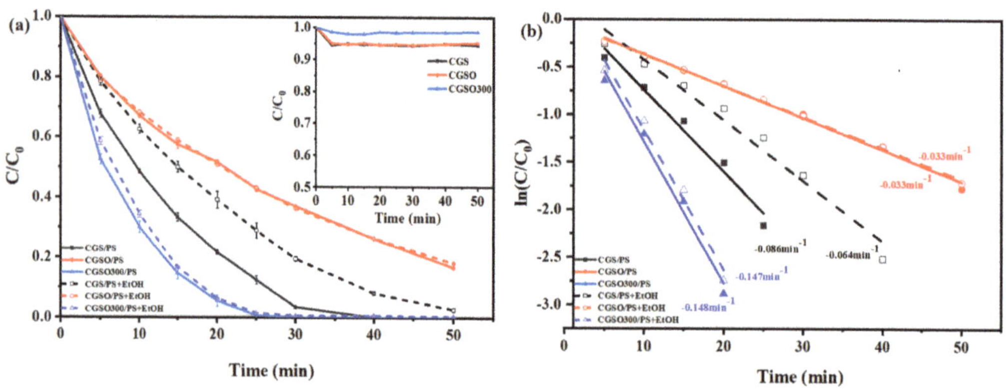

Figure 6. (**a**) Oxidation and adsorption curves and (**b**) kinetic rate of APAP with and without quenching ((APAP) = 50 mgL^{-1}, (CGS) = 0.1 gL^{-1}, (PS) = 2.5 mM, (EtOH/(PS) = 1000).

Ethanol (EtOH) is often used as a trapping agent to quench radicals. Because EtOH had a high reaction rate with both ·OH ($k_{Ethanol/OH} = (1.2 - 2.8) \times 10^9\ M^{-1}s^{-1}$) and $SO_4^{\bullet-}$ ($k_{Ethanol/SO4-} = (1.6 - 7.8) \times 10^7\ M^{-1}s^{-1}$), the reaction rates of CGSO/PS and CGSO300/PS systems were almost unaffected after ethanol was added. This indicated that there are few radicals in CGSO/PS and CGSO300/PS systems and that PS was activated through nonradical pathways. The APAP degradation was inhibited by EtOH in the CGS/PS system with the decreasing kinetic rate constant of 0.086 min^{-1} to 0.064 min^{-1}. Based on Equations (1) and (2), the contributions of nonradical pathways on APAP degradation in CGS/PS, CGSO/PS, and CGSO300/PS systems were calculated as 74%, 99%, and 100%, respectively. It can be found that the nonradical pathway was the dominant activation pathway of PS with modified gasification slag as the catalyst, which benefits the APAP degradation in high-salt wastewater.

$$R_{nonradical} = \frac{k_{nonradical}}{k_{APAP}} \times 100\% \quad (1)$$

$$R_{radical} = 100\% - R_{nonradical} \quad (2)$$

where k_{APAP} and $k_{nonradical}$ represent the total reaction rate constant of APAP oxidation degradation and the rate constant of nonradical oxidation reaction (min^{-1}), respectively.

$R_{radical}$ and $R_{nonradical}$ are the ratios of the radical reaction and the nonradical reaction, respectively.

As is well known, carbonyl and graphitic N are the active sites of carbon materials for persulfate activation, and the ratio of carbonyl and graphitic N can better reflect the catalytic performance of the catalyst. As shown in Figure 7, the reaction activity of the modified gasification slag increased gradually with the increase in the graphitic N and carbonyl content. The coefficient of determinations of graphitic N and carbonyl groups with the reaction rate constant are 0.978 and 0.875, respectively (Figure 7). This indicates that graphitic N and the carbonyl group have an obvious linear relationship with K, and the linear relationship between graphitic N and K is more obvious. The R^2 in Figure 7 is the coefficient of determination of the linear relationship between graphite N, carbonyl, and the kinetic rate constant. The higher R^2 of graphite N indicates that graphite N has a more significant linear relationship with the constant kinetic rate. This may reflect that graphite N has a more significant effect on the catalytic activity. It is indicated that graphitic N may be the main activated site for modified gasification slag due to its higher coefficient of determination.

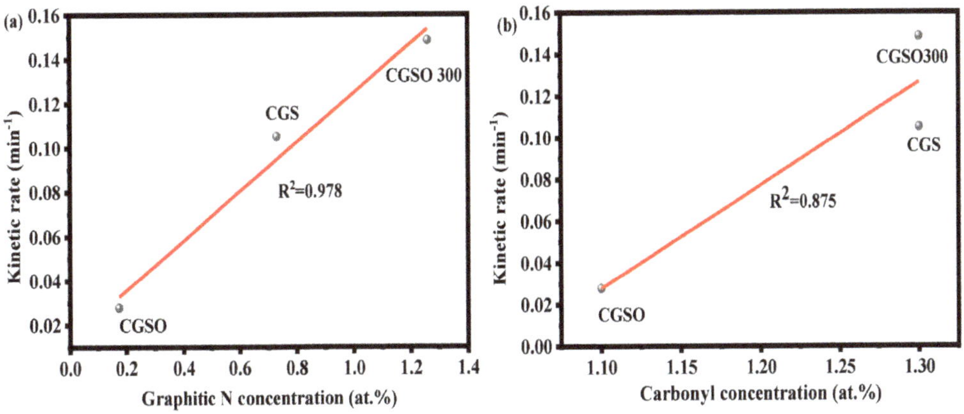

Figure 7. (**a**) Correlation between the kinetic rate and the graphitic N content and (**b**) the correlation between the kinetic rate and the carbonyl content ((APAP) = 50 mgL^{-1}, (CGS) = 0.1 gL^{-1}, (PS) = 2.5 mM).

To reflect the intrinsic activity of the modified gasification slag, the turnover frequency (TOF) values were calculated using Equation (3). As graphitic N and carbonyl are the activity sites, the ratio of graphitic N to the carbonyl group was related to TOF (Figure 8). It can be observed that the ratio of graphitic N to the carbonyl group and TOF has a linear correlation, with a coefficient of determination of 0.921. This shows that graphitic N and carbonyl work together in the activation of PS. As the ratio of graphitic N to carbonyl group increased, the site activity (TOF) increased significantly. This suggests that the graphitic group has a significant influence on the activity of active sites, and the presence of the graphitic group near the carbonyl group is favorable to enhancing the activity of the modified gasification slag.

$$\mathrm{TOF} = \frac{k}{(\text{Graphitic N} + \text{Carbonyl}) \times S_{BET}} \tag{3}$$

where TOF is the APAP turnover frequency ($min^{-1}gm^{-2}$) normalized by carbonyl and graphitic N sites, and k represents the apparent rate constant (min^{-1}). Graphitic represents the atomic ratio of graphitic N. Carbonyl represents the atomic ratio of carbonyl. S_{BET} is the total specific surface area (m^2g^{-1}) obtained by the BET test.

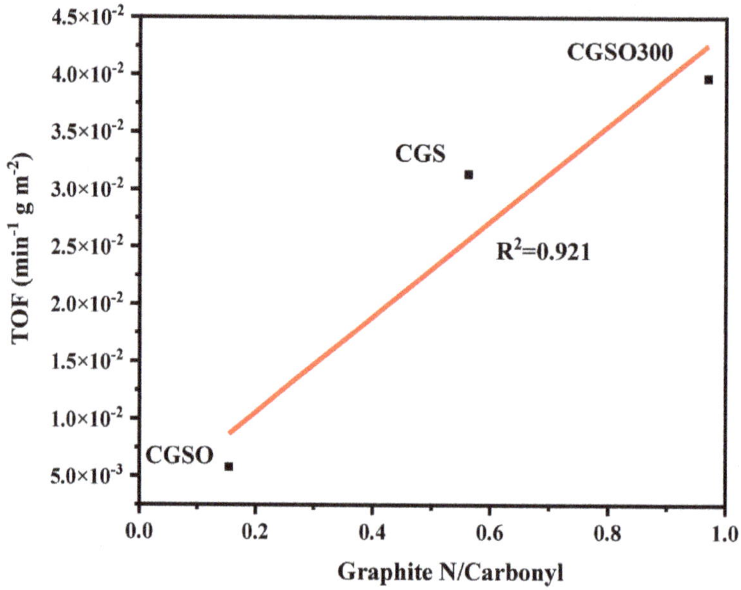

Figure 8. Relationship between the ratio of graphitic N to the carbonyl group and TOF.

2.3. Tests of Catalyst Reusability

To evaluate the reusability of the catalyst for the sustainable operation of the CGSO300/PS system, the catalyst was repeated at room temperature. As shown in Figure 9, the experimental data showed that the catalytic activity of CGSO300 decreased after use. This may be due to the inactivity of the active site on the surface of CGSO300 by oxidation. After calcination and regeneration at 300 °C (after the Used CGSO300 was placed in a tube furnace and heated to 300 °C at 5 °C per minute under a N_2 atmosphere), the activity of CGSO300 was improved, which may be due to the increased active sites such as graphitic N. The states of the N elements in CGSO300 were analyzed (Figure 10 and Table 5), which revealed that the proportion of graphitic N and pyridinic N functional groups on the surface of the Used CGSO300 decreased significantly after use, which may be due to their oxidation by persulfate to form electron-deficient nitrogen oxide functional groups. This makes it difficult for persulfate to continue to activate on its surface, resulting in a significant decrease in catalytic activity. After regeneration at 300 °C, the content of graphitic N functional groups increased, and the catalytic activity partially increased. However, due to the difficulty of converting pyrrole functional groups to active graphite N functional groups at 300 °C, the full recovery of activity may require a higher regeneration temperature. The study found that the content of graphite N decreased and nitrogen oxide increased in CGSO300 after calcination and recovery.

Table 5. Nitrogen-containing functional group proportion of fresh CGSO300 used CGSO300 and regenerated CGSO300.

Samples	Pyridine N (at.%)	Pyrrole N (at.%)	Graphitic N (at.%)	Nitrogen Oxide (at.%)
Fresh CGSO300	0.49	0.24	12.6	0.08
Used CGSO300	0.11	0.25	0.78	1.57
Regenerated CGSO300	0.17	0.28	0.94	1.93

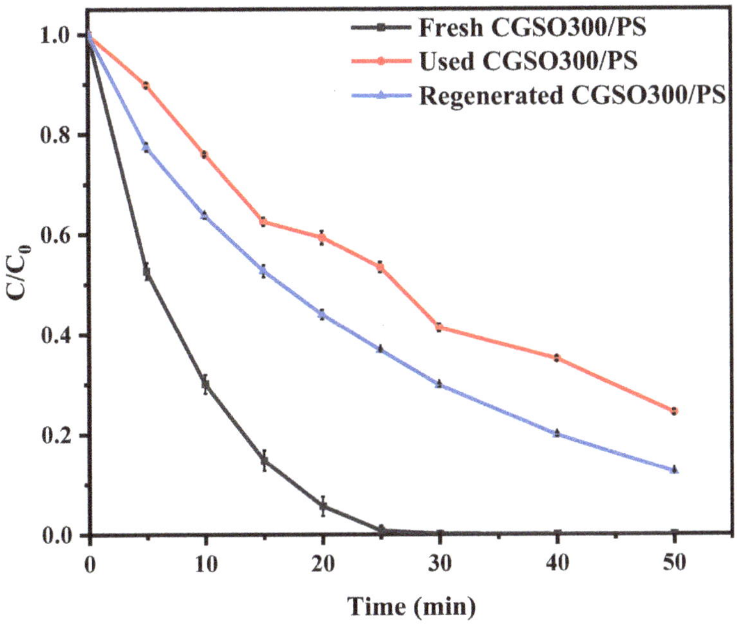

Figure 9. Fresh CGSO300, Used CGSO300, and Regenerated CGSO300 oxidative degradation curves of APAP ((APAP) = 50 mgL^{-1}, (CGSO300) = 0.1 gL^{-1}, (PS) = 2.5 mM).

As shown in Table 6, comparing the data from APAP degradation using different catalytic PS activation approaches shows that biochar has the best catalytic effect on activated PS, with the shortest degradation time and the highest degradation rate. The metal catalyst has a higher preparation cost, longer degradation time, and lower degradation rate.

Table 6. Comparison between CGSO300 and the previously reported different catalysts that activated PS to degrade APAP [36–41].

Catalyst Dosage (gL^{-1})	PS (mM)	APAP (mgL^{-1})	Removal Rate (%)	Pseudo First-Order Kinetics (min^{-1})	References
CGSO300 (0.1)	2.5	50.0	100.0 (25 min)	0.148	This study
Biochar (0.1)	0.5	50.0	100.0 (15 min)	\	[36]
Fe, Cu@g-C_3N_4 (0.01)	1.0	4.0	100.0 (40 min)	0.069	[37]
3DOMFeCo (0.2)	2.0	10.0	100.0 (60 min)	\	[38]
$CuFe_2O_4$ (0.3)	3.4	100.0	91.0 (60 min)	\	[39]
Fe^0 (2.0)	2.0	30.0	98.5 (35 min)	0.129	[40]
Fe_3O_4 (0.8)	0.2	10.0	75.0 (120 min)	0.012	[41]

2.4. Theoretical Calculation

DFT calculation was further used to analyze the interaction of $S_2O_8{}^{2-}$ with different sites. Figure 11 simulates the adsorption of $S_2O_8{}^{2-}$ on modified gasification slags with different ratios of graphitic N to the carbonyl group, as well as the derivations of various functional groups and the distribution of carbonyl groups in the new theoretical model. The adsorption energy and transfer charge of the different ratios of graphitic N to a carbonyl group are shown in Table 7. All adsorption energies are negative, which means that the adsorption of PS is spontaneous. The adsorption energy of PS (E_{ads}) was defined as follows:

$$E_{ads} = E_{total} - E_{subs} - E_{PS} \quad (4)$$

The E_{total}, E_{subs}, and E_{PS} were defined as the energy of catalysts with PS, catalysts, and free PS, respectively. It was found that the model with a higher ratio of graphitic N to carbonyl group had the higher adsorption energy for $S_2O_8^{2-}$. A higher adsorption energy means a stronger interaction between the catalyst and PS. The results of the transfer charge verified that the electron transfer occurred between PS and the catalyst and that CGS-N_3 had the largest transfer charge. The above results further indicate that catalysts with higher graphitic N and carbonyl are more conducive to PS activation, which is consistent with the trend of TOF values.

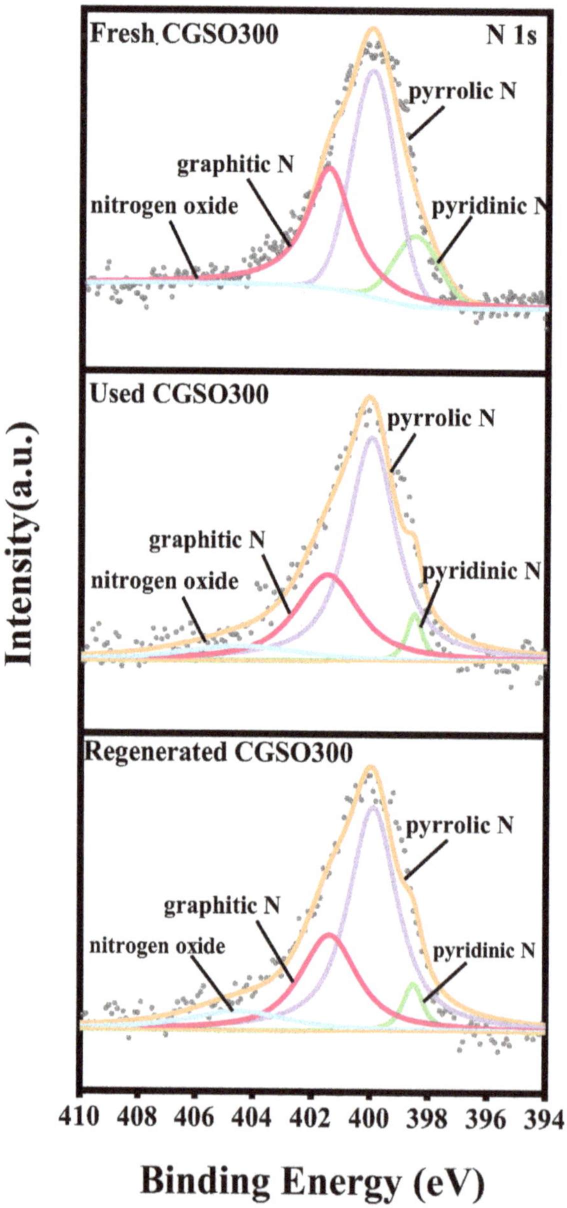

Figure 10. N 1s spectra of Fresh CGSO300, Used CGSO300, and Regenerated CGSO300.

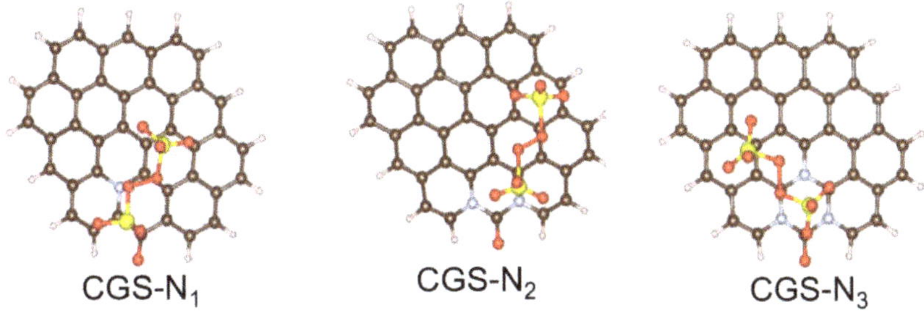

Figure 11. Structure of $S_2O_8^{2-}$ adsorbed on different sites.

Table 7. Adsorption energy and the transferred charge of different N-doped CGSs.

Catalysts	Ratio of Graphitic N to Carbonyl Group	Adsorption Energy (Ea, eV)	Transfer Charge (Q, e)
CGS-N_1	1:1	−1.982	−1.330
CGS-N_2	2:1	−2.979	−1.027
CGS-N_3	3:1	−3.870	−1.638

3. Experimental Procedure

3.1. Materials and Reagents

CGS was provided by Shaanxi Future Energy Chemical Co. (Yulin, China), methanol was supplied by Tianjin Tianli Chemical Reagent Co. (Tianjin, China), nitric acid (HNO_3) was provided by Tianjin Zhiyuan Chemical Reagent Co., Ltd. (Tianjin, China), and 4-acetaminophen (APAP, 99.0%) and potassium persulfate ($K_2S_2O_8$, 99.0%) were supplied by Macklin (Shanghai, China) and Aladdin (Shanghai, China). Ultrapure water was employed throughout the experiments.

3.2. Preparation of Modified Gasification Slag

During the nitric acid oxidation treatment, CGS was soaked in nitric acid (34%) at room temperature for 2 h. After filtering and washing to a neutral pH, the samples were dried in an oven at 80 °C to a constant weight, and then placed in a dryer for use, labeled as CGSO. The CGSO was placed in a quartz tube, calcined in N_2 at 300 °C for 1 h, and labeled as CGSO300.

3.3. Catalyst Characterization Methods

A Raman spectrometer (LabRAM HR800, Horiba, NJ, USA) with a 532 nm argon ion laser was used to detect the molecular structure of the material. The composition and chemical states of the elements on the surface of the carbon catalysts were determined by X-ray photoelectron spectroscopy (XPS, AXIS ULTRA DLD, Kratos, Manchester, UK) under irradiation with Al Kα X-ray (hv = 1486.6 eV). The textural properties of the gasification slag were determined using nitrogen sorption isotherms at 77 K using a TriStar II 3020. All samples were degassed at 80 °C for 8 h under vacuum before measurement. The specific surface area of the gasification slag adopted the Brunauer–Emmett–Teller (BET) method. The Barrett–Joyner–Halenda (BJH) calculation model for microporous carbon was used to measure the micropore surface area and volume of the samples. The pore volume and pore size distribution were determined from the volume adsorbed at a relative pressure of 0.99 and by using Density Functional Theory (DFT), respectively.

3.4. Measurement of Catalytic Activity

In this study, CGS, CGSO, and CGSO300 catalysts were used to activate persulfate for the catalytic degradation of acetaminophen. Firstly, 100 mL of acetaminophen solution (50 mgL^{-1}) with persulfate (2.5 mM) was continuously shanked in a thermostatic shaker. Then, 100 mgL^{-1} of CGS was added to the system. The reaction was initiated by adding the catalyst to the solution. Following this, 1 mL of the solution was withdrawn at regular intervals and filtered through a 0.45 μm aqueous needle filter. The concentration of acetaminophen was analyzed by high-performance liquid chromatography (UHPLC, UltiMate3000, Dionex, Waltham, MA, USA) with a diode array detector (DAD) at λ = 280 nm. The chromatographic column was a SunFire™ C18 (4.6 × 150 mm, 5.0 μm) and the mobile phase was 30% methanol and 70% ultrapure water with a flow rate of 0.8 $mLmin^{-1}$. The column temperature was set as 30 °C.

3.5. Theoretical Calculation

Spin polarization density functional theory (DFT) was calculated based on plane waves using the Vienna ab simulation package [42]. All calculations were performed using generalized gradient approximation (GGA) with the Perdew–Burke–Ernzerhof (PBE) exchange-correlation function [43]. The Bader charge analysis method was applied to calculate charge transfers [44].

4. Conclusions

In this work, modified gasification slag was successfully synthesized for use as an activator for PS in the APAP degradation process. The catalytic activity of the modified gasification slag was enhanced by O/N doping combined with calcination. It was found that GCSO calcinated at 300 °C had the highest catalytic activity for PS activation, with 100% APAP removal in 30 min. O/N doping accelerated the electron transfer process, in which carbonyl and graphitic N were demonstrated to be the main active sites. We reported that the increase in carbonyl and graphitic N not only enhanced the catalytic performance towards PS activation but also effectively facilitated the nonradical pathway. The contributions of nonradical pathways on APAP degradation in the CGS/PS, CGSO/PS, and CGSO300/PS systems were calculated as 74%, 99%, and 100%, respectively. This results in the high anti-interference potential of the PS activation system by CGS. It was also found that a higher ratio of graphitic N to carbonyl group led to the higher catalytic activity of the modified gasification slag, which was also proven through the DFT calculation of the adsorption energy for $S_2O_8^{2-}$. This study provided a new feasible way for the degradation of organic pollutants using modified gasification slag to activate persulfate. Benefiting from its simple preparation process, modified gasification slag is anticipated to be a novel and cheap carbon catalyst for wastewater treatment.

Author Contributions: Investigation, W.S. and Y.N.; writing—original draft, W.S.; methodology, F.Q.; project administration, Z.Z.; software, Q.W.; writing—review and editing, F.Q., Z.Z. and Z.H.; supervision, K.W. and Z.H.; resources, Z.H. All authors have read and agreed to the published version of the manuscript.

Funding: This research was funded by the Fundamental Research Program of Shanxi Province (202103021224443), ICC CAS (SCJC-WRW-2022-19), the Youth Innovation Promotion Association of the Chinese Academy of Sciences (2022174, Z.Z.Q.), the Science and Technology Major Project of Shanxi Province (202005D121002), and Yulin University and the Dalian National Laboratory for Clean Energy (Grant YLU-DNL, Fund 2021003).

Data Availability Statement: The data presented in this study are available in the article.

Conflicts of Interest: The authors declare no conflict of interest.

References

1. Santos, J.L.; Aparicio, I.; Alonso, E.; Callejón, M. Simultaneous determination of pharmaceutically active compounds in wastewater samples by solid phase extraction and high-performance liquid chromatography with diode array and fluorescence detectors. *Anal. Chim. Acta* **2005**, *550*, 116–122. [CrossRef]
2. Zhang, H.; Cao, B.; Liu, W.; Lin, K.; Feng, J. Oxidative removal of acetaminophen using zero-valent aluminum-acid system: Efficacy, influencing factors, and reaction mechanism. *J. Environ. Sci.* **2012**, *24*, 314–319. [CrossRef]
3. Do, Q.C.; Kim, D.G.; Ko, S.O. Nonsacrificial Template Synthesis of Magnetic-Based Yolk-Shell Nanostructures for the Removal of Acetaminophen in Fenton-like Systems. *ACS Appl. Mater. Interfaces* **2017**, *9*, 28508–28518. [CrossRef]

4. Lu, H.; Zhu, Z.; Zhang, H.; Zhu, J.; Qiu, Y.; Zhu, L.; Kuppers, S. Fenton-Like Catalysis and Oxidation/Adsorption Performances of Acetaminophen and Arsenic Pollutants in Water on a Multimetal Cu-Zn-Fe-LDH. *ACS Appl. Mater. Interfaces* **2016**, *8*, 25343–25352. [CrossRef]
5. Basha, S.; Keane, D.; Nolan, K.; Oelgemoller, M.; Lawler, J.; Tobin, J.M.; Morrissey, A. UV-induced photocatalytic degradation of aqueous acetaminophen: The role of adsorption and reaction kinetics. *Environ. Sci. Pollut. Res. Int.* **2015**, *22*, 219–230. [CrossRef]
6. Jallouli, N.; Elghniji, K.; Trabelsi, H.; Ksibi, M. Photocatalytic degradation of paracetamol on TiO_2 nanoparticles and TiO_2/cellulosic fiber under UV and sunlight irradiation. *Arab. J. Chem.* **2017**, *10*, S3640–S3645. [CrossRef]
7. Ding, Y.; Fu, L.; Peng, X.; Lei, M.; Wang, C.; Jiang, J. Copper catalysts for radical and nonradical persulfate based advanced oxidation processes: Certainties and uncertainties. *Chem. Eng. J.* **2022**, *427*, 131776. [CrossRef]
8. Wen, Q.; Wang, Y.; Zeng, Z.; Qi, F.; Gao, P.; Huang, Z. Covalent organic frameworks-derived hierarchically porous N-doped carbon for 2,4-dichlorophenol degradation by activated persulfate: The dual role of graphitic N. *J. Hazard. Mater.* **2022**, *426*, 128065. [CrossRef]
9. Tian, K.; Shi, F.; Cao, M.; Zheng, Q.; Zhang, G. A Review of Persulfate Activation by Magnetic Catalysts to Degrade Organic Contaminants: Mechanisms and Applications. *Catalysts* **2022**, *12*, 1058. [CrossRef]
10. Gao, Y.; Wang, Q.; Ji, G.; Li, A. Degradation of antibiotic pollutants by persulfate activated with various carbon materials. *Chem. Eng. J.* **2022**, *429*, 132387. [CrossRef]
11. Xie, X.; Liu, Y.; Li, Y.; Tao, J.; Liu, C.; Feng, J.; Feng, L.; Shan, Y.; Yang, S.; Xu, K. Nitrogen-doped Fe-MOFs derived carbon as PMS activator for efficient degradation of tetracycline. *J. Taiwan Inst. Chem. Eng.* **2023**, *146*, 104891. [CrossRef]
12. Wu, L.; Wu, T.; Liu, Z.; Tang, W.; Xiao, S.; Shao, B.; Liang, Q.; He, Q.; Pan, Y.; Zhao, C.; et al. Carbon nanotube-based materials for persulfate activation to degrade organic contaminants: Properties, mechanisms and modification insights. *J. Hazard. Mater.* **2022**, *431*, 128536. [CrossRef]
13. Luo, J.; Yi, Y.; Fang, Z. Nitrogen-rich magnetic biochar prepared by urea was used as an efficient catalyst to activate persulfate to degrade organic pollutants. *Chemosphere* **2023**, *339*, 139614. [CrossRef]
14. Qi, F.; Zeng, Z.; Wen, Q.; Huang, Z.; Wang, Y.; Xu, Y. Asymmetric enhancement of persulfate activation by N-doped carbon microelectrode: Electro-adsorption and activation pathway regulation. *Sep. Purif. Technol.* **2022**, *301*, 121916. [CrossRef]
15. Xu, J.; Zhou, P.; Shi, P.; Min, Y.; Xu, Q. Insights into the multiple mechanisms of chlorophenols oxidation via activating peroxymonosulfate by 3D N-doped porous carbon. *J. Environ. Chem. Eng.* **2021**, *9*, 106545. [CrossRef]
16. Manz, K.E.; Kulaots, I.; Greenley, C.A.; Landry, P.J.; Lakshmi, K.V.; Woodcock, M.J.; Hellerich, L.; Bryant, J.D.; Apfelbaum, M.; Pennell, K.D. Low-temperature persulfate activation by powdered activated carbon for simultaneous destruction of perfluorinated carboxylic acids and 1,4-dioxane. *J. Hazard. Mater.* **2023**, *442*, 129966. [CrossRef]
17. Yu, W.; Zhang, H.; Wang, X.; Rahman, Z.U.; Shi, Z.; Bai, Y.; Wang, G.; Chen, Y.; Wang, J.; Liu, L. Enrichment of residual carbon from coal gasification fine slag by spiral separator. *J. Environ. Manag.* **2022**, *315*, 115149. [CrossRef]
18. Liu, X.; Jin, Z.; Jing, Y.; Fan, P.; Qi, Z.; Bao, W.; Wang, J.; Yan, X.; Lv, P.; Dong, L. Review of the characteristics and graded utilization of coal gasification slag. *Chin. J. Chem. Eng.* **2021**, *35*, 92–106. [CrossRef]
19. Ren, L.; Ding, L.; Guo, Q.; Gong, Y.; Yu, G.; Wang, F. Characterization, carbon-ash separation, and resource utilization of coal gasification fine slag: A comprehensive review. *J. Clean. Prod.* **2023**, *398*, 136554. [CrossRef]
20. Yuan, N.; Zhao, A.; Hu, Z.; Tan, K.; Zhang, J. Preparation and application of porous materials from coal gasification slag for wastewater treatment: A review. *Chemosphere* **2022**, *287 Pt 2*, 132227. [CrossRef]
21. Li, X.; Liu, X.; Huang, X.; Lin, C.; He, M.; Ouyang, W. Activation of peroxymonosulfate by WTRs-based iron-carbon composites for atrazine removal: Performance evaluation, mechanism insight, and byproduct analysis. *Chem. Eng. J.* **2021**, *421*, 127811. [CrossRef]
22. Long, Y.; Yang, P.; Wang, C.; Wu, W.; Chen, X.; Liu, W.; Cao, Z.; Zhan, X.; Liu, D.; Huang, W. Peroxymonosulfate activation by iron-carbon composite derived from coal gasification slag for sulfamethoxazole removal: Performance evaluation and mechanism insight. *Chem. Eng. J.* **2023**, *456*, 140996. [CrossRef]
23. Liu, B.; Lv, P.; Wu, R.; Bai, Y.; Wang, J.; Su, W.; Song, X.; Yu, G. Coal gasification fine slag based multifunctional nanoporous silica microspheres for synergistic adsorption of Pb(II) and Congo red. *Sep. Purif. Technol.* **2023**, *323*, 124478. [CrossRef]
24. Palaniselvam, T.; Aiyappa, H.B.; Kurungot, S. An efficient oxygen reduction electrocatalyst from graphene by simultaneously generating pores and nitrogen-doped active sites. *J. Mater. Chem.* **2012**, *22*, 23799–23805. [CrossRef]
25. Dresselhaus, M.S.; Jorio, A.; Hofmann, M.; Dresselhaus, G.; Saito, R. Perspectives on carbon nanotubes and graphene Raman spectroscopy. *Nano Lett.* **2010**, *10*, 751–758. [CrossRef]
26. Demiral, İ.; Samdan, C.; Demiral, H. Enrichment of the surface functional groups of activated carbon by modification method. *Surf. Interfaces* **2021**, *22*, 100873. [CrossRef]
27. He, H.; Jiang, B.; Yuan, J.; Liu, Y.; Bi, X.; Xin, S. Cost-effective electro generation of H_2O_2 utilizing HNO_3 modified graphite/polytetrafluoroethylene cathode with exterior hydrophobic film. *J. Colloid Interface Sci.* **2019**, *533*, 471–480. [CrossRef]
28. Wang, X.; Zhu, W.; Wang, N.; Zuo, X. Nitrogen-doped oxidized activated carbon as efficient metal-free bifunctional material for bisphenol removal: A tuning of surface chemistry. *J. Environ. Chem. Eng.* **2022**, *10*, 108509. [CrossRef]
29. Forouzesh, M.; Ebadi, A.; Aghaeinejad-Meybodi, A.; Khoshbouy, R. Transformation of persulfate to free sulfate radical over granular activated carbon: Effect of acidic oxygen functional groups. *Chem. Eng. J.* **2019**, *374*, 965–974. [CrossRef]

30. Yang, W.; Jiang, Z.; Hu, X.; Li, X.; Wang, H.; Xiao, R. Enhanced activation of persulfate by nitric acid/annealing modified multi-walled carbon nanotubes via non-radical process. *Chemosphere* **2019**, *220*, 514–522. [CrossRef]
31. Duan, X.; Ao, Z.; Sun, H.; Indrawirawan, S.; Wang, Y.; Kang, J.; Liang, F.; Zhu, Z.H.; Wang, S. Nitrogen-doped graphene for generation and evolution of reactive radicals by metal-free catalysis. *ACS Appl. Mater. Interfaces* **2015**, *7*, 4169–4178. [CrossRef]
32. Gao, Y.; Zhu, Y.; Lyu, L.; Zeng, Q.; Xing, X.; Hu, C. Electronic Structure Modulation of Graphitic Carbon Nitride by Oxygen Doping for Enhanced Catalytic Degradation of Organic Pollutants through Peroxymonosulfate Activation. *Environ. Sci. Technol.* **2018**, *52*, 14371–14380. [CrossRef]
33. Li, H.; Li, N.; Zuo, P.; Qu, S.; Shen, W. Efficient adsorption-reduction synergistic effects of sulfur, nitrogen and oxygen heteroatom co-doped porous carbon spheres for chromium(VI) removal. *Colloids Surf. A Physicochem. Eng. Asp.* **2021**, *618*, 126502. [CrossRef]
34. Meng, Y.; Li, Z.; Tan, J.; Li, J.; Wu, J.; Zhang, T.; Wang, X. Oxygen-doped porous graphitic carbon nitride in photocatalytic peroxymonosulfate activation for enhanced carbamazepine removal: Performance, influence factors and mechanisms. *Chem. Eng. J.* **2022**, *429*, 130860. [CrossRef]
35. Pham, X.N.; Nguyen, H.T.; Pham, T.N.; Nguyen, T.T.-B.; Nguyen, M.B.; Tran, V.T.-T.; Doan, H.V. Green synthesis of H-ZSM-5 zeolite-anchored O-doped g–C_3N_4 for photodegradation of Reactive Red 195 (RR 195) under solar light. *J. Taiwan Inst. Chem. Eng.* **2020**, *114*, 91–102. [CrossRef]
36. Alvandi, M.; Nourmoradi, H.; Nikoonahad, A.; Aghayani, E.; Abbas Mirzaee, S. LED visible light assisted photo-oxidation of acetaminophen using *a* one-step synthesis of Cu, Fe@g-C_3N_4 nanosheet—Activated persulfate system in aqueous solutions. *Arab. J. Chem.* **2023**, *16*, 105251. [CrossRef]
37. An, S.; Zhang, G.; Wang, T.; Zhang, W.; Li, K.; Song, C.; Miller, J.T.; Miao, S.; Wang, J.; Guo, X. High-Density Ultra-small Clusters and Single-Atom Fe Sites Embedded in Graphitic Carbon Nitride (g-C_3N_4) for Highly Efficient Catalytic Advanced Oxidation Processes. *ACS Nano* **2018**, *12*, 9441–9450. [CrossRef]
38. Li, W.; Nie, C.; Wang, X.; Ye, H.; Li, D.; Ao, Z. Alkaline lignin-derived N-doped biochars as peroxymonosulfate activators for acetaminophen degradation: Performance and catalytic bridging mediated Electron-Transfer mechanism. *Sep. Purif. Technol.* **2023**, *323*, 124418. [CrossRef]
39. Tan, C.; Gao, N.; Deng, Y.; Deng, J.; Zhou, S.; Li, J.; Xin, X. Radical induced degradation of acetaminophen with Fe_3O_4 magnetic nanoparticles as *a* heterogeneous activator of peroxymonosulfate. *J. Hazard. Mater.* **2014**, *276*, 452–460. [CrossRef]
40. Zhang, Y.; Zhang, Q.; Dong, Z.; Wu, L.; Hong, J. Structurally modified $CuFe_2O_4$/persulfate process for acetaminophen scavenging: High efficiency with low catalyst addition. *J. Chem. Technol. Biotechnol.* **2018**, *94*, 785–794. [CrossRef]
41. Zhu, Z.; Zhang, Q.; Xu, M.; Xue, Y.; Zhang, T.; Hong, J. Highly active heterogeneous FeCo metallic oxides for peroxymonosulfate activation: The mechanism of oxygen vacancy enhancement. *J. Environ. Chem. Eng.* **2023**, *11*, 109071. [CrossRef]
42. Morales-Garcia, A.; Rhatigan, S.; Nolan, M.; Illas, F. On the use of DFT+U to describe the electronic structure of TiO_2 nanoparticles: $(TiO_2)35$ as a case study. *J. Chem. Phys.* **2020**, *152*, 244107. [CrossRef]
43. Samat, M.H.; Ali AM, M.; Taib MF, M.; Hassan, O.H.; Yahya, M.Z.A. Structural and electronic properties of TiO_2 polymorphs with effective on-site coulomb repulsion term: DFT+U approaches. *Mater. Today Proc.* **2019**, *17*, 472–483. [CrossRef]
44. Qi, F.; Wang, Q.; Zeng, Z.; Wen, Q.; Huang, Z. Insight into the roles of microenvironment and active site on the mechanism regulation in metal-free persulfate activation process coupling with an electric field. *J. Hazard. Mater.* **2022**, *439*, 129673. [CrossRef]

catalysts

MDPI

Article

Assessing the Efficacy of A Mo_2C/Peroxydisulfate System for Tertiary Wastewater Treatment: A Study of Losartan Degradation, *E. coli* Inactivation, and Synergistic Effects

Alexandra A. Ioannidi [1], Maria Vlachodimitropoulou [1], Zacharias Frontistis [2], Athanasia Petala [3], Eleni Koutra [1], Michael Kornaros [1] and Dionissios Mantzavinos [1,*]

[1] Department of Chemical Engineering, University of Patras, GR-26504 Patras, Greece; alex.ioannidi@chemeng.upatras.gr (A.A.I.); m.vlachodimitropoulou@gmail.com (M.V.); ekoutra@chemeng.upatras.gr (E.K.); kornaros@chemeng.upatras.gr (M.K.)

[2] Department of Chemical Engineering, University of Western Macedonia, GR-50132 Kozani, Greece; zfronistis@uowm.gr

[3] Department of Environment, Ionian University, GR-29100 Zakynthos, Greece; apetala@ionio.gr

* Correspondence: mantzavinos@chemeng.upatras.gr

Abstract: This work examines the use of pristine Mo_2C as an intriguing sodium persulfate (SPS) activator for the degradation of the drug losartan (LOS). Using 500 mg/L Mo_2C and 250 mg/L SPS, 500 μg/L LOS was degraded in less than 45 min. LOS decomposition was enhanced in acidic pH, while the apparent kinetic constant decreased with higher LOS concentrations. According to experiments conducted in the presence of scavengers of reactive species, sulfate radicals, hydroxyl radicals, and singlet oxygen participated in LOS oxidation, with the latter being the predominant reactive species. The presence of competitors such as bicarbonate and organic matter reduced the observed efficiency in actual matrices, while, interestingly, the addition of chloride accelerated the degradation rate. The catalyst showed remarkable stability, with complete LOS removal being retained after five sequential experiments. The system was examined for simultaneous LOS decomposition and elimination of *Escherichia coli*. The presence of *E. coli* retarded LOS destruction, resulting in only 30% removal after 3 h, while the system was capable of reducing *E. coli* concentration by 1.23 log. However, in the presence of simulated solar irradiation, *E. coli* was reduced by almost 4 log and LOS was completely degraded in 45 min, revealing a significant synergistic effect of the solar/Mo_2C/SPS system.

Keywords: 2D material; peroxydisulfate; pharmaceutical; pathogen; process synergy; solar irradiation

1. Introduction

Following the "technological revolution", also known as the second industrial revolution, aquatic environmental pollution escalated due to numerous compounds, the effects of which on aquatic ecosystems and humans were unknown. Only in the last half-century, prompted by the threat of drinking water shortages, did the World Health Organization, in collaboration with nations worldwide, establish detection limits for many substances and heavy metals in aquatic environments that are harmful to all organisms [1,2]. However, industrial, agricultural, and sewage wastewater treatment plants have proven to be inadequate for certain types of pollutants (e.g., pharmaceuticals, endocrine disruptors, and pesticides). Many researchers have detected these in surface and groundwater at extremely low concentrations (ng/L–μg/L) [3]. These micropollutants have been found to be toxic to many microorganisms and may cause metabolic disorders and various types of cancer [3,4].

Antihypertensive drugs, which are widely prescribed each year, are a significant type of pharmaceutical [5,6]. In the United States alone, the number of adults with hypertension was approximately 37 million in 2017 [7,8]. Overconsumption of antihypertensive medications has resulted in their detection in surface waters, wastewater effluents, and environments like wetlands, rivers, and seawaters, as well as in hospital wastewater. Their concentrations ranged from 0.6 ng/L to 17.7 mg/L [9–16].

Losartan (LOS), an antihypertensive medication, reduces the risk of strokes and heart attacks [17,18]. It has also been proven to be a beneficial agent in preventing kidney damage caused by diabetes [17,18]. However, LOS has been found in effluents at concentrations from 19.7 to 2760 ng/L [17,19], indicating the incomplete degradation of LOS using biological wastewater treatment [20–22]. Also, Cortez et al. [9] noted the occurrence of LOS in a Brazilian coastal region with concentrations reaching up to 32 ng/L. Furthermore, Cortez et al. [9], Ladhari et al. [20], Osorio et al. [21], and Adams et al. [22] revealed that despite LOS's numerous benefits for human health, it can be toxic to various organisms, including humans, fish, crustaceans, algae, *Daphnia magna*, and *Desmodesmus subspicatus*. Its byproducts can be more harmful and persistent than the parent compound (LOS). Thus, there is a pressing need to develop alternative methods to efficiently degrade antihypertensive drugs in aqueous media.

Over the past fifteen years, numerous attempts have been made to replace traditional disinfection processes like ozonation [23–25], chlorination [25,26], and UVC irradiation [25,26] due to the potential carcinogenicity of their byproducts or the increased cost. As a result, many scientists have turned their focus towards advanced oxidation processes (AOPs) to find a technology with high efficiency for degrading micropollutants and eradicating pathogenic microorganisms [27–29].

AOPs include a large set of processes (e.g., photocatalysis, Fenton, electrochemical oxidation, and persulfate activation), all of which share the common characteristic of producing strong reactive species such as hydroxyl radicals ($HO^•$), superoxide radicals, singlet oxygen, and sulfate radicals ($SO_4^{•-}$) [30]. Technologies based on persulfate activation have garnered considerable attention from the research community for treating hazardous micropollutants in aquatic environments. Persulfate can be activated by heat, solar irradiation, ultrasound, alkaline conditions, and transition metals to form $SO_4^{•-}$ with a redox potential (E = 2.5–3.1 V) similar to the redox potential of $HO^•$ (E = 1.8–2.7 V) [31]. In addition, $SO_4^{•-}$ has advantages over hydroxyl radicals in terms of activity over a large pH range, lifespan, and selectivity [32–34]. Furthermore, in the past decade, heterogeneous persulfate activation has gained interest due to the possible reuse of catalytic materials.

Recently, two-dimensional (2D) materials such as graphene [35] and MoS_2 [36] have gained recognition in the field of wastewater treatment. Up until now, 2D materials have primarily found applications in renewable and sustainable energy production [36,37] because of their high electrical conductivity, thermal stability, high hardness, and adsorption capacity. Considering that 2D materials are characterized by high electron mobility, they are potential persulfate activators. Bekris et al. [35] showed that graphene is an effective persulfate (SPS) activator for propyl paraben degradation, while Zhou et al. [36] examined the activation of peroxymonosulfate (PMS) and SPS with MoS_2 for the degradation of carbamazepine. However, an excessive use of MoS_2 after persulfate activation may lead to sulfuric leaching and the formation of H_2S, which is toxic [38].

Molybdenum carbide (Mo_2C) can be considered an alternative choice for persulfate activation since Mo is the primary active site in Mo-based materials, and given that Mo_2C has two Mo atoms, there is a high probability of it being more reactive. Additionally, it does not exhibit the drawbacks associated with the use of MoS_2 [39–41].

The exploration of Mo_2C as a persulfate activator for micropollutant degradation has only recently begun. Yang et al. [38] treated Mo_2C with 5% Cu to enhance its ability to activate PMS for tetracycline degradation, while Chen et al. [31] and Bao et al. [42] synthesized Mo_2Ga_2C and Mo_2C/C respectively, to activate PMS for bisphenol A and carbamazepine removal. However, there are no studies that investigate solely the potential of pristine Mo_2C to activate SPS, a less expensive oxidant than PMS, for micropollutant degradation and pathogen disinfection without increasing catalyst preparation costs through material modification.

To date, the oxidation of LOS has been explored through various methods, including electrochemical oxidation, heat-activated persulfate, acoustic cavitation, UV/H_2O_2, UV/Fe^{2+}/H_2O_2, and photoelectro-Fenton, all of which have shown promising yields, as can be seen in Table 1. However, these processes require energy (and thus a relatively high operating cost), necessitating an alternative approach to degrade pharmaceuticals such as LOS. This alternative could be the use of heterogeneous processes. For instance, Andrade et al. [43] used N-doped porous carbon to activate PMS and achieved complete LOS removal after 240 min of reaction time. However, there is no data on LOS degradation or *E. coli* inactivation using a Mo_2C/SPS process.

Table 1. Degradation of LOS using various AOPs.

Process	Conditions	LOS Removal	Reference
Electrochemical oxidation	$[LOS]_o$ = 377 µM, $[Na_2SO_4]$ = 0.05 M, BDD anode, SS cathode, J = 80 mA/cm^2, pH = 7	100% in 180 min (UPW) -	[44]
Ultrasound	$[LOS]_o$ = 40 µM, P = 88 W/L, pH = 6.5, 20 °C	100% in 120 min (UPW) 100% in 120 min (hospital wastewater)	[16]
UVC/Fe^{2+}/H_2O_2	$[LOS]_o$ = 40 µM, $[Fe^{2+}]$ = 40 µM, $[H_2O_2]$ = 400 µM, pH = 6.1, 23 °C	76% TOC in 120 min (UPW) 30% TOC in 120 min (groundwater)	[17]
N-doped porous carbon and PMS	$[LOS]_o$ = 40 µM, [catalyst] = 26 mg/L, [PMS] = 6.5 mM, pH = 5.2, 25 °C	100% (240 min) 96% in 240 min (20 mg/L humic acid)	[43]
UVC/H_2O_2	$[LOS]_o$ = 43.4 µM, $[H_2O_2]$ = 500 µM pH 6.1	65.7% in 20 min (distilled water) ≈27% in 20 min (fresh urine)	[45]
Photoelectron Fenton	$[LOS]_o$ = 45 µM, $[Na_2SO_4]$ = 0.05 M, $[Fe^{2+}]$ = 36 µM, BDD anode, SS cathode, j = 5 mA/cm^2, UVA = 1.4 W/m^2, pH = 3	≈95% (30 min) -	[46]
Heat/US/SPS	$[LOS]_o$ = 1.18 µM, [SPS] = 200 µM, US = 36 W/L, 20 kHz, 50 °C, pH = 5.25	100% in 15 min (UPW) 60% in 45 min (WW)	[47]
Biochar from spent malt rootlet treated with acid and SPS	$[LOS]_o$ = 0.59 µM, [SPS] = 1.05 mM, [Acid-C] = 90 mg/L pH = 5.6, 25 °C	91% in 90 min (UPW) 20% in 90 min (WW)	[48]
Mo_2C and SPS	$[LOS]_o$ = 1.18 µM, [SPS] = 1.05 mM, $[Mo_2C]$ = 500 mg/L pH = 5.5, 25 °C	100% in 45 min (UPW) 34% in 45 min (WW)	This study

Furthermore, a promising strategy to enhance treatment performance is the concurrent use of multiple processes. Although some studies have investigated the use of treated molybdenum carbide to activate PMS, a combination of these systems with renewable energy provided by solar irradiation is an intriguing strategy towards enhanced efficiency. At the same time, most published studies are limited to the examination of the efficiency over a probe compound, while the use of the proposed hybrid system as a tertiary treatment (i.e., including disinfection) has not been explored.

In summary, to the best of the authors' knowledge, there is no data on losartan degradation via the Mo_2C/SPS process or the synergistic effect of the Mo_2C/SPS process combined with solar irradiation for the simultaneous elimination of losartan and inactivation of *E. coli*. Therefore, this study aims at (i) investigating the effects of different oxidants, catalysts, SPS and LOS concentrations, initial solution pH, synthetic and real water matrices on LOS oxidation and the degradation reaction mechanism using various scavengers, as well as Mo_2C reusability; (ii) testing the ability of the Mo_2C/SPS process to eliminate pathogenic microorganisms; and (iii) examining the synergistic effects of a hybrid process for simultaneous losartan elimination and *E. coli* inactivation by coupling Mo_2C/SPS with simulated solar irradiation.

2. Results and Discussion

2.1. Mo_2C Characterization

Figure 1 shows the X-ray diffraction pattern of the Mo_2C used in the present study. The typical crystal planes (100), (002), (101), (102), (110), (103), (112), and (201) of Mo_2C (JCPDS No 01-1188) are clearly discernible [49–51]. In addition, its primary crystallite size, as estimated using the Debye–Scherrer formula, was found to be equal to 35 nm.

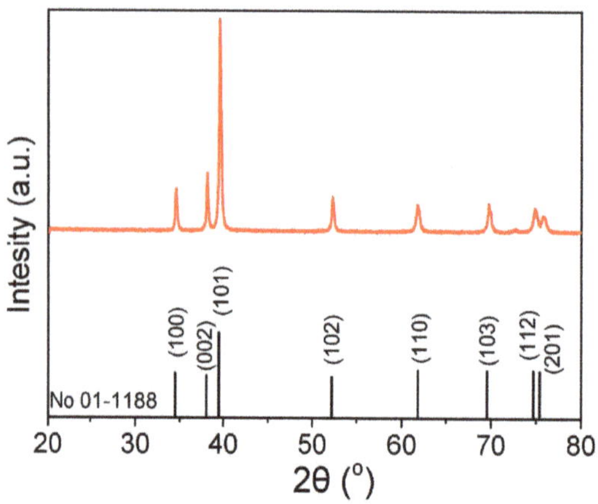

Figure 1. XRD pattern of Mo_2C.

The zeta potential of Mo_2C was measured in ultrapure water at various pH values to determine its isoelectric point. The calculated pH_{zpc} (pH at zero-point charge) was found to be equal to 3.7, which is consistent with previous studies [42].

The morphology of Mo_2C was studied by means of SEM/EDS, and characteristic images are shown in Figure 2. It was observed that Mo_2C consisted of almost spherical particles agglomerated with each other homogeneously with an estimated average diameter of approximately 1.41 μm (Figure 2D). Moreover, it could be stated that particle size distribution (Figure 2D) was rather broad, including particles from 0.4 to ca. 2.8 μm [52]. In addition, EDS spectra confirmed the presence of Mo, O and C alone without impurities.

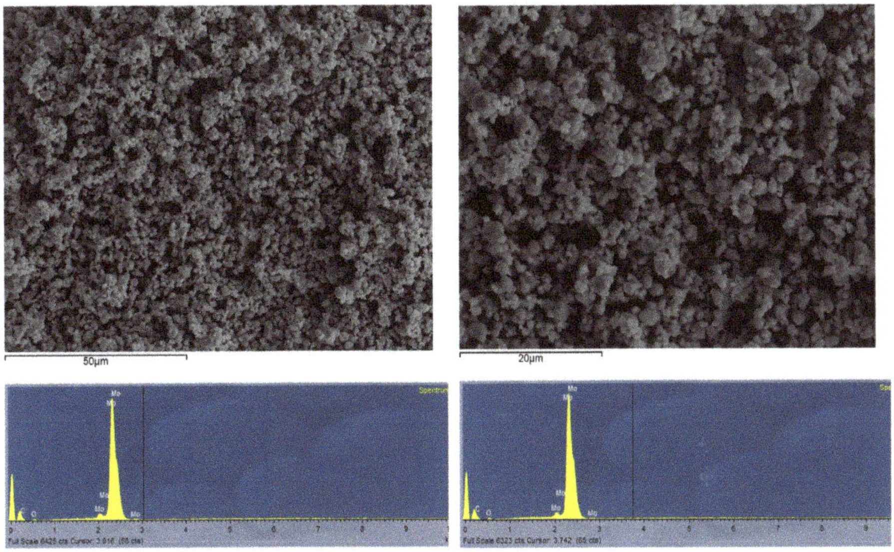

(A) (B)

Element	C	Mo
Atomic (%)	27.77	72.23

(C)

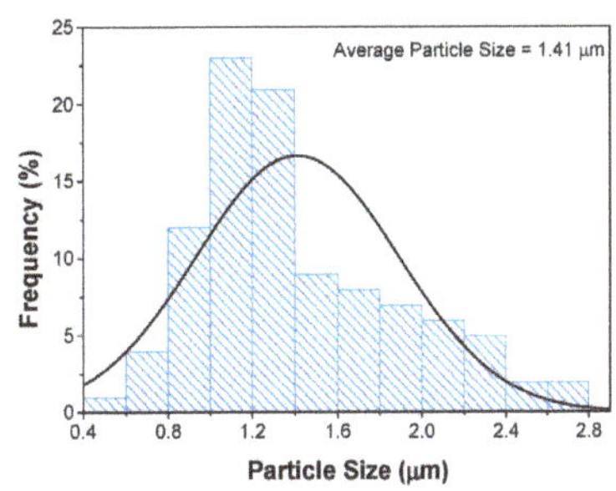

(D)

Figure 2. (**A**,**B**) Characteristic SEM images of Mo_2C at different magnifications with the corresponding EDS spectra. (**C**) Atomic percentage of Mo_2C elements. (**D**) Particle size distribution of Mo_2C.

2.2. Catalytic Results

2.2.1. Activation of Various Types of Oxidants

The decomposition of 500 μg/L LOS using 500 mg/L Mo_2C to activate sodium metaperiodate ($NaIO_4$), sodium percarbonate ($Na_2H_3CO_6$), H_2O_2, and SPS was examined, and the results are displayed in Figure 3. The concentration of oxidants used was 1.05 mM. A complete degradation of LOS was achieved using Mo_2C-SPS after 45 min of reaction, whereas in the cases of Mo_2C-$NaIO_4$, Mo_2C-H_2O_2, and Mo_2C-$Na_2H_3CO_6$, only 58%, 39%, and 17% of LOS removal were achieved after 45 min, respectively. These results clearly demonstrate the superiority of SPS over $NaIO_4$, H_2O_2 and $Na_2H_3CO_6$ when Mo_2C is used as the activator since the decomposition rate, as demonstrated by the computed k_{app} values, is 1–2 orders of magnitude greater. Similar results were reported by Bao et al. [42] and Yang et al. [38]. Bao et al. [42] investigated the degradation of 5 mg/L carbamazepine with 300 mg/L Mo_2C/C and 0.75 mM PMS. They found that the Mo_2C/C-PMS process could degrade carbamazepine after 75 min of reaction. Meanwhile, Yang et al. [38] studied the degradation of 40 mg/L tetracycline with 300 mg/L 5% Cu-Mo_2C and 300 mg/L PMS. They reported a complete degradation of tetracycline after 20 min of reaction.

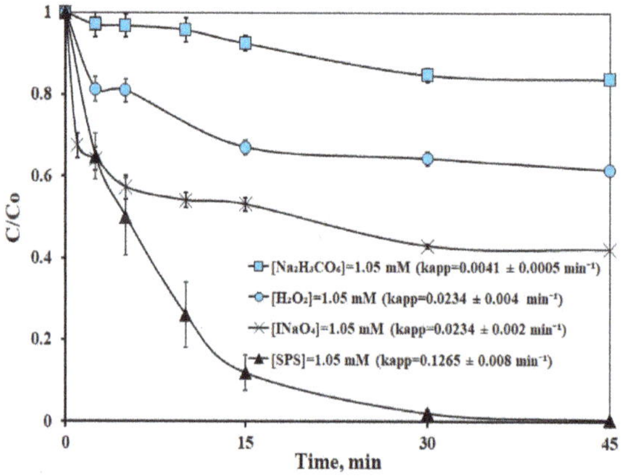

Figure 3. Performance of Mo_2C in activating various oxidants for 500 µg/L LOS degradation in UPW and inherent pH ≈ 5.5. Experimental conditions: [Mo_2C] = 500 mg/L, [$NaIO_4$] = [$Na_2H_3CO_6$] = [H_2O_2] = [SPS] = 1.05 mM.

Thus, both pure Mo_2C and modified Mo_2C-based catalysts are capable of activating SPS and PMS, respectively. However, it should be noted that none of the aforementioned studies thoroughly examined the ability of pure, unmodified Mo_2C to activate SPS, which is a less expensive oxidant than PMS.

Interestingly, the Mo_2C/persulfate system has the capability to generate different reactive species, including sulfate and hydroxyl radicals and singlet oxygen. More researchers are attributing this high efficiency to the redox potential of the formed radicals [53], as well as to higher electrostatic adsorption and collision efficiency. Although the activation of hydrogen peroxide can generate hydroxyl radicals with high redox potential, they are also characterized by no selectivity and a very short lifetime. Conversely, the percarbonate system also produces carbonate radicals with higher selectivity but reduced redox potential [53]. Meanwhile, $NaIO_4$ has a redox potential of 1.60 V. However, oxidation reactions initiated by this ion are known to be selective and significantly slower than those involving hydroxyl and sulfate radicals [54]. Therefore, in this study, Mo_2C was chosen as the activator and SPS as the oxidant for the degradation of LOS for further study.

2.2.2. Effect of Operating Parameters (Initial Concentration of Catalyst, Persulfate, Losartan, and pH)

It is well known that parameters such as the pH of the solution and the concentrations of catalysts, oxidants, and micropollutants can all affect catalytic performance.

Figure 4A shows that by increasing Mo_2C concentration, both the degradation rate and the adsorption of 500 µg/L LOS increased. Specifically, 41%, 63%, 88%, and 93% of LOS removal were achieved with 125, 250, 500, and 750 mg/L Mo_2C, respectively, after 15 min of reaction, while LOS adsorption rose from 4% for 125 mg/L Mo_2C to 30% for 750 mg/L Mo_2C during the same reaction time. (The respective k_{app} values are shown in Figure S1.) Increasing the amount of Mo_2C meant that more active sites were available for LOS adsorption and SPS activation, leading to faster degradation rates. However, as seen in Figure 4A, LOS degradation with 750 mg/L Mo_2C was not significantly faster than that for 500 mg/L Mo_2C; thus, the chosen catalyst concentration was 500 mg/L. Moreover, an additional experiment was carried out in the absence of Mo_2C to investigate the potential of SPS to oxidize LOS as a mild oxidant; only 20% LOS removal was achieved after 45 min of reaction.

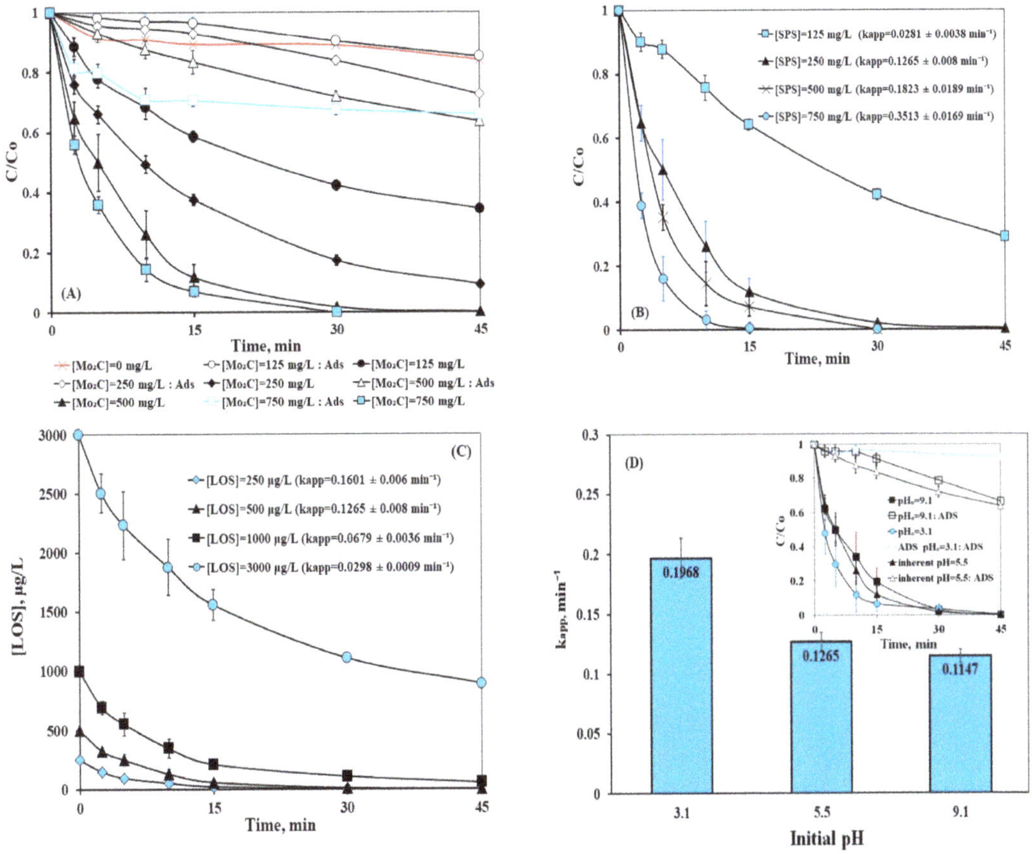

Figure 4. Effect of (**A**) initial concentration of catalyst on 500 μg/L LOS degradation with 250 mg/L SPS, (**B**) initial concentration of persulfate on 500 μg/L LOS degradation with 500 mg/L Mo_2C, (**C**) initial concentration of LOS on its degradation with 500 mg/L Mo_2C and 250 mg/L SPS, and (**D**) initial pH on 500 μg/L LOS degradation with 500 mg/L Mo_2C and 250 mg/L SPS in UPW.

Consequently, several SPS concentrations (125, 250, 500, 750 mg/L) were tested; the results are displayed in Figure 4B, showing that the yield increased as SPS concentration increased. k_{app} values increased 4.5-fold when SPS concentration rose from 125 to 250 mg/L. Further increases in SPS concentration led only to 1.44-fold and 2.78-fold higher k_{app} for 500 mg/L, and 750 mg/L SPS, respectively, using the k_{app} value for 250 mg/L SPS as a reference. Taking into consideration the environmental impact that the generated sulfate anions would have on aquatic systems, as well as the yield of the present process, 250 mg/L SPS was selected for further experiments [55].

Figure 4C presents concentration–time profiles for different initial LOS concentrations, as well as their respective k_{app} values. Although the rate constant decreased as the initial LOS concentration increased (i.e., over five-fold, from 250 to 3000 μg/L LOS), the Mo_2C/SPS process was capable of degrading efficiently relatively high concentrations of LOS; such concentrations are likely to occur in hospital wastewaters but not in secondary treated effluents or surface waters [56]. After 15 min of reaction, the remaining quantity of LOS was 1559, 211, 59, and 23 μg/L for 3000, 1000, 500, and 250 μg/L initial LOS concentration, respectively. It is worth noting that, despite the 48% degradation observed after 45 min of oxidation for 3000 μg/L of LOS, the TOC removal after 45 min was only

14%. This indicates the production of transformation products, as already demonstrated in other advanced oxidation processes [47].

Solution pH may be an important operating parameter since real water matrices have different pH values than that of UPW, which is ≈5.5. Therefore, adsorption and oxidation experiments were conducted at pH = 3.1 and pH = 9.1. As seen in Figure 4D, LOS degradation was favored under acidic conditions (94% LOS removal in 15 min) and slightly inhibited as the pH value increased (88% and 80% LOS removal after 15 min at pH 5.5 and 9.1, respectively). The k_{app} value decreased only 1.72 times as pH increased from 3.1 to 9.1. In contrast to oxidation, LOS adsorption was favored at an inherent pH ≈ 5.5; there was a slight reduction in LOS adsorption at pH = 9.1, while at pH = 3.1 it was almost completely inhibited. The zero-point charge of Mo_2C is 3.7, while the pKa value of LOS is 5.05 [47]. Thus, at pH = 3, Mo_2C is positively charged, and LOS exists in neutral form, while at inherent pH and pH = 9, both Mo_2C and LOS are negatively charged. As a result, the adsorption mechanism cannot be claimed to be due to repulsive/electrostatic forces but to π–π interactions [57–59], which are not favored below pH = 3.

Therefore, the high degradation efficiency at pH = 3 is likely attributed to the higher adsorption of $S_2O_8^{2-}$ on the positively charged surface of Mo_2C, resulting in the formation of more sulfate radicals.

2.2.3. Effect of Scavengers

Molybdenum is renowned for its various oxidation states (II, IV, VI), indicating a high electron mobility and leading to the generation of reactive species with high redox potential, such as $SO_4^{\bullet-}$, $HO^{\bullet}$, and $^1O_2$.

To evaluate the contribution of singlet oxygen, sulfate, and hydroxyl radicals, 9.24 mM of NaN_3, MeOH, and t-BuOH were used as scavengers, respectively. As depicted in Figure 5, the introduction of each scavenger reduced process efficiency, with NaN_3 resulting in the most severe inhibition, wherein the k_{app} value decreased almost 22 times. On the other hand, the k_{app} value decreased almost 2.5 and 5 times after the addition of t-BuOH and MeOH, respectively. These results suggest that all examined reactive species contribute to the oxidation of LOS via the multivalent characteristics of Mo, with singlet oxygen being the predominant reactive species. A plausible mechanism of SPS activation by Mo_2C can be described by Equations (1)–(10).

$$Mo^{2+} + 2S_2O_8^{2-} \rightarrow Mo^{4+} + 2SO_4^{\bullet-} + 2SO_4^{2-} \tag{1}$$

$$Mo^{4+} + 2S_2O_8^{2-} \rightarrow Mo^{6+} + 2SO_4^{\bullet-} + 2SO_4^{2-} \tag{2}$$

$$Mo^{4+} + 2S_2O_8^{2-} \rightarrow Mo^{2+} + 2S_2O_8^{\bullet-} \tag{3}$$

$$Mo^{6+} + 2S_2O_8^{2-} \rightarrow Mo^{4+} + 2S_2O_8^{\bullet-} \tag{4}$$

$$S_2O_8^{2-} + 2H_2O \rightarrow HO_2^- + 2SO_4^{2-} + 3H^+ (\text{SPS hydrolysis catalyzed by } M_2C) \tag{5}$$

$$HO_2^- + S_2O_8^{2-} \rightarrow SO_4^{\bullet-} + SO_4^{2-} + H^+ + O_2^{\bullet-} \tag{6}$$

$$2SO_4^{\bullet-} + H_2O \rightarrow H^+ + SO_4^{2-} + HO^{\bullet} \tag{7}$$

$$HO^{\bullet} + O_2^{\bullet-} \rightarrow 1O_2 + HO^- \tag{8}$$

$$2H^+ + 2O_2^{\bullet-} \rightarrow H_2O_2 + 1O_2 \tag{9}$$

$$HO^{\bullet}/SO_4^{\bullet-}/1O_2 + LOS \rightarrow by - product......... \rightarrow CO_2 + H_2O \quad (10)$$

The contributions of $HO^{\bullet}$, $SO_4^{\bullet-}$, and $1O_2$ were also reported by Bao et al. [42], who studied the degradation of carbamazepine using Mo_2C/C and PMS, and Yang et al. [38], who investigated the degradation of tetracycline using 5% Cu/Mo_2C and PMS. However, in their studies, sulfate radicals were reported to be the dominant reactive species, unlike in the present work.

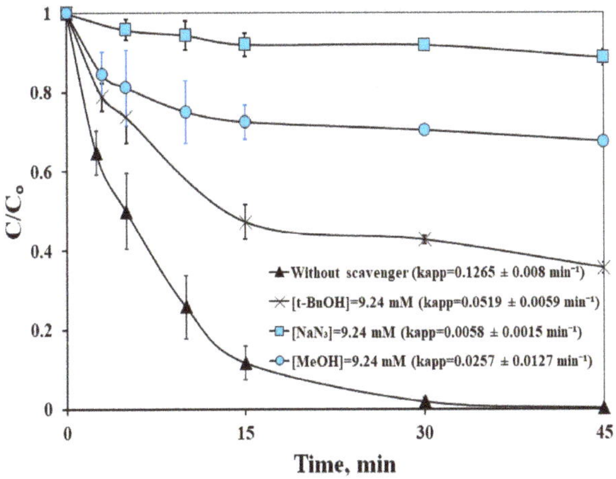

Figure 5. Effect of reactive species scavengers on 500 μg/L LOS degradation with 500 mg/L Mo_2C and 250 mg/L SPS in UPW.

2.2.4. Effect of Water Matrix

Figure 6A illustrates how LOS degradation was affected by the addition of 10 mg/L HA or 250 mg/L NaCl, $NaNO_3$, and $NaHCO_3$, respectively. Considering the findings from Section 2.2.2 (i.e., pH effect) and 2.2.3, the negative impact of HCO_3^-, HA and, to a lesser extent, NO_3^- was probably not related to pH alteration since alkaline conditions did not delay LOS degradation. Consequently, it may have been be due to LOS adsorption being hindered by the presence of NO_3^-, HCO_3^-, and HA, and the competition between LOS and inorganic and organic matter for the generated reactive species ($SO_4^{\bullet-}$, $HO^{\bullet}$, $^1O_2$, $O_2^{\bullet-}$) [60]. NO_3^- as well as HCO_3^- reacted with the formed reactive species, and radicals with smaller redox potential were produced, as described by Equations (11)–(15) [38,42,44,60].

$$SO_4^{\bullet-} + NO_3^- \rightarrow SO_4^{2-} + NO_3^{\bullet} \quad (11)$$

$$HO^{\bullet} + NO_3^- \rightarrow HO^- + NO_3^{\bullet} \quad (12)$$

$$SO_4^{\bullet-} + HCO_3^- \rightarrow SO_4^{2-} + HCO_3^{\bullet} \quad (13)$$

$$HO^{\bullet} + HCO_3^- \rightarrow H_2O + CO_3^{\bullet-} \quad (14)$$

$$HCO_3^- + O_2^{\bullet-} \rightarrow CO_3^{2-} + HO_2^{\bullet} \quad (15)$$

Interestingly, the addition of NaCl facilitated the degradation of LOS, leading to complete LOS removal after 150 min (Figure 6A); hence, the effect of NaCl concentration was studied further. As depicted in Figure 6B, increasing NaCl concentration from 0 to 500 mg/L led to a 5-fold degradation rate increase. Chloride anions may have reacted with hydroxyl and sulfate radicals to form active chlorine or hypochlorous reactive species, as shown in Equations (16)–(20):

$$SO_4^{\bullet -} + Cl^- \rightarrow SO_4^{2-} + Cl^\bullet \quad (16)$$

$$Cl^- + HO^\bullet \rightarrow HOCl^{\bullet -} \quad (17)$$

$$Cl^- + HO^\bullet \rightarrow HO^- + Cl^\bullet \quad (18)$$

$$Cl^- + Cl^\bullet \rightarrow Cl_2^{\bullet -} \quad (19)$$

$$Cl_2^{\bullet -}/Cl^\bullet/HOCl^{\bullet -} + LOS \rightarrow by - products \quad (20)$$

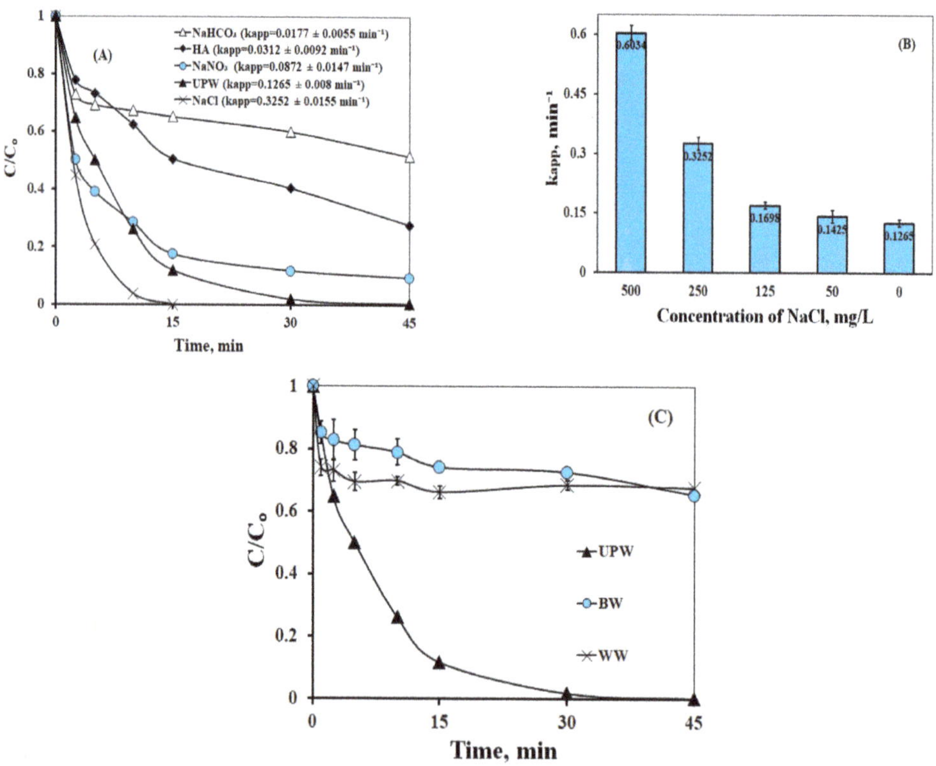

Figure 6. (**A**) Effect of 10 mg/L HA and 250 mg/L NaCl, $NaHCO_3$, and $NaNO_3$, respectively, on LOS degradation. (**B**) Effect of NaCl on LOS degradation. (**C**) Effect of real water matrices (BW, WW) on LOS degradation. Experimental conditions: [LOS] = 500 μg/L, [Mo_2C] = 500 mg/L, and [SPS] = 250 mg/L.

These results seem contradictory when compared to other studies, which documented that the addition of Cl^- had either no effect on their processes [38,60] or a detrimental effect on the yield of other systems [44,61]. However, these results align with the works of Bao et al. [42] and Chen et al. [31], who studied the degradation of carbamazepine and bisphenol A using Mo_2C/C and Mo_2Ga_2C for PMS activation, respectively. Both studies reported an enhancement in carbamazepine and bisphenol A degradation in the presence of Cl^- and attributed it to the direct reaction of chlorine or hypochlorous reactive species with the unsaturated bonds of micropollutants.

In addition, real water matrices were used (BW and WW) and the results are shown in Figure 6C. Only 25% and 34% of LOS removal were achieved after 15 min in BW and WW, respectively, while in UPW, more than 90% of LOS removal was achieved. This decrease in performance in real matrices was attributed to the presence of inorganic and organic matter, with HCO_3^- having the highest concentration among them (Table S1). However, it should be noted that the aim of the present work was not to optimize the system, but to study how the system was affected by various operating parameters. Therefore, the optimization of the Mo_2C/SPS system should be undertaken under representative conditions if the proposed process is to be scaled up from the lab to a pilot unit.

2.2.5. Reuse

Eventually, the Mo_2C/SPS system was evaluated in terms of catalyst reusability according to the following procedure: The mixture was allowed to react for 45 min. After this period, LOS conversion was measured, the mixture was centrifuged, the Mo_2C sample was collected, dried for 12 h, weighed, and then used for the degradation of another LOS solution and repetition of the experiment. This cycle was run five times (including the fresh sample) and, in all cases, a complete LOS conversion was achieved, indicating a high stability of Mo_2C, as seen in Figure S3. Moreover, the stability of the reused Mo_2C was corroborated by XRD analysis, as shown in Figure S2. It was observed that Mo_2C remained practically intact after exposure to reaction conditions for 225 min, with the primary crystallite size remaining unchanged and equal to 35 nm.

2.3. Disinfection and Synergistic Effects with Simulated Solar Irradiation

In a final set of experiments, the Mo_2C/SPS process was assessed as a disinfectant agent. Adsorption, blank, and degradation experiments were performed in the presence of 500 μg/L LOS and $\approx 10^6$–10^7 CFU/mL *E. coli*. Figure 7A illustrates the degradation of LOS, while Figure 7B shows the reduction in *E. coli* population. The presence *of E. coli* significantly impacted the degradation and adsorption of LOS. Only 17% LOS removal and 10% LOS adsorption were observed after 60 min, while the *E. coli* population declined by 1.23 ± 0.11 log and 0.9 ± 0.12 log at the same time during the oxidation and adsorption experiment, respectively. After 60 min, no further *E. coli* inactivation and adsorption were observed, whereas in the case of LOS, slow oxidation was noted, with 30% LOS removal after 180 min of reaction. These results suggest that both LOS and *E. coli* competed for the active centers on the Mo_2C surface and for the generated reactive species.

To accelerate the degradation of LOS and *E. coli*, a coupling of the Mo_2C/SPS process with simulated solar irradiation was conducted. The choice of simulated solar radiation was made on the basis that in real-world conditions, solar irradiation is a cost-free, green energy source that is abundant on Earth.

As observed in Figure 7, the solar/SPS process exhibited a high efficiency for LOS degradation, while as for *E. coli* inactivation, the yield of the solar/SPS process was relatively poor. After 180 min of reaction, 95% LOS oxidation and 1.30 ± 0.13 log *E. coli* elimination were achieved. Regarding the inactivation of *E. coli* by the solar/SPS process, similar results were reported by Wang et al. [62], who studied *E. coli* inactivation by visible light persulfate activation. Specifically, they mentioned that a system using visible light with 1 mM persulfate was not able to reduce the population of *E. coli*. However, when the persulfate concentration increased to 4 mM, they reported a 7 log *E. coli* elimination after 80 min [62].

The way reactive species inactivate/eliminate *E. coli* is similar to the mechanism proposed by Wang et al. [62] and Wang et al. [63], who studied the hybrid processes of heat/visible light/persulfate and visible light/hydrochar/persulfate for *E. coli* inactivation, respectively. They reported that reactive species cause severe damage to the cell membrane. The reactive species then pass through the membrane, and the cell's defense system tries to protect itself by producing a high level of intracellular antioxidant enzymes. But as the reactive species continually penetrate through the membrane, the defense systems become incapacitated, leading to the destruction of the cell.

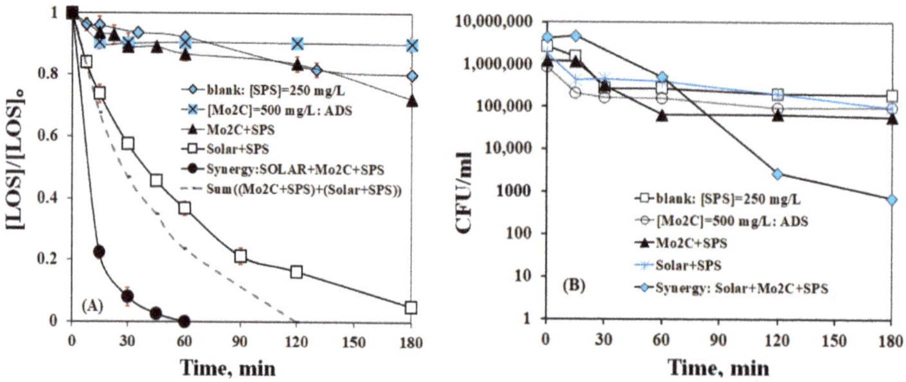

Figure 7. Simultaneous degradation of 500 µg/L LOS (**A**) and inactivation of $\approx 10^6$–10^7 CFU/mL *E. coli* (**B**) in UPW with 500 mg/L Mo_2C and 250 mg/L SPS and the synergistic effects with simulated solar irradiation.

However, the results of the synergistic solar/Mo_2C/SPS process were promising for both pollutant and microorganism elimination. Complete LOS removal was achieved in 60 min, comparable to LOS degradation (100% removal in 45 min) in the absence of *E. coli*. Regarding *E. coli* inactivation performance, a reduction by 3.98 log $\pm$ 0.21 was observed, which is significantly higher than that achieved by the Mo_2C/SPS and solar/SPS processes alone.

The hybrid system with $k_{(Solar/Mo2C/SPS)} = 0.083 \pm 0.005\ min^{-1}$ exhibited a higher yield than the sum of the individual processes with $k_{(Mo2C/SPS)} = 0.0024 \pm 0.003\ min^{-1}$ and $k_{(solar/SPS)} = 0.0165 \pm 0.004\ min^{-1}$, as seen in Figure 7A (curve with dotted lines). The extent of synergy, S, for LOS decomposition may be computed as follows:

$$S = \frac{k_{Solar/Mo_2C/SPS}}{k_{Mo_2C/SPS} + k_{solar/SPS}} \quad \text{where, } \begin{array}{ll} S > & 1(\text{synergistic effect}) \\ S = & 1(\text{additive effect}) \\ S < & 1(\text{antagonistic effect}) \end{array} \tag{21}$$

Subsequently, the value of S from the data in Figure 7A is 4.4.

To examine the observed synergy, an additional experiment was conducted for the simultaneous removal of LOS and *E. coli* using the Mo_2C/solar system in the absence of the oxidant. Despite the fact that, according to the literature, the energy gap of Mo_2C is 1.22 V [64], the obtained results were similar to adsorption (i.e., absence of light and oxidant). A possible explanation may be the position of the conduction and valence bands of Mo_2C, which are -0.41 V and 0.81 V, respectively [64]. Therefore, the photogenerated holes have very low oxidation potential, while the Mo_2C/solar system is incapable of producing hydroxyl radicals. On the other hand, SPS can trap the photogenerated electrons, thus increasing the separation of photoproduced holes and electrons and producing additional reactive species into the system through oxidant activation by the photoproduced electrons. Therefore, the observed synergy can be justified by the higher concentration of reactive species and the different activation mechanisms in the combined system. Thus, the role of solar irradiation is associated with an acceleration of the production of additional $SO_4^{\bullet-}$ and $HO^{\bullet}$ through SPS activation, which consequently leads to the formation of more $O_2^{\bullet-}$ and $^1O_2$ (Equations (6), (8), and (9)), which are available for the degradation of LOS and *E. coli*. In summary, in the hybrid system, persulfate is activated in three ways: (i) homogeneously by solar irradiation, which allows a direct attack on LOS molecules and on the cell membranes of *E. coli,* (ii) heterogeneously by Mo_2C, and (iii) photocatalytically by the photogenerated electrons.

3. Materials and Methods

3.1. Materials

Molybdenum carbide ($Mo_2C \geq 99.5\%$, CAS: 12069-89-5), sodium (meta)periodate ($NaIO_4 \geq 99.0\%$, CAS: 7790-28-5), sodium nitrate ($NaNO_3 \geq 99.0\%$, CAS: 7631-99-4), sodium persulfate (SPS $\geq 99.5\%$), losartan potassium (LOS), humic acid (HA, technical grade), Petri dishes (D = 9 cm), tert-butyl alcohol (t-BuOH), sodium azide (NaN_3), sodium bicarbonate ($NaHCO_3$), ortho-phosphoric acid ($H_3PO_4 \geq 85\%$), and sulfuric acid (H_2SO_4) were obtained from Sigma Aldrich (St. Louis, MO, USA). Further details for the above reagents and LOS can be found in the study of Ioannidi et al. (2022). Sodium percarbonate ($Na_2CO_3{\cdot}1.5H_2O_2$, CAS: 15630-89-4), and hydrogen peroxide solution 30% (H_2O_2, CAS: 7722-84-1) were bought from Acros organics (Geel, ANTWERP, Belgium) and Carlo Ebra reagents (Denzlinger Str., Emmendingen, Germany), respectively, while sodium hydroxide (NaOH, CAS: 1310-73-2), and sodium chloride (NaCl $\geq 99.5\%$, CAS: 16610-31000) were purchased from Penta (Radiová, Prague, Czechia). Methanol (MeOH, hplc-grade, CAS: 67-56-1) was purchased from Fischer chemicals (Riesbachstrasse, Zürich, Switzerland). All chemicals were used without any further purification. Furthermore, microbiological yeast extract (CAS:1702.00) and tryptone (CAS:1612.00) were purchased from Condalab (C. Forja, Torrejón de Ardoz, Madrid, Spain), while agar (CAS: 9002-18-0) was purchased from Serva (Carl-benz-str., Heidelberg, Baden-Wuerttemberg, Germany). *E. coli* (DSM 1103) was obtained by the Leibniz Institute DSMZ (German Collection of Microorganisms and Cell Cultures).

Most experiments were performed in ultrapure water (UPW, 18.2 MΩ·cm and pH $\approx$ 5.5), while commercial bottled water (BW) and secondary effluent from the University of Patras campus wastewater treatment plant (WW) were also used. Details of the water matrices can be found in Table S1.

3.2. Procedure of Degradation Experiments

For LOS oxidation experiments, a batch reactor with a maximum volume capacity of 150 mL was used. The working volume of the reactor with the desired LOS concentration was 100 mL. The reaction started after the simultaneous addition of the preferred amount of Mo_2C and SPS under continuous magnetic stirring. At fixed time intervals, samples of 1.2 mL were collected, quenched with 0.3 mL MeOH, and filtered through 0.22 μm PVDF filter.

To study the role of pH, 1 M NaOH and 1 M H_2SO_4 were used. pH adjustment to 9 or 3 was performed prior to the addition of the catalyst and oxidant. During the reaction, solution pH was monitored but left uncontrolled. To investigate the contribution of several kinds of reactive species, MeOH, t-BuOH, and NaN_3 were used as scavengers.

3.3. Experimental Procedure of Disinfection

To evaluate the potential of the Mo_2C/SPS system for disinfection, *E. coli* was used as a representative Gram-negative bacterium. All experiments were conducted in the presence of 500 μg/L LOS to investigate the simultaneous degradation of xenobiotics and pathogens typically existing in secondary effluent. The necessary equipment (e.g., glass vessels, Eppendorf tubes, pipette tips, stirring magnets, beakers, nutrient medium, ultrapure water, and 0.8% *w*/*v* NaCl aqueous solution) for carrying out all inactivation experiments was sterilized in an autoclave. Details of bacteria cultivation, concentration estimation, and the execution of inactivation experiments can be found in the supplementary material [65–67].

3.4. Analytical Methods

The concentration of LOS was monitored using a high-performance liquid chromatograph (HPLC, waters alliance 2695). Details of the analytical method can be obtained in the study of Ioannidi et al. [47]. Total organic carbon (TOC) was measured using a Shimadzu TOC-L_{CSH} analyzer (Kyoto, Japan).

3.5. Mo_2C Characterization

Scanning electron microscopy (JEOL 6300) (Tokyo, Japan) equipped with an energy-dispersive spectrometer (EDS) was adopted to study the surface morphology and chemical composition of Mo_2C, while the crystal structure of the sample was obtained by X-ray diffraction (XRD) (Bruker D8-Advanced diffractometer) (Billerica, MA, USA). Its primary crystallite size was calculated by means of the Debye–Scherrer equation [68]. A zeta potential analyzer (Zetasizer Nano Z (Malvern), UK) was employed to assess the electrical characteristics of the catalyst's solid-liquid interface within the reaction solution. Additional information can be found in former publications from our group [69].

3.6. Data Analysis

LOS degradation obeys a pseudo-first-order kinetic rate with $R^2 > 0.98$ in all cases. The apparent rate constants (k_{app}, min^{-1}) were calculated according to Equation (22):

$$\text{rate} = -\frac{d[\text{LOS}]}{dt} = k_{app}[\text{LOS}] \rightarrow \ln\left(\frac{[\text{LOS}]}{[\text{LOS}]_o}\right) = -k_{app}t \quad (22)$$

where $[\text{LOS}]_o$, and [LOS] refer to LOS concentration at times zero and t, respectively.

4. Conclusions

In this work, Mo_2C, a 2D material, demonstrated notable efficiency in persulfate activation and LOS decomposition. The primary conclusions of this work can be summarized as follows:

- The investigated system was capable of eliminating 500 μg/L of LOS in less than 45 min. This concentration is well over the upper limit typically found in surface waters and/or secondary treated effluents, which implies that milder treatment conditions would suffice to deal with environmentally relevant concentrations.
- Oxidation adhered to pseudo-first-order kinetics, and the apparent kinetic constant decreased with increasing LOS concentration and increased in acidic pH.
- The presence of organic matter and carbonate impeded LOS degradation in the experiments conducted in secondary effluent and bottled water; moreover, the presence of *E. coli* slowed down LOS decomposition, while *E. coli* removal was only 1.23 log after 180 min. This clearly highlights the critical role of the environmentally relevant water matrix.
- Sulfate radicals, hydroxyl radicals, and singlet oxygen participated in LOS destruction, but singlet oxygen emerged as the predominant species. In this respect, the possible interplay between catalyst, oxidants, and the target and non-target species may be quite complicated.
- Experiments conducted in the presence of simulated solar irradiation demonstrated a significant synergy and efficiency improvement, leading to complete LOS elimination in 60 min and nearly a 4-log reduction of *E. coli* in 180 min. This implies that process coupling may be a step in the right direction in terms of enhancing treatment efficiency; nonetheless, this should be complemented by cost-efficient strategies that can be offered by the application of renewable energy sources.
- Though the use of 2D materials in environmental remediation appears as a promising strategy, further research is needed to examine the operation of similar systems under continuous flow, and scaling up is required. Future work must also delve deeper into the mechanism and transformation products, as well as investigate toxicity and Mo leaching during the treatment.

Supplementary Materials: The following supporting information can be downloaded at: https://www.mdpi.com/article/10.3390/catal13091285/s1, Figure S1: Apparent rate constants at several initial concentrations of Mo_2C for LOS degradation and adsorption in UPW. Experimental conditions: [LOS] = 500 µg/L, and [SPS] = 250 mg/L; Figure S2: XRD patterns of the "fresh" Mo_2C, and after exposure to reaction conditions for 225 min (500 mg/L Mo_2C, 250 mg/L SPS and 500 µg/L LOS); Figure S3. Reuse experiments of Mo_2C for LOS removal in UPW. Experimental conditions: [Mo_2C] = 500 mg/L, [SPS] = 250 mg/L, and [LOS] = 500 µg/L at inherent pH $\approx$ 5.5.; Table S1: Main properties of various water matrices used in this study.

Author Contributions: Conceptualization, A.A.I.; Z.F. and D.M.; methodology, A.A.I.., A.P., E.K., M.K. and D.M.; formal analysis, A.A.I., M.V., Z.F. and A.P.; investigation, A.A.I., M.V., E.K. and A.P.; resources, Z.F.,M.K. and D.M.; data curation, A.A.I., M.V. and A.P.; writing—original draft preparation, A.A.I., Z.F. and A.P.; writing—review and editing, A.A.I., A.P., E.K., M.K. and D.M.; visualization, A.A.I., M.V. and A.P.; supervision, D.M.; funding acquisition, D.M. All authors have read and agreed to the published version of the manuscript.

Funding: Alexandra Ioannidi acknowledges that the implementation of the doctoral thesis was co-financed by Greece and the European Union (European Social Fund-ESF) through the Operational Programme "Human Resources Development, Education and Lifelong Learning" in the context of the Act "Enhancing Human Resources Research Potential by undertaking a Doctoral Research" Sub-action 2: IKY Scholarship Program for PhD candidates in the Greek Universities.

Data Availability Statement: Not applicable.

Conflicts of Interest: The authors declare no conflict of interest.

References

1. Ekramul Mahmud, H.N.M.; Huq, A.K.O.; Yahya, R.B. The removal of heavy metal ions from wastewater/aqueous solution using polypyrrole-based adsorbents: A review. *RSC Adv.* **2016**, *6*, 14778–14791. [CrossRef]
2. Priya, A.K.; Gnanasekaran, L.; Rajendran, S.; Qin, J.; Vasseghian, Y. Occurrences and removal of pharmaceutical and personal care products from aquatic systems using advanced treatment- A review. *Environ. Res.* **2022**, *204*, 112298. [CrossRef]
3. Asghar, M.A.; Zhu, Q.; Sun, S.; Peng, Y.; Shuai, Q. Suspect screening and target quantification of human pharmaceutical residues in the surface water of Wuhan, China, using UHPLC-Q-Orbitrap HRMS. *Sci. Total Environ.* **2018**, *635*, 828–837. [CrossRef]
4. Busch, W.; Schmidt, S.; Kühne, R.; Schulze, T.; Krauss, M.; Altenburger, R. Micropollutants in European rivers: A mode of action survey to support the development of effect-based tools for water monitoring: Micropollutants in European rivers: A mode-of-action. Environ. Toxicol. *Chem.* **2016**, *35*, 1887–1899. [CrossRef] [PubMed]
5. Kane, S.P. Valsartan, ClinCalc DrugStats Database, Version 2022.08. ClinCalc. Available online: https://clincalc.com/DrugStats/Drugs/Valsartan (accessed on 29 July 2023).
6. Statista: Number of Losartan Potassium Prescriptions in the U.S. from 2004 to 2020 (in Million). Available online: https://www.statista.com/statistics/781681/losartan-potassium-prescriptions-number-in-the-us/ (accessed on 29 July 2023).
7. Centers for Disease Control and Prevention. Hypertension Cascade: Hypertension Prevalence, Treatment and Control Estimates Among U.S. Adults Aged 18 Years and Older Applying the Criteria from the American College of Cardiology and American Heart Association's 2017 Hypertension Guideline—NHANES 2017–2020. Atlanta, GA: 12 May 2023. Available online: https://www.cdc.gov/bloodpressure/facts.htm#:~:text=This%20includes%2037%20million%20U.S.%20adults.&text=About%2034%20million%20adults%20who,140%2F90%20mmHg%20or%20higher (accessed on 6 July 2023).
8. Rouette, J.; McDonald, E.G.; Schuster, T.; Brophy, J.M.; Azoulay, L. Treatment and prescribing trends of antihypertensive drugs in 2.7 million UK primary care patients over 31 years: A population-based cohort study. *BMJ Open* **2022**, *12*, e057510. [CrossRef] [PubMed]
9. Cortez, F.S.; Souza, L.S.; Guimarães, L.L.; Almeida, J.E.; Pusceddu, F.H.; Maranho, L.A.; Mota, L.G.; Nobre, C.R.; Moreno, B.B.; Abessa, D.M.S.; et al. Ecotoxicological effects of losartan on the brown mussel Perna perna and its occurrence in seawater from Santos Bay (Brazil). *Sci. Total Environ.* **2018**, *637–638*, 1363–1371. [CrossRef]
10. Mandaric, L.; Diamantini, E.; Stella, E.; Cano-Paoli, K.; Valle-Sistac, J.; Molins- Delgado, D.; Bellin, A.; Chiogna, G.; Majone, B.; Diaz-Cruz, M.S.; et al. Contamination sources and distribution patterns of pharmaceuticals and personal care products in Alpine rivers strongly affected by tourism. *Sci. Total Environ.* **2017**, *590–591*, 484–494. [CrossRef]
11. Nuel, M.; Laurent, J.; Bois, P.; Heintz, D.; Wanko, A. Seasonal and ageing effect on the behaviour of 86 drugs in a full-scale surface treatment wetland: Removal efficiencies and distribution in plants and sediments. *Sci. Total Environ.* **2018**, *615*, 1099–1109. [CrossRef]
12. Ashfaq, M.; Li, Y.; Wang, Y.; Chen, W.; Wang, H.; Chen, X.; Wu, W.; Huang, Z.; Yu, C.-P.; Sun, Q. Occurrence, fate, and mass balance of different classes of pharmaceuticals and personal care products in an anaerobic-anoxic-oxic wastewater treatment plant in Xiamen, China. *Water Res.* **2017**, *123*, 655–667. [CrossRef]

13. Botero-Coy, A.M.; Martínez-Pachon, D.; Boix, C.; Rincon, R.J.; Castillo, N.; Arias- Marín, L.P.; Manrique-Losada, L.; Torres-Palma, R.; Moncayo-Lasso, A.; Hernandez, F. An investigation into the occurrence and removal of pharmaceuticals in Colombian wastewater'. *Sci. Total Environ.* **2018**, *642*, 842–853. [CrossRef]
14. Castro, G.; Rodríguez, I.; Ramil, M.; Cela, R. Selective determination of sartan drugs in environmental water samples by mixed-mode solid-phase extraction and liquid chromatography tandem mass spectrometry. *Chemosphere* **2019**, *224*, 562–571. [CrossRef] [PubMed]
15. Wang, J.; Wang, S. Activation of persulfate (PS) and peroxymonosulfate (PMS) and application for the degradation of emerging contaminants. *Chem. Eng. J.* **2018**, *334*, 1502–1517. [CrossRef]
16. Serna-Galvis, E.A.; Isaza-Pineda, L.; Moncayo-Lasso, A.; Hernández, F.; Ibáñez, M.; Torres-Palma, R.A. Comparative degradation of two highly consumed antihypertensives in water by sonochemical process. Determination of the reaction zone, primary degradation products and theoretical calculations on the oxidative process. *Ultrason. Sonochem.* **2019**, *58*, 104635. [CrossRef] [PubMed]
17. Kaur, B.; Dulova, N. UV-assisted chemical oxidation of antihypertensive losartan in water. *J. Environ. Manag.* **2020**, *261*, 110170. [CrossRef]
18. Matsoukas, J.; Apostolopoulos, V.; Zulli, A.; Moore, G.; Kelaidonis, K.; Moschovou, K.; Mavromoustakos, T. From Angiotensin II to Cyclic Peptides and Angiotensin Receptor Blockers (ARBs): Perspectives of ARBs in COVID-19 Therapy. *Molecules* **2021**, *26*, 618. [CrossRef]
19. Gurke, R.; R¨oßler, M.; Marx, C.; Diamond, S.; Schubert, S.; Oertel, R.; Fauler, J. Occurrence and removal of frequently prescribed pharmaceuticals and corresponding metabolites in wastewater of a sewage treatment plant. *Sci. Total Environ.* **2015**, *532*, 762–770. [CrossRef]
20. Ladhari, A.; La Mura, G.; Di Marino, C.; Di Fabio, G.; Zarrelli, A. Sartans: What they are for, how they degrade, where they are found and how they transform. *Sustain. Chem. Pharm.* **2021**, *20*, 100409. [CrossRef]
21. Osorio, V.; Larrañaga, A.; Aceña, J.; P'erez, S.; Barcel'o, D. Concentration and risk of pharmaceuticals in freshwater systems are related to the population density and the livestock units in Iberian Rivers. *Sci. Total Environ.* **2016**, *540*, 267–277. [CrossRef]
22. Adams, E.; Neves, B.B.; Prola, L.D.T.; De Liz, M.V.; Martins, L.R.R.; Ramsdorf, W.A.; De Freitas, A.M. Ecotoxicity and genotoxicity assessment of losartan after UV/H_2O_2 and UVC/photolysis treatments. *Environ. Sci. Pollut. Res.* **2021**, *28*, 23812–23821. [CrossRef]
23. Czekalski, N.; Imminger, S.; Salhi, E.; Veljkovic, M.; Kleffel, K.; Drissner, D.; Hammes, F.; Bürgmann, H.; Von Gunten, U. Inactivation of Antibiotic Resistant Bacteria and Resistance Genes by Ozone: From Laboratory Experiments to Full-Scale Wastewater Treatment. *Environ. Sci. Technol.* **2016**, *50*, 11862–11871. [CrossRef]
24. Iakovides, I.C.; Michael-Kordatou, I.; Moreira, N.F.F.; Ribeiro, A.R.; Fernandes, T.; Pereira, M.F.R.; Nunes, O.C.; Manaia, C.M.; Silva, A.M.T.; Fatta-Kassinos, D. Continuous ozonation of urban wastewater: Removal of antibiotics, antibiotic-resistant Escherichia coli and antibiotic resistance genes and phytotoxicity. *Water Res.* **2019**, *159*, 333–347. [CrossRef] [PubMed]
25. Nnadozie, C.F.; Odume, O.N. Freshwater environments as reservoirs of antibiotic resistant bacteria and their role in the dissemination of antibiotic resistance genes. *Environ. Pollut.* **2019**, *254*, 113067. [CrossRef] [PubMed]
26. Yoon, S.-H.; Ha, S.-M.; Kwon, S.; Lim, J.; Kim, Y.; Seo, H.; Chun, J. Introducing EzBioCloud: A taxonomically united database of 16S rRNA gene sequences and whole-genome assemblies. *Int. J. Syst. Evol. Microbiol.* **2017**, *67*, 1613–1617. [CrossRef] [PubMed]
27. Kokkinos, P.; Venieri, D.; Mantzavinos, D. Advanced Oxidation Processes for Water and Wastewater Viral Disinfection. A Systematic Review. *Food Environ. Virol.* **2021**, *13*, 283–302. [CrossRef] [PubMed]
28. García-Espinoza, J.D.; Robles, I.; Durán-Moreno, A.; Godínez, L.A. Photo-assisted electrochemical advanced oxidation processes for the disinfection of aqueous solutions: A review. *Chemosphere* **2021**, *274*, 129957. [CrossRef]
29. Chen, Y.; Duan, X.; Zhou, X.; Wang, R.; Wang, S.; Ren, N.; Ho, S.-H. Advanced oxidation processes for water disinfection: Features, mechanisms and prospects. *J. Chem. Eng.* **2021**, *409*, 128207. [CrossRef]
30. Ma, D.; Yi, H.; Lai, C.; Liu, X.; Huo, X.; An, Z.; Li, L.; Fu, Y.; Li, B.; Zhang, M.; et al. Critical review of advanced oxidation processes in organic wastewater treatment. *Chemosphere* **2021**, *275*, 130104. [CrossRef]
31. Chen, X.; Gudda, F.O.; Hu, X.; Waigi, M.G.; Gao, Y. Degradation of bisphenol A in an oxidation system constructed from Mo_2C MXene and peroxymonosulfate. *npj Clean Water* **2022**, *5*, 66. [CrossRef]
32. Guo, S.; Wang, H.; Yang, W.; Fida, H.; You, L.; Zhou, K. Scalable synthesis of Ca-doped α-Fe_2O_3 with abundant oxygen vacancies for enhanced degradation of organic pollutants through peroxymonosulfate activation. *Appl. Catal. B* **2020**, *262*, 118250. [CrossRef]
33. Ghanbari, F.; Moradi, M. Application of peroxymonosulfate and its activation methods for degradation of environmental organic pollutants: Review. *Chem. Eng. J.* **2017**, *310*, 41–62. [CrossRef]
34. Forouzesh, M.; Ebadi, A.; Aghaeinejad-Meybodi, A. Degradation of metronidazole antibiotic in aqueous medium using activated carbon as a persulfate activator. *Sep. Purif. Technol.* **2019**, *210*, 145–151. [CrossRef]
35. Bekris, L.; Frontistis, Z.; Trakakis, G.; Sygellou, L.; Galiotis, C.; Mantzavinos, D. Graphene: A new activator of sodium persulfate for the advanced oxidation of parabens in water. *Water Res.* **2017**, *126*, 111–121. [CrossRef] [PubMed]
36. Zhou, H.; Lai, L.; Wan, Y.; He, Y.; Yao, G.; Lai, B. Molybdenum disulfide (MoS_2): A versatile activator of both peroxymonosulfate and persulfate for the degradation of carbamazepine. *Chem. Eng. J.* **2020**, *384*, 123264. [CrossRef]

37. Alaba, P.A.; Abbas, A.; Huang, J.; Daud, W.M.A.W. Molybdenum carbide nanoparticle: Understanding the surface properties and reaction mechanism for energy production towards a sustainable future. *Renew. Sust. Energ. Rev.* **2018**, *91*, 287–300. [CrossRef]
38. Yang, L.; Chen, H.; Jia, F.; Peng, W.; Tian, X.; Xia, L.; Wu, X.; Song, S. Emerging Hexagonal Mo_2C Nanosheet with (002) Facet Exposure and Cu Incorporation for Peroxymonosulfate Activation Toward Antibiotic Degradation. ACS Appl. *Mater. Interfaces* **2021**, *13*, 14342–14354. [CrossRef] [PubMed]
39. Chen, P.; Liang, Y.; Yang, B.; Jia, F.; Song, S. In Situ Reduction of Au(I) for Efficient Recovery of Gold from Thiosulfate Solution by the 3D MoS_2/Chitosan Aerogel. *ACS Sustain. Chem. Eng.* **2020**, *8*, 3673–3680. [CrossRef]
40. Zhu, L.; Ji, J.; Liu, J.; Mine, S.; Matsuoka, M.; Zhang, J.; Xing, M. Designing 3D-MoS_2 Sponge as Excellent Cocatalysts in Advanced Oxidation Processes for Pollutant Control. *Angew. Chem. Int. Ed.* **2020**, *59*, 13968–13976. [CrossRef]
41. Ji, J.; Aleisa, R.M.; Duan, H.; Zhang, J.; Yin, Y.; Xing, M. Metallic Active Sites on MoO_2(110) Surface to Catalyze Advanced Oxidation Processes for Efficient Pollutant Removal. *Iscience* **2020**, *23*, 100861. [CrossRef]
42. Bao, Y.; Chen, T.; Zhu, Z.; Zhang, H.; Qiu, Y.; Yin, D. Mo_2C/C catalyst as efficient peroxymonosulfate activator for carbamazepine degradation. *Chemosphere* **2022**, *287*, 132047. [CrossRef]
43. De Andrade, J.R.; Vieira, M.G.A.; Da Silva, M.G.C.; Wang, S. Oxidative degradation of pharmaceutical losartan potassium with N-doped hierarchical porous carbon and peroxymonosulfate. *J. Chem. Eng.* **2020**, *382*, 122971. [CrossRef]
44. Salazar, C.; Contreras, N.; Mansilla, H.D.; Yáñez, J.; Salazar, R. Electrochemical degradation of the antihypertensive losartan in aqueous medium by electro-oxidation with boron-doped diamond electrode. *J. Hazard. Mater.* **2016**, *319*, 84–92. [CrossRef] [PubMed]
45. Guateque-Londoño, J.F.; Serna-Galvis, E.A.; Ávila-Torres, Y.; Torres-Palma, R.A. Degradation of losartan in fresh urine by sonochemical and photochemical advanced oxidation processes. *Water* **2020**, *12*, 3398. [CrossRef]
46. Martínez-Pach'on, D.; Espinosa-Barrera, P.; Rinc'on-Ortíz, J.; Moncayo-Lasso, A. Advanced oxidation of antihypertensives losartan and valsartan by photo-electro- Fenton at near-neutral pH using natural organic acids and a dimensional stable anode-gas diffusion electrode (DSA-GDE) system under light emission diode (LED) lighting. *Environ. Sci. Pollut. Res.* **2019**, *26*, 4426–4437. [CrossRef]
47. Ioannidi, A.; Arvaniti, O.S.; Nika, M.-C.; Aalizadeh, R.; Thomaidis, N.S.; Mantzavinos, D.; Frontistis, Z. Removal of drug losartan in environmental aquatic matrices by heat-activated persulfate: Kinetics, transformation products and synergistic effects. *Chemosphere* **2022**, *287*, 131952. [CrossRef] [PubMed]
48. Ioannidi, A.A.; Vakros, J.; Frontistis, Z.; Mantzavinos, D. Tailoring the Biochar Physicochemical Properties Using a Friendly Eco-Method and Its Application on the Oxidation of the Drug Losartan through Persulfate Activation. *Catalysts* **2022**, *12*, 1245. [CrossRef]
49. García, G.; Guillén-Villafuerte, O.; Rodríguez, J.L.; Arévalo, M.C.; Pastor, E. Electrocatalysis on metal carbide materials. *Int. J. Hydrog. Energy.* **2016**, *41*, 19664–19673. [CrossRef]
50. Ma, L.; Ting, L.R.L.; Molinari, V.; Giordano, C.; Yeo, B.S. Efficient hydrogen evolution reaction catalyzed by molybdenum carbide and molybdenum nitride nanocatalysts synthesized via the urea glass route. *J. Mater. Chem. A* **2015**, *3*, 8361–8368. [CrossRef]
51. García, G.; Roca-Ayats, M.; Guillén-Villafuerte, O.; Rodríguez, J.L.; Arévalo, M.C.; Pastor, E. Electrochemical performance of α-Mo2C as catalyst for the hydrogen evolution reaction. *J. Electroanal. Chem.* **2017**, *793*, 235–241. [CrossRef]
52. Kim, J.; Lee, D.H.; Yang, Y.; Chen, K.; Liu, C.; Kang, J.; Li, O.L. Hybrid Molybdenum Carbide/Heteroatom-Doped Carbon Electrocatalyst for Advanced Oxygen Evolution Reaction in Hydrogen Production. *Catalysts* **2020**, *10*, 1290. [CrossRef]
53. Han, M.; Wang, H.; Jin, W.; Chu, W.; Xu, Z. The performance and mechanism of iron-mediated chemical oxidation: Advances in hydrogen peroxide, persulfate and percarbonate oxidation. *J. Environ. Sci.* **2023**, *128*, 181–202. [CrossRef]
54. Lee, C.; Yoon, J. Application of photoactivated periodate to the decolorization of reactive dye: Reaction parameters and mechanism. *J. Photochem. Photobiol., A* **2004**, *165*, 35–41. [CrossRef]
55. Ioannidi, A.; Frontistis, Z.; Mantzavinos, D. Destruction of propyl paraben by persulfate activated with UV-A light emitting diodes. J. Environ. *Chem. Eng.* **2018**, *6*, 2992–2997. [CrossRef]
56. Serna-Galvis, E.A.; Botero-Coy, A.M.; Rosero-Moreano, M.; Lee, J.; Hernández, F.; Torres-Palma, R.A. An Initial Approach to the Presence of Pharmaceuticals in Wastewater from Hospitals in Colombia and Their Environmental Risk. *Water* **2022**, *14*, 950. [CrossRef]
57. Avramiotis, E.; Frontistis, Z.; Manariotis, I.D.; Vakros, J.; Mantzavinos, D. Oxidation of Sulfamethoxazole by Rice Husk Biochar-Activated Persulfate. *Catalysts* **2021**, *11*, 850. [CrossRef]
58. Tang, H.; Zhao, Y.; Shan, S.; Yang, X.; Liu, D.; Cui, F.; Xing, B. Theoretical insight into the adsorption of aromatic compounds on graphene oxide. *Environ. Sci. Nano* **2018**, *5*, 2357–2367. [CrossRef]
59. Mrozik, W.; Minofar, B.; Thongsamer, T.; Wiriyaphong, N.; Khawkomol, S.; Plaimart, J.; Vakros, J.; Karapanagioti, H.; Vinitnantharat, S.; Werner, D. Valorisation of agricultural waste derived biochars in aquaculture to remove organic micropollutants from water—Experimental study and molecular dynamics simulations. *J. Environ. Manag.* **2021**, *300*, 113717. [CrossRef] [PubMed]
60. Jawad, A.; Lang, J.; Liao, Z.; Khan, A.; Ifthikar, J.; Lv, Z.; Long, S.; Chen, Z.; Chen, Z. Activation of persulfate by CuO_x@Co-LDH: A novel heterogeneous system for contaminant degradation with broad pH window and controlled leaching. *J. Chem. Eng.* **2018**, *335*, 548–559. [CrossRef]
61. Shahzad, A.; Ali, J.; Ifthikar, J.; Aregay, G.G.; Zhu, J.; Chen, Z.; Chen, Z. Non-radical PMS activation by the nanohybrid material with periodic confinement of reduced graphene oxide (rGO) and Cu hydroxides. *J. Hazard. Mater.* **2020**, *392*, 122316. [CrossRef]

62. Wang, W.; Wang, H.; Li, G.; An, T.; Zhao, H.; Wong, P.K. Catalyst-free activation of persulfate by visible light for water disinfection: Efficiency and mechanisms. *Water Res.* **2019**, *157*, 106–118. [CrossRef]
63. Wang, W.; Wang, H.; Li, G.; Wong, P.K.; An, T. Visible light activation of persulfate by magnetic hydrochar for bacterial inactivation: Efficiency, recyclability and mechanisms. *Water Res.* **2020**, *176*, 115746. [CrossRef]
64. Dong, J.; Shi, Y.; Huang, C.; Wu, Q.; Zeng, T.; Yao, W. A New and stable Mo-Mo_2C modified g-C_3N_4 photocatalyst for efficient visible light photocatalytic H_2 production. *Appl. Catal. B.* **2019**, *243*, 27–35. [CrossRef]
65. Cold Spring Harbor Protocols. Available online: http://cshprotocols.cshlp.org/content/2006/1/pdb.rec8141.full?text_only=true (accessed on 24 May 2023).
66. Sanders, E.R. Aseptic Laboratory Techniques: Plating Methods. *J. Vis. Exp.* **2012**, *63*, e3064. [CrossRef]
67. Venieri, D.; Karapa, A.; Panagiotopoulou, M.; Gounaki, I. Application of activated persulfate for the inactivation of fecal bacterial indicators in water. *J. Environ. Manage.* **2020**, *261*, 110223. [CrossRef] [PubMed]
68. Lin, K.-Y.A.; Chen, Y.-C.; Lin, Y.-F. $LaMO_3$ perovskites (M = Co, Cu, Fe and Ni) as heterogeneous catalysts for activating peroxymonosulfate in water. *Chem. Eng. Sci* **2017**, *160*, 96–105. [CrossRef]
69. Gkika, C.; Petala, A.; Frontistis, Z.; Bampos, G.; Hela, D.; Konstantinou, I.; Mantzavinos, D. Heterogeneous activation of persulfate by lanthanum strontium cobaltite for sulfamethoxazole degradation. *Catal. Today* **2021**, *361*, 130–138. [CrossRef]

catalysts

MDPI

Article

The Influence of Au Loading and TiO_2 Support on the Catalytic Wet Air Oxidation of Glyphosate over TiO_2+Au Catalysts

Gregor Žerjav [1,*], Alen Albreht [2] and Albin Pintar [1]

1 Department of Inorganic Chemistry and Technology, Laboratory for Environmental Sciences and Engineering, National Institute of Chemistry, Hajdrihova 19, SI-1001 Ljubljana, Slovenia; albin.pintar@ki.si
2 Department of Analytical Chemistry, Laboratory for Food Chemistry, National Institute of Chemistry, Hajdrihova 19, SI-1001 Ljubljana, Slovenia; alen.albreht@ki.si
* Correspondence: gregor.zerjav@ki.si

Abstract: This study aimed to explore the impact of varying amounts of added Au (0.5 to 2 wt.%) and the structural characteristics of anatase TiO_2 supports (nanoparticles (TP, S_{BET} = 88 m^2/g) and nanorods (TR, S_{BET} = 105 m^2/g)) on the catalytic efficiency of TiO_2+Au catalysts in eliminating the herbicide glyphosate from aqueous solutions via the catalytic wet air oxidation (CWAO) process. The investigation was conducted using a continuous-flow trickle-bed reactor. Regardless of the TiO_2 support and the amount of Au added, the addition of Au has a positive effect on the glyphosate degradation rate. Regarding the amount of Au added, the highest catalytic activity was observed with the TP + 1% Au catalyst, which had a higher Schottky barrier (SB) than the TP + 2% Au catalyst, which helped the charge carriers in the TiO_2 conduction band to increase their reduction potential by preventing them from returning to the Au. The role of glyphosate degradation product adsorption on the catalyst surface is crucial for sustaining the long-term catalytic activity of the investigated TiO_2+Au materials. This was particularly evident in the case of the TR + 1% Au catalyst, which had the highest glyphosate degradation rate at the beginning of the CWAO experiment, but its catalytic activity then decreased over time due to the adsorption of glyphosate degradation products, which was favoured by the presence of strong acidic sites. In addition, the TR + 1% Au solid had the smallest average Au particle size of all analyzed materials, which were more easily deactivated by the adsorption of glyphosate degradation products. The analysis of the degradation products of glyphosate shows that the oxidation of glyphosate in the liquid phase involves the rupture of C–P and C–N bonds, as amino-methyl-phosphonic acid (AMPA), glyoxylic acid and sarcosine were detected.

Keywords: glyphosate; catalytic wet air oxidation; noble metals; gold nanoparticles; trickle-bed reactor; water treatment

1. Introduction

The utilization of the wet air oxidation (WAO) process is highly appealing for treating wastewater that is either too toxic for biological purification or too diluted for incineration [1]. The use of WAO in large-scale wastewater treatment is limited due to energy demanding processes (high temperature and high pressure), which cause high operating costs. The addition of a suitable catalyst mitigates the WAO process by increasing the reaction rate, shortening the reaction time, and reducing the operating costs [2,3]. In the catalytic WAO (CWAO) process, organic impurities are oxidised to form biodegradable intermediates or mineralised to water, CO_2, and corresponding inorganic salts, facilitated by activated O_2 species in the presence of catalysts. This takes place in a temperature range of 130 to 250 °C and at a pressure of 10 to 50 bar [4]. Mixed metal oxide systems, metal oxides, noble metal catalysts supported on materials like titanium oxide (TiO_2) and cerium-based composite oxides have been studied for catalytic wet air oxidation (CWAO) of diverse organic pollutants [4–8].

The widely used broad-spectrum herbicide glyphosate (*N*-(phosphono-methyl)glycine) is potentially carcinogenic and can cause endocrine disruption in humans even at trace levels (<0.02 mg/L) [9–12]. In Europe, glyphosate accounts for a third of all herbicides used. The discussion concerning a possible ban on the use of glyphosate was triggered by concerns about its possible indirect and direct effects on the environment and human health. Several European countries have announced that they will ban or severely restrict the use of glyphosate in the future [13–16]. The U.S. Environmental Protection Agency has set the permitted level for glyphosate in water at 700 μg/L [17], while EU regulations have set this level at 0.1 μg/L [18,19]. Complete mineralization of glyphosate by conventional wastewater treatment processes is difficult due to the high chemical stability of the glyphosate molecule, which is due to the presence of C–N and C–P bonds [20,21]. The widespread use of glyphosate, improper handling of herbicide-containing containers, and its good solubility in water (11.6 g/L, 25 °C) led to an increase in herbicide concentrations in groundwater or surface water, resulting in reported glyphosate concentrations of up to 76 mg/L [22–24]. Gupt et al. [25] studied the mineralization of glyphosate in CWAO using carbon nanofibers (CNF) with iron nanoparticles (Fe NP) deposited on activated carbon beads (ACB) as a catalyst (0.75 g/L catalyst dose) with high thermal stability (25–1100 °C) and high S_{BET} value (~296 m^2/g). After 2 h at 220 °C and 25 bar atmospheric pressure, ~70% degradation of aqueous glyphosate (100 mg/L) was achieved. Complete degradation of glyphosate was achieved after 6 h under the same experimental conditions when Fe-CNF/ACB with higher specific surface area (~338 m^2/g), Fe loading of 4 mg/g, increased exposure of Fe-NPs to the surrounding liquid, and graphitic properties (ID/IG = 0.946) of CNFs were used [26]. Xing et al. [27] carried out CWAO of water-dissolved glyphosate (500–2500 mg/L) at 10 bar and 130 °C using activated carbons modified by H_2O_2 oxidation and thermal treatment with ammonia or melamine as catalysts and obtained ~97% removal of glyphosate in the continuous 55-day test.

The objective of this study was to thoroughly examine how varying amounts of added Au and the structural characteristics of anatase TiO_2 supports (nanoparticles (TP) and nanorods (TR)) influence the catalytic performance of TiO_2+Au catalysts in the degradation of the water-dissolved herbicide glyphosate. This investigation was conducted using a continuous-flow trickle-bed reactor operating under low-interaction conditions (scheme of the reactor system in [28]). Liquid chromatography and UV spectrophotometry coupled with mass spectrometry were employed to explore and quantify potential transformation products of glyphosate, aiming to identify predominant by-products and elucidate a plausible oxidation pathway of glyphosate via the CWAO process. To our knowledge, this is the first application of TiO_2+Au catalysts for the CWAO of glyphosate dissolved in water.

2. Results and Discussion

2.1. *Catalyst Characterization*

The XRD diffraction patterns of the investigated solids illustrated in Figure 1 show the main diffraction peaks of anatase TiO_2 in all cases. No diffraction peaks were detected that would belong to other TiO_2 polymorphs. For the Ti_2O+Au catalysts, no diffraction peaks belonging to gold were observed, which could be due to several facts such as: (i) low gold loading, (ii) presence of small gold particles, (iii) good distribution of gold on the surface of the TiO_2 supports, etc. [29–32]. The Scherrer equation, utilizing the main diffraction peak of anatase at $2\theta = 25.3°$, was used to determine the average anatase crystallite sizes, which are shown in Table 1. We can see that there is a difference between the average anatase crystallite sizes of the bare TiO_2 supports of about 5 nm and that the TP sample has the highest anatase crystallite size of 21.3 nm. Furthermore, we can conclude that the wet impregnation process used had no influence on the average anatase crystallite size, regardless of which TiO_2 support was used.

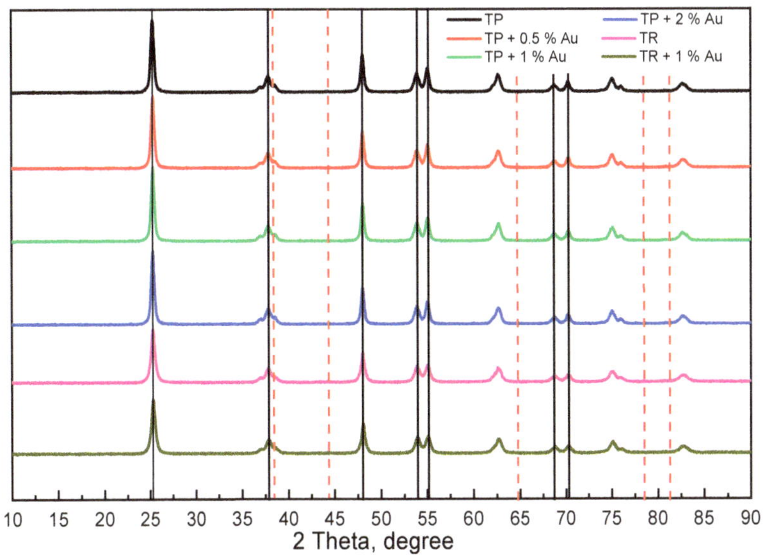

Figure 1. XRD patterns of both pure TiO_2 supports and TiO_2+Au catalysts with varying Au loadings are presented. In these patterns, solid vertical lines indicate anatase TiO_2 (JCPDS 00-021-1272), while dotted vertical lines denote fcc Au (JCPDS 01-1174).

The SEM images, N_2 adsorption–desorption isotherms and the corresponding BJH pore size distributions of the investigated solids are shown in Figures 2 and S2. Table 1 shows the results of the N_2 adsorption–desorption measurements (S_{BET}, V_{pore} and d_{pore}). The obtained SEM and TEM images of the TiO_2+Au catalysts reveal that there is a difference in the morphology of the synthesized catalysts depending on which TiO_2 support was used. In the case of the TP sample, we can observe elongated TiO_2 particles with a length of 40 nm and a diameter of about 30 nm, while in the case of the TR solid as the TiO_2 support, we can clearly observe TiO_2 in the form of nanorods with a length of 80–100 and a diameter of about 20 nm. The results of the N_2 adsorption–desorption measurements performed on the investigated materials (Table 1 and Figure S1) reflect well the differences in the TiO_2 support shapes, where the TR-based materials (S_{BET}(TR+Au) = 94–105 m^2/g) have a higher specific surface area (of about 20%) compared to the TP-based materials (S_{BET}(TP+Au) = 71–88 m^2/g). The results also show that impregnating TiO_2 supports with gold did not significantly affect the average pore diameter and the total pore volume compared to the pure TiO_2 supports. This fact indicates that the gold particles were mainly distributed over the surface of the TiO_2 supports and not incorporated into the TiO_2 supports [33]. Scanning electron microscopy was also used to obtain data on the actual gold loading in the fabricated TiO_2+Au catalysts compared to the nominal gold loading. The results of the SEM–EDS analysis can be found in Table S1, which shows that the actual gold loading is almost equal to the nominal for all TiO_2+Au catalysts investigated. The SEM–EDS elemental mapping images of the investigated TiO_2+Au catalysts in Figure S2 also show a good distribution of gold in the synthesized TiO_2+Au catalysts. The diagrams of the gold particle size distribution illustrated in Figure S4 were prepared using transmission electron microscopy (Figure S3). The results in Figure S4 and Table 1 show that the size distribution of the gold particles and the average gold particle size depend on the used TiO_2 substrate (d_{Au}(TR+Au) = 9.2 nm and d_{Au}(TP+Au) = 37–54 nm). The TR+Au catalysts show a narrow size distribution of Au particles compared to the TP+Au catalysts, where a broader size distribution of Au particles is observed. This could be due to the fact that the TiO_2 supports have different BET specific surface areas and a different shape of the curve representing the dependence of the zeta potential on the pH of the solution, as shown in Figure S5. The pH value of the $HAuCl_4 \times 3H_2O$ aqueous solution used to prepare TiO_2+Au catalysts with 1 wt.% gold content was 3.4. In Figure S5, we can see that, at this pH value, the surfaces of the TiO_2 supports were positively charged, which promoted the adsorption of $(Au(OH)_3Cl)^-$ by an electrostatic effect that was stronger in the case of the TP support (20 mV) than in the case of the TR support (10 mV). For the TP+Au catalysts, we can observe that, by adding more $HAuCl_4 \times 3H_2O$ to the aqueous solution to obtain catalysts with a higher Au content, the pH value of the $HAuCl_4 \times 3H_2O$ solution also decreased. This led to an increase in the surface charge of the TP support, which had an additional positive effect on the formation of larger gold clusters in the TP + 2% Au catalyst, apart from the fact that, with the addition of more $HAuCl_4 \times 3H_2O$ to the solution, the so-called "clustering" effect also occurs, as has already been observed by other authors [34].

Table 1. Comparison of specific surface area (S_{BET}), total pore volume (V_{pore}), average pore diameter (d_{pore}), Au particle size, anatase TiO_2 crystallite size, TiO_2 support diameter (d_{TiO2}) and length (l_{TiO2}), and Schottky barrier height (SBH) in the examined solids.

Sample	S_{BET}	V_{pore}	d_{pore}	d_{Au} *	Crystallite Size	d_{TiO2}	l_{TiO2}	SBH **
	m^2/g	cm^3/g	nm	nm	nm	nm	nm	eV
TP	88	0.30	14.8	/	21.3	30	40	/
TP + 0.5% Au	71	0.28	14.6	37.0	21.3			0.26
TP + 1% Au	74	0.29	14.2	36.9	21.3			0.22
TP + 2% Au	75	0.27	13.7	54.0	21.3			0.15
TR	105	0.57	19.3	/	16.7	20	80–100	/
TR + 1% Au	94	0.44	18.9	9.2	16.5			0.16

* Au particle size was calculated by counting more than 100 particles from acquired TEM images. ** Calculated from the determined values of VBM in Figure 3.

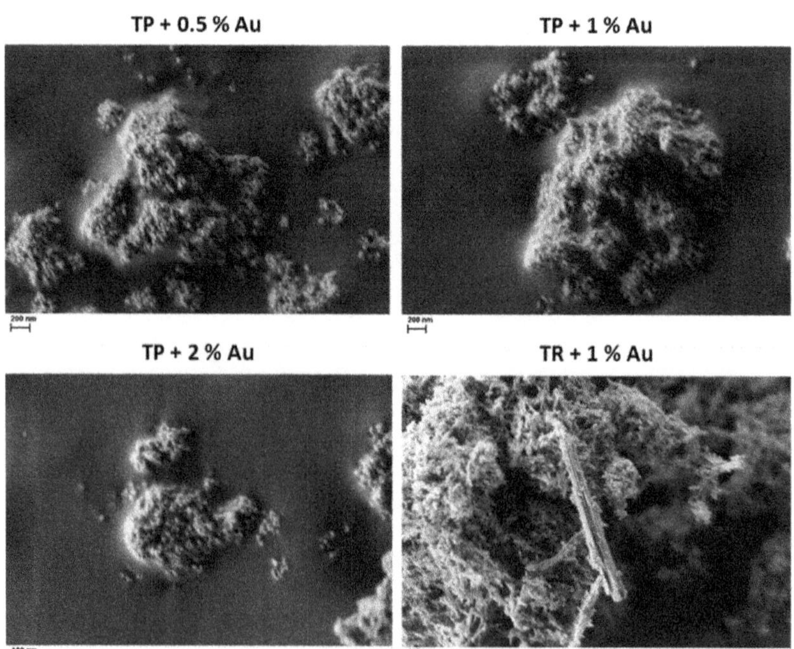

Figure 2. SEM images of the TiO_2+Au catalysts investigated.

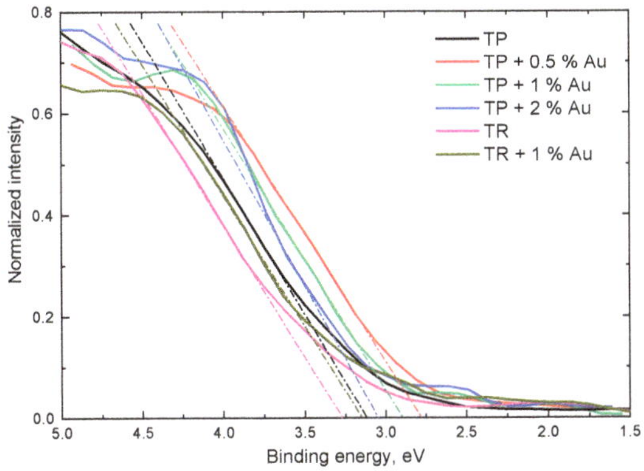

Figure 3. Determination of VBM of bare TiO_2 supports and TiO_2+Au catalysts by XPS analysis.

Figure S6 illustrates the UV–Vis DR spectra of the bare TiO_2 supports and the prepared TiO_2+Au. Across all materials, the UV–Vis DR spectra exhibit pronounced absorption in the wavelength range below 400 nm, attributable to the band gap of the TiO_2 supports [35]. The incorporation of Au resulted in heightened light absorption within the range of 450 to 600 nm. This enhancement can be ascribed to the existence of metallic Au nanoparticles within the TiO_2+Au catalysts and the consequent localized surface plasmon resonance effect induced by the plasmonic characteristics of Au [36,37]. A slight shift of the absorption edge in the UV light region of the TiO_2+Au catalysts compared to bare TiO_2 supports indicates that Au particles are deposited only on the surface of the TiO_2 supports [38], which is in agreement with the results of N_2 physisorption measurements. The outcomes of the solid-state PL measurements (depicted in Figure S7) show that all TiO_2+Au catalysts have lower solid-state PL intensities than the pure TiO_2 supports, which means that the catalysts have a lower charge carrier recombination rate than bare TiO_2. The lower charge carrier recombination rates in the TiO_2+Au catalysts indicate that there is a junction between the TiO_2 supports and the Au particles. At the interface between TiO_2 and Au, a barrier forms, i.e., the Schottky barrier (SB), which is an obstacle for the charge carriers when they want to migrate either from Au to TiO_2 or vice versa. Furthermore, the results of the PL measurements confirm the results of the XRD analysis, namely that the TiO_2 is present in the anatase form, as a characteristic anatase TiO_2 peak occurs at 3.17 eV [39,40]. In the solid state TiO_2+Au PL spectra a blue shift of the 3.17 eV peak to about 3.2 eV can be observed, as the gold particles suppress indirect phonon-assisted transitions in the anatase TiO_2 [41,42]. The peak at 2.9 eV indicates the lowest indirect transition $\Gamma 1b \rightarrow X1a$. The peaks at 2.7, 2.55 and 2.34 eV can be attributed to the existence of TiO_2 defects, oxygen vacancies and shallow trap levels [43,44].

The high-resolution XPS spectra for O 1s, Ti 2p, and Au 4f of the analyzed materials are presented in Figure S8a–c. The O 1s binding energies for all analyzed materials are approximately 530.5 eV, corresponding to bulk oxide (O^{2-}) in the TiO_2 lattice (Figure S8a). In the high-resolution O 1s spectra of TP+Au catalysts, a high-energy shoulder appears about 0.9–1.5 eV above the main O^{2-} peak, attributed to weakly adsorbed species [45] or low-coordinated oxygen ions (O^-) indicating oxygen vacancies [46,47]. Figure S8 shows that the intensity of this high-energy shoulder (~532.4 eV) increases with the amount of Au in the TP+Au catalysts, suggesting an increase in oxygen vacancies and adhesion between Au and TiO_2. These oxygen vacancies can act as electrostatic field centres at the TiO_2–Au interface, enhancing the adhesion between Au and TiO_2 and forming a network of Au-O-Ti [48]. In the Ti 2p spectra of all materials (Figure S8), two peaks at ~464.9 eV (Ti $2p_{1/2}$) and ~459.2 eV (Ti $2p_{3/2}$) are observed, typical for Ti in the +4 oxidation state [49]. These results indicate that the chemical bonding between Ti and O remains consistent across all materials studied, as the Ti $2p_{3/2}$ binding energy is ~459.2 eV in all samples (Figure S8b).

The Au 4f spectra for the investigated TiO_2+Au catalysts are shown in Figure S8c. The Au $4f_{7/2}$ and Au $4f_{5/2}$ binding energies at 83.9 and 87.6 eV, respectively, match the reported values for fully reduced Au [50]. As already noted by other researchers [51], the intensity of the Au 4f peak increases with higher Au content in the TP+Au catalysts. The presence of a peak at 86.6 eV in the spectrum of the TNR + 1% Au catalyst suggests that amorphous Au_2O_3 (Au $4f_{7/2}$ = 86.9 eV) or other Au^{3+}-containing species (e.g., $Au(OH)_3$) could coexist with metallic Au on the surface of the TNR nanorods [52]. Figure 3 shows the results of the measurement of the valence band maxima (VBM) of the analyzed solids. The differences between the VBMs of the TiO_2 supports and the TiO_2+Au catalysts represent the height of the SBs [53] (Table 2). It is noticeable that, for TP-based solids, the Schottky barrier height decreases with the augmentation of Au content in the catalysts. Furthermore, the catalyst TR + 1% Au has a lower height of SB than the catalyst TP + 1% Au. For the charge carriers formed in Au, a high SB is not preferred as they need more time and energy to move to TiO_2, which causes them to bunch and recombine at SB. For the charge carriers moving from the TiO_2 conduction band to Au, a high SB is positive as they cannot move to Au and remain in the conduction band of TiO_2. This allows them to achieve higher energy levels, thereby increasing the reduction potential [54].

Table 2. Amount and density of acidic sites determined with TPD of pyridine for the catalysts studied.

Sample	Amount of Acidic Sites	Density of Acidic Sites	Peak of Pyridine Desorption
	mmol/g	mmol/(m^2/g)	°C
TP	0.174	0.0019	312, 454, 554, above 700
TP *	0.183	0.0021	310, 450, 550
TP + 0.5% Au	0.121	0.0017	320, 441, 619
TP + 1% Au	0.139	0.0018	320, 453, 614
TP + 1% Au *	0.140	0.0018	318, 451, 613
TP + 2% Au	0.068	0.0009	376, above 700
TR	0.210	0.0020	310, 630
TR *	0.204	0.0019	310, 630
TR + 1% Au	0.190	0.0020	310, 629, above 750
TR + 1% Au *	0.190	0.0020	308, 625, above 750

* Samples were calcined at 150 °C for 3 h.

Besides the specific surface area, the separation of charge carriers, and the degree of crystallinity, the surface acidity of the investigated catalysts also significantly influences their behavior in the CWAO process [28,55]. To measure the surface acidic properties of the investigated catalysts, the temperature-programmed desorption (TPD) method was used with pyridine as the probe molecule (Figure 4). The temperatures at which the peaks of pyridine desorption appear during the TPD measurements, the density of the acidic sites and the number of acidic sites present in the studied solids are presented in Table 2. The positions of the peaks of pyridine desorption in high- or low-temperature ranges are due to the desorption of pyridine from strong or weak acidic sites. The presence of strong acidic sites on the catalyst surface could trigger either electrostatic interactions or side reactions, which in turn could lead to the accumulation of carbonaceous deposits on the catalyst surface (e.g., coking) and ultimately to the deactivation of the catalyst [28,56]. For the catalysts studied, it can be said that there is a linear trend between the decreasing number of acidic sites (from 0.210 mmol/g for the TR to 0.190 mmol/g for the TR + 1% Au) and the decreasing BET specific surface area of the solids studied (from 105 m^2/g for the TR to 94 m^2/g for the TR + 1% Au). For the two TiO_2+Au series produced, we can observe that, compared to the pure TiO_2 supports, the number of acidic sites decreases with the addition of gold and that, for the TP+Au series, the amount of acidic surface sites decreases drastically when the gold loading increases from 1 (0.139 mmol/g) to 2 wt.% (0.068 mmol/g). The distribution of the strength of the acidic sites, expressed by the temperatures of the peaks of pyridine desorption (Table 2 and Figure 4), shows that bare TP and TR supports exhibit the presence of medium and strong acidic sites, although the ratio between them is different. In the case of the TP sample, more strong acidic sites are present than weak acidic sites, while for the TR sample the trend is just the opposite. In addition, we must emphasize that we have also observed the desorption of pyridine from the TP sample in the temperature range above 700 °C (Figure 4). For the TR + 1% Au catalyst, it can be observed that the addition of gold did not alter the ratio of weak to strong acidic sites in comparison to the pure TR support. However, desorption of pyridine is still evident within the temperature range exceeding 750 °C. For the TP+Au series, the peak for the strong acidic site was shifted to lower temperatures in comparison to the bare TP support (from 550 °C for TP sample to 450 °C for TP + 0.5% Au and TP + 1% Au catalysts). In contrast, the TP + 2% Au sample showed a completely different distribution of the strength of the acidic sites than the pure TP support and the catalysts with 0.5 and 1 wt.% Au loadings. This could be due to the fact that the TP + 2% Au catalyst also showed the highest average gold particle size among all TP+Au catalysts investigated. It is evident that the presence of small gold nanoparticles in the TiO_2+Au catalysts has no effect on the characteristics of the acidic-basic properties compared to bare TiO_2 supports, in contrast to larger gold nanoparticles, which is due to the change in metal–support interaction. Furthermore, in the case of the TP + 2% Au catalyst, we observed the desorption of pyridine in the temperature range above 700 °C, indicating the presence of strong acidic sites. This could lead to a decrease in the catalytic activity of the TP + 2% Au catalyst during the CWAO reaction, as the presence of strong acidic sites favours the accumulation of carbonaceous deposits [28]. We also performed the TPD analysis for the pure TiO_2 supports and the catalysts with 1 wt.% Au loading, which were calcined at 150 °C for 3 h prior to the measurements to simulate the effects of reaction temperature on the catalyst properties studied (Figure S9). The analysis reveals that calcination has no discernible impact on the quantity, density, or strength of the acidic sites in the investigated materials. This suggests that the properties of the studied catalysts remain unaffected by the operating conditions of the CWAO process.

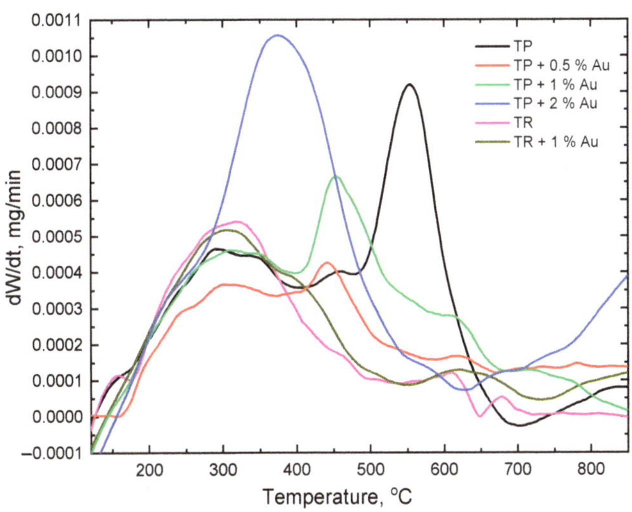

Figure 4. TPD of pyridine from the surface of the catalysts studied.

2.2. CWAO of Glyphosate

The CWAO operating and reaction conditions were chosen based on the results of preliminary experiments, where we used 500 mg of inert SiC and different experimental conditions, such as different reaction temperatures (120 to 180 °C) and different flow rates of the 10.0 mg/L glyphosate aqueous solution (0.5–1 mL/min). The data presented in

Figure S10 suggest that, under conditions of 180 °C and a flow rate of 1 mL/min, approximately 75% of the glyphosate undergoes degradation through non-catalytic oxidation reactions within the liquid phase. The rate of glyphosate degradation decreases to about 35, 10 and 5% when the reaction temperature decreases to 150, 135 and 120 °C, respectively. Decreasing the rate of aqueous glyphosate solution from 1 mL/min to 0.5 mL/min at 120 °C exhibits only a minimal effect on the glyphosate degradation rate. Based on these results, we decided to start the CWAO experiment using the synthesized catalysts at 135 °C and a flow rate of 1 mL/min and operate at these conditions for 48 h. After this, we increased the temperature to 150 °C and continued with the experiment for another 48 h.

Figure 5 shows the results of glyphosate and TOC conversions over time in a three-phase trickle-bed reactor packed with the synthesized catalysts. The data were used to calculate glyphosate degradation and TOC conversion over the 96-h CWAO experiment and are listed in Table 3. We also performed CHNS elemental analysis of fresh and 96-h operated catalysts to calculate TOC accumulation (TOC_{accu}) and TOC mineralization (TOC_{miner}, Table 3). After 35 h at 135 °C, all materials achieved steady-state performance. After 48 h, both TiO_2 supports had similar glyphosate degradation rates. The TR support started at nearly 100% degradation, dropping to ~25% in 30 h, while the TP support started at 30% and remained constant. These differences are due to varying S_{BET} values and the nature of acidic surface sites. The TP support's strong acidic sites and lower surface area led to lower initial activity as glyphosate degradation products blocked its surface. The TR support showed slow deactivation over time due to higher initial activity. CHNS analysis (Table 3) confirmed higher TOC_{accu} on the TR support, almost twice that of the TP support, due to its high initial catalytic activity generating more degradation products.

Table 3. Glyphosate conversions obtained in the presence of the investigated catalysts during the whole 96 h on stream. The carbon content, as determined by CHNS elemental analysis, was assessed on the surfaces of newly prepared solids (TC_{fresh}) and catalysts employed in the CWAO process (TC_{spent}). Additionally, the total organic carbon (TOC) conversions (96 h on stream) were correlated with the percentage of TOC deposition observed on the catalyst surfaces (TOC accumulation on the surface of spent catalyst (TOC_{accu}) and TOC mineralization to CO_2 and H_2O (TOC_{miner})).

Sample	Glyphosate Conversion 0–96 h	TOC Conversion 0–96 h	TC_{fresh}	TC_{spent}	TOC_{accu}	TOC_{miner}
	%		mg/g		%	
TP	35	19	0.16	0.27	2.7	16.3
TP + 0.5% Au	51	41	0.13	0.46	8.1	33.9
TP + 1% Au	67	48	0.11	0.36	6.1	41.9
TP + 2% Au	45	30	0.17	0.50	8.1	21.9
TR	51	29	0.13	0.33	4.9	24.1
TR + 1% Au	65	42	0.15	0.57	10.3	31.7

The results in Figure 5 show that materials containing Au exhibit higher catalytic activity than the bare TiO_2 supports. Heating the catalysts in the CWAO process generates charge carriers, similar to those produced in heterogeneous photocatalysis, by illuminating the catalysts with light of suitable energy [57]. As noted, charge carriers and reactive oxygen species oxidize glyphosate even with only bare TiO_2 supports (Figure 5). Adding Au enhances the separation of electron-hole pairs, reducing signal intensities in the photoluminescence (PL) spectra of TiO_2+Au catalysts compared to bare TiO_2 (Figure S7). UV–Vis DR measurements (Figure S6) suggest that metallic Au in TiO_2+Au catalysts can absorb visible light and near-infrared radiation. It can be tentatively proposed that, in the CWAO process, near-infrared radiation absorbed by Au particles may trigger the Au surface plasmon resonance effect, forming electrons that are injected into the TiO_2 support, thus increasing the catalytic activity of TiO_2+Au catalysts compared to bare TiO_2. For the investigated materials, the surface area of the TiO_2 supports in the TiO_2+Au catalysts is significantly larger than that of the Au particles, which leads to the conclusion that the activation of oxygen in the CWAO process mainly occurs at the surface of the TiO_2 supports. This means that the interface between TiO_2 and Au as well as the surface of Au particles should not be covered by the glyphosate degradation products, so that the Au–TiO_2 junction and the Au particles can participate in the separation and generation of the charge carriers. We see (Figure 5) that the adsorption of glyphosate degradation products influences the glyphosate degradation rates of the used TiO_2+Au catalysts regardless of which TiO_2 support was used. At 48 h, the highest degradation of glyphosate, when considering only the TP based materials, was achieved by that with 1 wt.% Au loading (~50%) and the lowest by that with 2 wt.% of Au (~35%). The TR + 1% Au catalyst exhibited a 5% higher glyphosate degradation rate after 48 h relative to the TP + 1% Au sample. After increasing the temperature to 150 °C, the glyphosate degradation rates also increased for all tested catalysts by ~15 to 20%, as glyphosate degradation over SiC also increased by ~15%. It should be noted that the desorption of some weakly bound glyphosate degradation products from the catalyst surface can also be achieved by increasing the reaction temperature. Almost all investigated catalysts, except the TR + 1% Au sample, reached a steady-state performance after 10 h on stream at 150 °C. After 48 h at 150 °C (all together 96 h), the highest glyphosate degradation rate was obtained in the presence of TP + 1% Au catalyst (~75%), followed by the TP + 0.5% Au (~62%) and TP + 2% Au (~52%) catalysts. Bare TR support and TR + 1% Au catalyst exhibited almost the same glyphosate degradation rates as the TP + 2% Au sample. In the case of the TR + 1% Au sample, we can see that the initial glyphosate degradation rate was around 75%, which then gradually decreased with the time on stream almost linearly until ~55%, where it reached the glyphosate degradation rate of the bare TR sample. The decline in the glyphosate degradation rate observed with the TR + 1% Au catalyst can be attributed to the accumulation of glyphosate degradation products on the material's surface during the CWAO process. Additionally, the TR + 1% Au catalyst generated a greater quantity of glyphosate degradation products compared to the pristine TR support, further contributing to this effect. Another possibility would be that the Au particles are washed off the TR surface during the CWAO process. For this reason, we also performed SEM–EDX analysis of the spent TiO_2+Au catalysts and, as the results in Table S1 show, the Au loading of the spent catalysts was almost the same as that of the fresh samples. It should be also emphasized that the TR + 1% Au catalyst exhibited the highest carbon accumulation (and highest TOC_{accu}) rate among all the materials investigated, as shown by the CHNS elemental analysis results provided in Table 3. We also measured the TOC conversions at 20, 45, 70 and 95 h on stream. The results displayed in Figure 5b show that the TOC conversion rate orders for the investigated catalysts at 45 and 95 h on stream are the same as the orders for the catalysts based on the glyphosate degradation rates obtained after 48 and 96 h on stream. Furthermore, regarding the TOC conversion rates, an enhancement in conversions is observed for all tested materials with the elevation of the reaction temperature to 150 °C. A comparison of the activity of the catalysts of the TP+Au series shows that a higher Au loading with the same average Au particle size and the same SB height is favourable for the long-term catalytic activity in the CWAO degradation of glyphosate, as the comparison of the samples TP + 0.5% Au and TP + 1% Au shows. This could be due to the fact that the glyphosate degradation products can cover the surface of the Au particles, which represent the catalytically active phase in the CWAO process, more quickly in the case of the 0.5 wt.% Au loading. It can also be said that a higher SB is advantageous, as the TP + 1% Au catalyst already has a higher catalytic activity at the beginning of the CWAO of glyphosate than the TP + 2% Au solid. A higher SB allows the charge carriers to remain in the TiO_2 conduction band and increase their reduction potential [54] for the generation of reactive oxygen species (ROS) as, for example, hydroxyl radicals ($^{\bullet}OH$) and superoxide anion radicals ($^{\bullet}O_2^-$). The presence of the latter in our investigated reactor system is evidenced by the detected degradation products of glyphosate (see Section 3.3), although we cannot clearly state whether they are formed by a direct or indirect process [58]. The low catalytic activity of the TP + 2% Au sample in the CWAO of glyphosate can also stem from the existence of strong acidic sites and glyphosate degradation products adsorption (high TOC_{accu} of 8.1%) [28]. When comparing the performance of TR + 1% Au (TOC_{accu} = 10.3%; TOC_{miner} = 31.7%) and TP + 1% Au (TOC_{accu} = 6.1%; TOC_{miner} = 41.9%) catalysts, it can also be said that a higher SB is favourable for the catalytic activity in the long-term CWAO run, but it must be taken into account here that smaller Au nanoparticles (d_{Au} = 9.2 nm, Table 1) are found on the TR + 1% Au surface, compared to the TP + 1% Au solid (d_{Au} = 36.9 nm, Table 1), which could be more easily covered with glyphosate degradation products, leading to the catalyst deactivation. In addition, the pyridine TPD profile of TR + 1% Au catalyst (Figure 4) shows the desorption of pyridine even in the temperature range above 750 °C and thus the presence of strong acidic sites that favour the adsorption of carbonaceous deposits during CWAO of glyphosate [28].

(**a**)

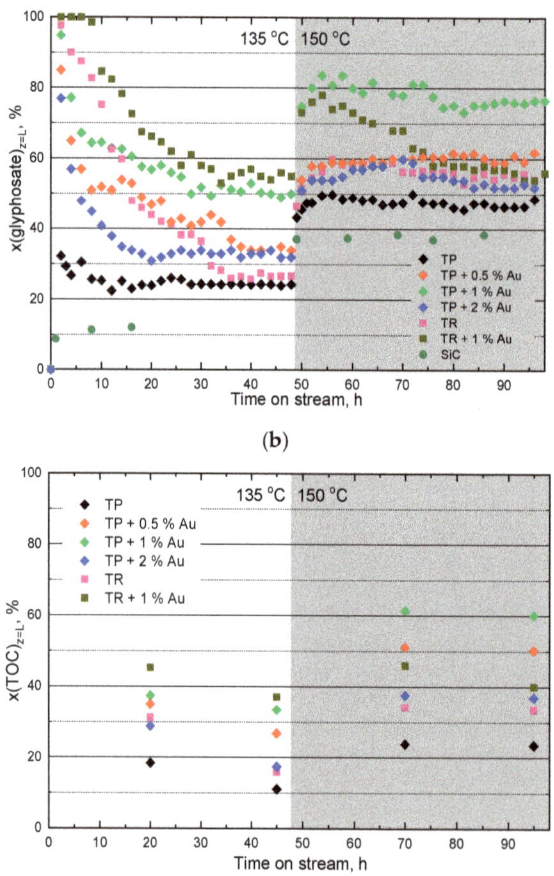

Figure 5. (**a**) Glyphosate and (**b**) TOC conversion as a function of time on the stream obtained in the trickle-bed reactor in the presence of prepared materials.

TPO analysis of the spent catalysts (results are shown in Figure 6) was employed to monitor the oxidation of the carbonaceous species adsorbed on the surface of the investigated catalyst. Comparing only the TiO_2 supports, the highest weight loss was measured for the TR support, confirming that a larger specific surface area favourably affects the adsorption of glyphosate degradation products, as the calculated weight loss per surface area is almost the same for TR and TP supports (0.0171% weight loss/(m^2/g) for TR and 0.0178% weight loss/(m^2/g) for TP). Further, the TiO_2+Au catalysts exhibited a higher weight loss compared to the bare TiO_2 supports as they also degraded more glyphosate and produced more degradation products. The highest weight loss was measured for the TR + 1% Au catalyst, which is to be expected considering the results of the CWAO runs (and the pyridine TPD and CHNS elemental analyses), where the catalyst showed a high affinity to adsorb glyphosate degradation products. Figure 6b shows the derivations of the mass as a function of temperature for the thermogravimetric measurements. According to the literature research, the peaks up to 410 °C belong to amorphous carbon, those between 410 and 520 °C to single-walled carbon nanotubes and the peaks between 700 and 800 °C to graphite [59]. In the materials we examined, most of the deposits belong to amorphous carbon. However, some products in the form of carbon nanotubes and graphite can also be observed. The latter in particular are known to have a strong adsorption affinity for organic compounds and can thus enhance the apparent removal of glyphosate degradation products from the liquid phase on the catalyst surface [60].

(**a**)

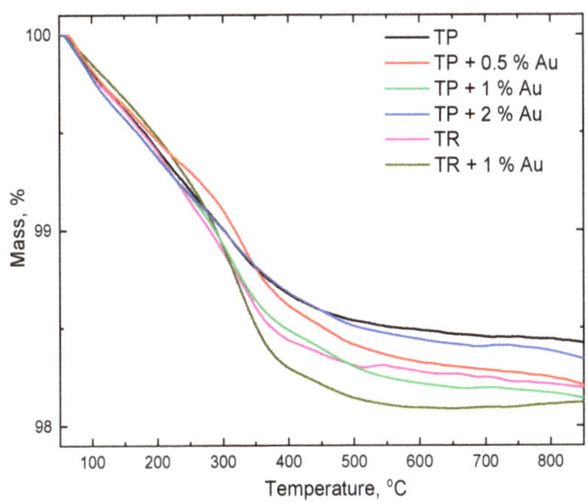

(**b**)

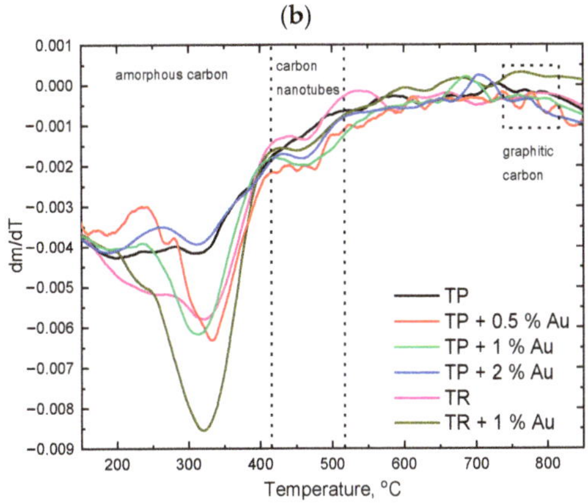

Figure 6. (**a**) TGA–TPO profiles of the TiO_2 and TiO_2+Au samples after use in the CWAO process. (**b**) Derivation of mass as a function of temperature.

2.3. Identification of Degradation Products

One of the goals of our study was also to identify the major degradation products generated during CWAO of glyphosate and, based on the data, propose a possible oxidation pathway characteristic of our reaction system. Multiple investigations posit two conceivable oxidation pathways for glyphosate under oxidative conditions [22,27,61–63]. The initial pathway involves glyphosate transforming into amino-methyl-phosphonic acid (AMPA) through the rupture of the C–N bond. Subsequently, AMPA can undergo further oxidation to yield methylamine, NH_4^+, CH_2O, NO_3^-, and PO_4^{3-}. In an alternative route, glyphosate is transformed into sarcosine via direct cleavage of the C–P bond, which can then be further oxidized to glycine, CH_2O, and NH_4^+. Xing et al. [27], based on their experimental findings regarding the evolution of primary intermediates and end products, proposed a preliminary degradation pathway for glyphosate in the CWAO process utilizing activated carbons as catalysts. Initially, the C–N bond cleavage of glyphosate takes place, yielding equivalent amounts of formaldehyde (CH_2O) and AMPA through reactive radicals as, for example, superoxide anion radicals ($^\bullet O_2^-$) and hydroxyl radicals ($^\bullet OH$). Subsequently, in the second step, cleavage of the C–N bond of AMPA ensues, leading to the formation of PO_4^{3-}, NH_4^+, and CH_2O. In addition, oxidation of CH_2O and NH_4^+ to HCOOH and NO_3^- takes place. Our qualitative and quantitative data on the catalytic degradation products of glyphosate when using the TP + 1% Au as a model catalyst are shown in Table 4. At t = 0, no degradation products of glyphosate, apart from AMPA, were found in the sample. The unexpected presence of this acid could be explained by a chemical impurity of the commercial glyphosate or by a spontaneous degradation of glyphosate in the aqueous solution, prior to its introduction to the reactor. More importantly, the results in Table 4 demonstrate that the oxidation of glyphosate must include the cleavage of the C–N bond, as both glyoxylic acid and AMPA were detected in samples withdrawn from the reactor after 20–90 h on stream. No traces of sarcosine were found in the tested samples; however, glycine, a downstream degradation product of sarcosine, was determined at similar concentration levels as glyoxylic acid, which indicates that glyphosate also degraded by following a secondary pathway that involves the C–P bond cleavage. Munner et al. [64] observed the same phenomenon, namely that no sarcosine was formed when $^\bullet OH$ radicals attacked the non-adsorbed glyphosate, leading to the rupture of the C–P bond and formation of glycine. Other downstream degradation products, such as oxalic acid and MPA, were not detected in the studied samples.

Table 4. Determination of the degradation products of glyphosate in the end-product solutions using the TP + 1% Au catalyst in the CWAO experiment.

Time on Stream	Glyoxylic Acid	Sarcosine	Glycine	Oxalic Acid	MPA	AMPA
h	mg/L					
0	n.d.	n.d.	n.d.	n.d.	n.d.	0.73
20	0.21	n.d.	0.18	n.d.	n.d.	1.20
40	0.25	n.d.	0.32	n.d.	n.d.	1.69
90	0.31	n.d.	0.50	n.d.	n.d.	2.30

n.d.—not detected.

3. Experimental

3.1. Catalyst Preparation

Two distinct TiO_2 substrates underwent wet impregnation to yield TiO_2+Au catalysts. The first, commercially sourced TiO_2 DT-51 (generously provided by CristalACTiV^{TM}, Thann, France), designated as TP, underwent a pre-calcination step in air for 2 h at 200 °C before the wet impregnation synthesis process. This pre-treatment ensured that the intrinsic properties of TP remained unaffected by the subsequent operating conditions of the CWAO experiment. TiO_2 nanorods (TR), prepared by a hydrothermal process [65] and using DT-51 as the starting material, were employed as the second TiO_2 support. TR underwent further calcination in air for 2 h at 500 °C to achieve complete transformation of the amorphous TiO_2 to anatase TiO_2. An amount of 0.5 g of the TiO_2 support was added to an aqueous solution of $HAuCl_4 \times 3H_2O$ (25 mL) and stirred for 20 h. The molarity of $HAuCl_4 \times 3H_2O$ was adjusted to obtain TiO_2+Au catalysts with Au weight percent (wt.%) between 0.5 and 2. The suspension was allowed to dry for 12 h at room temperature before undergoing calcination at 300 °C for 2 h, with a heating ramp of 300 °C per hour. The samples were designated as TP(TR) + x% Au, where x indicates the nominal weight percent of Au (from 0.5 to 2).

3.2. Catalyst Characterization

Images of the catalysts' morphology and elemental composition were obtained using a field emission scanning electron microscope (SEM, Carl Zeiss SUPRA 35 VP Oberkochen, Germany) equipped with an energy dispersive X-ray detector spectrometer (EDS) Inca 400 from Oxford Instruments. Additionally, transmission electron microscopy (TEM) analysis was performed using a JEM-2100 instrument from Jeol Inc. (Tokyo, Japan), operating at 200 keV, to generate gold particle size distribution diagrams for the materials under investigation.

The phase composition of the materials under investigation was determined using X-ray powder diffractometry (XRD) with a PANalytical X'pert PRO MPD instrument (Almero, The Netherlands), employing Cu Kα1 radiation (wavelength of 1.54056 Å) in a reflection geometry covering an angular range from 10 to 90° with steps of 0.0341°. Crystalline phases were identified using PDF standards from the International Centre for Diffraction Data (ICDD). Prior to N_2 adsorption–desorption isotherm measurements at −196 °C using the Micromeritics TriStar II 3020 analyzer, the materials underwent pretreatment in an N_2 stream (purity 6.0, Linde, Munich, Germany,) for 1 h at 90 °C and for 4 h at 180 °C using the Micromeritics SmartPrep degasser.

A Perkin Elmer Lambda 650 UV–Vis spectrophotometer equipped with a Harrick (New York, NY, USA) Scientific Praying Mantis DRP-SAP accessory was used to obtain the UV–Vis diffuse reflectance (UV–Vis DR) spectra of the catalysts.

The LS-55 UV–Vis fluorescence spectrometer from Perkin Elmer (Waltham, MA, USA), was used to obtain the photoluminescence (PL) emission spectra of the analyzed materials (excitation wavelength 315 nm).

The valence band maxima (VBM) of the analyzed catalysts were determined using the PHI–TFA XPS spectrometer (Physical Electronics Inc., Chanhassen, MN, USA). The surface of the analyzed materials was excited with X-rays emitted from a monochromatic Al-Kα source with an energy of 1468.6 eV. An energy analyzer with a pass energy of 29 eV and a resolution of 0.6 eV was used to obtain high energy resolution spectra.

The acidic properties of the studied catalysts were measured using the Perkin Elmer Pyris 1 TGA instrument. The studied catalysts were heated for 10 min at 200 °C in air and then cooled to 120 °C. After reaching 120 °C, a stream of N_2 and pyridine was turned on and the catalyst surface was saturated with pyridine for 10 min. To remove the excess pyridine from the catalyst surface, the solids were purged with N_2 for another 90 min. Then, temperature-programmed desorption (TPD) of pyridine was performed by increasing the temperature to 750 °C with a heating ramp of 20 °C/min.

The CHNS elemental analyzer employing the Perkin Elmer 2400 series II was used to determine the extent of carbonaceous deposits on the catalyst surfaces after the CWAO experiments.

The nature of the carbonaceous species accumulated on the surface of the studied catalysts after their use in the CWAO process was determined by the TGA measurements performed with the Perkin Elmer Pyris 1 TGA instrument. The spent catalyst samples were heated from 50 to 850 °C in air (10 °C/min).

A thermogravimetric analysis (TGA-TPO) of the spent catalysts in air (50 mL/min) was performed with the Perkin Elmer Pyris 1 TGA instrument, which allowed us to monitor the oxidation of the carbonaceous species adsorbed on the catalyst surface during the oxidation of the aqueous glyphosate solution.

3.3. Glyphosate Oxidation Runs in a Trickle-Bed Reactor

A computer-controlled and fully automated Microactivity-Reference (MA-Ref) reactor system from PID Eng and Tech (Madrid, Spain) was used to perform the CWAO experiments. The MA-Ref unit can be defined as a continuous-flow, three-phase trickle-bed reactor with the concurrent down-flow of gas and liquid phase. Oxidation of the contaminant dissolved in water was carried out in a Hastelloy C-276 tubular reactor (Autoclave Engineers, Erie, PA, USA) with a length of 305 mm and inner diameter of 9 mm. The tubular reactor was located in the central part of the reactor unit. An HPLC positive alternative displacement pump (Gilson, Middleton, WI, USA, model 307) was used to pump the 10.0 mg/L aqueous glyphosate solution into the reactor unit. The Bronkhorst EL-FLOW HI-TECH mass flow controller was used to supply pure O_2 (Messer, Ruše, Slovenia, USA, purity 5.0) to the reactor system at a prescribed flow rate. Before entering the reactor, the preheated gas and liquid streams were combined in a T-joint. Sintered stainless steel (SS 316, 10 μm) filters were installed at the bottom and top of the reactor to protect the reactor assembly from contamination. A total of 300 mg of a catalyst was placed on a sintered 2 μm frit located in the center of the reactor tube. The reaction temperature was monitored with a K-type thermocouple installed through the top of the reactor tube and in contact with the catalyst bed. The PID temperature controller (TOHO, Sagamihara-Shi, Japan, model TTM-005) was used to control the reaction temperature within ±1.0 °C. The liquid and gas phases were separated at the outlet of the reactor unit using a high-pressure liquid-gas separator cooled by a Peltier cell. The operating conditions for the CWAO experiments are listed in Table 5.

Table 5. The operational parameters of the continuous-flow, three-phase trickle-bed reactor during the CWAO experiments.

Parameter	Value
Mass of catalyst in bed, g	0.3
Reaction temperature, °C	135–150
Total operating pressure, bar	13.1–14.7
Oxygen partial pressure, bar	10.0
Gas flow rate, mL/min	60.0
Liquid flow rate, mL/min	1.0
Glyphosate feed concentration, mg/L	10.0

3.4. Analysis of the End-Product Solutions

Ion chromatography (IC) with pulsed amperometric detection was used to measure glyphosate concentration in liquid-phase samples collected during the CWAO experiments. Measurements were conducted using an Metrohm IC system consisting of a 930 Compact IC Flex with an amperometric detector with working electrode made of gold (3 mm diameter) and reference electrode made of palladium. The working temperature of the amperometric detector was 35 °C. Samples (100 μL) were injected automatically using an 858 Professional Sample Processor. Glyphosate was separated at 30 °C on the Metrosep Carb 2–150/4.0 high-capacity anion separation column. The sodium acetate eluent contained 300 mmol/L sodium acetate and 10 mmol/L sodium hydroxide. The eluent flow rate was 0.4 mL/min from 0 to 16 min and 0.8 mL/min from 16 to 30 min. The IC system was controlled by MagIC Net Compact software (version 3.2).

A Shimadzu TOC-L analyzer, paired with an ASI-L autosampler, was employed for the analysis of total organic carbon (TOC) levels in both untreated and treated glyphosate solutions. High-temperature catalytic oxidation of the organics contained in the injected samples was carried out at 680 °C in synthetic air (Messer, purity 5.0). The amount of CO_2 formed was quantified using a calibrated NDIR detector. The observed error for three replicates was within ±1%.

The main intermediates and final products of the glyphosate degradation were identified and quantified by hydrophilic interaction liquid chromatography (HILIC). Chromatographic analyses were carried out on a Thermo Scientific Accela 1250 UHPLC system (Waltham, MA, USA), which was coupled to a Thermo Scientific LTQ Velos mass spectrometer (MS). The separation of compounds was achieved by using a Waters Acquity UPLC BEH HILIC 1.7 μm column (100 × 2.1 mm i.d., Milford, MA, USA). The autosampler and column oven were kept at temperatures of 25 °C and 40 °C, respectively, throughout the experiment. The mobile phase comprised solvent A (40 mM ammonium acetate) and solvent B (10 mM ammonium acetate in a water-acetonitrile mixture at a ratio of 1:9, *v*/*v*). A solvent gradient was employed at a flow rate of 0.6 mL/min as follows: from 0 to 6 min (100% to 60% B), from 6 to 7 min (60% to 100% B), and from 7 to 13 min (maintaining 100% B). The data were acquired by using a PDA detector at 220 nm and by a mass spectrometer scanning in the range m/z = 60–300. Electrospray probe was operated in negative ion mode and the MS parameters were set as follows: spray voltage = 3 kV, T heater = 400 °C, T transfer capillary = 300 °C, sheath gas flow rate = 70 arbitrary units, and auxiliary gas flow rate = 20 arbitrary units. For each analysis, 10 μL of sample was injected, which was prepared in the following manner: the solvent from the aqueous solutions collected from the reactor (1.5 mL) at predetermined time intervals was removed under reduced pressure and the solid residue was reconstituted in 0.15 mL water-acetonitrile (3:7, *v*/*v*) which contained 1 mM EDTA. The resulting solution was then subjected to LC–UV–MS analysis. The samples were analyzed in duplicates. Quantitation of seven analytes, i.e., glyoxylic acid, sarcosine, glycine, oxalic acid, methyl-phosphonic acid (MPA), glyphosate, and amino-methyl-phosphonic acid (AMPA), was carried out by using an external standard calibration method.

4. Conclusions

The objective of this study was to use TiO_2+Au catalysts and CWAO for the degradation of the herbicide glyphosate and to examine the impact of Au loading and the structural characteristics of the anatase TiO_2 supports (nanoparticles (TP, S_{BET} = 88 m^2/g) and nanorods (TR, S_{BET} = 105 m^2/g)) on the catalytic activity of the TiO_2+Au catalysts. The results show that the addition of Au has a positive effect on the glyphosate degradation rate independent of the TiO_2 support and the amount of Au added and that, at the beginning of the CWAO experiment (reaction temperature 135 °C), the TR + 1% Au sample with the largest specific surface area and the lowest SB height has the highest activity. In the TP+Au series, where different Au loadings from 0.5 to 2 wt.% were investigated, the highest catalytic activity was observed for the TP + 1% Au catalyst, which had a higher SB, helping the charge carriers in the TiO_2 conduction band to increase their reduction potential by preventing them from returning to the Au particles. Compared to the TP + 0.5% Au sample, the higher Au loading in the case of the TP + 1% Au catalyst and the presence of more Au particles improved the catalytic activity, as the adsorption and blocking of Au particles by glyphosate degradation products is less detrimental. The detrimental effects of the adsorption of glyphosate degradation products on the catalytic activity of the TiO_2+Au catalysts studied became particularly evident when we increased the reaction temperature to 150 °C. The higher specific surface area of the TR + 1% Au sample and the presence of highly acidic sites favouring the adsorption of glyphosate degradation products, together with the smallest average size of Au particles under the examined TiO_2+Au catalysts, led to an almost linear decrease of the catalytic activity of the TR + 1% Au sample with time on stream. The main degradation products of glyphosate were glyoxylic acid and AMPA, indicating that the glyphosate C–N bond was ruptured. No traces of sarcosine were found, but glycine was determined in similar concentrations as glyoxylic acid, indicating that the glyphosate C–P bond was also cleaved. The results of the present study indicate that high long-term catalytic activity of TiO_2+Au catalysts in the CWAO process for the degradation of glyphosate dissolved in water is a combination of the appropriate specific surface area, SB height, acidic-basic properties, Au loading, and average Au particle size, as obtained in the case of the TP + 1% Au catalyst.

Supplementary Materials: The following supporting information can be downloaded at: https://www.mdpi.com/article/10.3390/catal14070448/s1, Table S1: The outcomes of the SEM-EDS elemental analysis were derived from scans conducted at 15 kV in area scan mode across more than ten randomly chosen locations. Figure S1. (a) N_2 adsorption-desorption isotherms of the fabricated materials and (b) corresponding BJH pore size distributions. Figure S2. SEM images of the TiO_2+Au catalysts under investigation are provided alongside corresponding elemental mapping. In these images, the colors cyan, green, and red are used to denote the elemental components of Ti, O, and Au, respectively. Figure S3. TEM images of the investigated TiO_2+Au catalysts, taken at different magnifications. The red arrows mark the Au nanoparticles detected by TEM elemental analysis. Figure S4. Au particle size distribution in the investigated TiO_2+Au catalysts. Figure S5. Zeta potential as a function of pH value for TP and TR supports. Figure S6. UV-Vis DR spectra of the analyzed TiO_2 supports and TiO_2+Au catalysts. Figure S7. The photoluminescence (PL) emission spectra in solid-state were captured for both pure TiO_2 supports and the synthesized TiO_2+Au catalysts. Figure S8. High resolution spectra for (a) O 1s, (b) Ti 2p and (c) Au 4f for the examined catalysts. Figure S9. TPD of pyridine from the surface of (a) fresh TP and TP + 1% Au solids and TP and TP + 1% Au samples calcined at 150 °C for 3 h in air (marked as TP calc@150 °C/3 h and TP + 1% Au calc@150 °C/3 h), and (b) fresh TR and TR + 1% Au solids and TR and TR + 1% Au samples calcined at 150 °C for 3 h in air (marked as TR calc@150 °C/3 h and TR + 1% Au calc@150 °C/3 h). Figure S10. Conversion of glyphosate as a function of temperature and flow rate obtained in the trickle-bed reactor in the presence of 500 mg SiC ($c_{glyphosate, feed}$ = 10.0 mg/L).

Author Contributions: Conceptualization, G.Ž. and A.P.; methodology, G.Ž. and A.A.; validation, G.Ž. and A.P.; formal analysis, G.Ž. and A.A.; investigation, G.Ž. and A.A.; resources, A.A. and A.P.; writing—original draft preparation, G.Ž. and A.A.; writing—review and editing, G.Ž., A.A. and A.P.; visualization, G.Ž. and A.A.; supervision, G.Ž.; funding acquisition, A.A. and A.P. All authors have read and agreed to the published version of the manuscript.

Funding: The authors acknowledge financial support from the Slovenian Research and Innovation Agency (research core funding nos. P2-0150 and P1-0005).

Data Availability Statement: Data will be made available on request.

Acknowledgments: The authors thank Janez Zavašnik and Janez Kovač (Jožef Stefan Institute, Ljubljana, Slovenia) for carrying out TEM and XPS analyses.

Conflicts of Interest: The authors declare that they have no known competing financial interests or personal relationships that could have appeared to influence the work reported in this paper.

References

1. Roy, S.; Vashishtha, M.; Saroha, A.K. Catalytic wet air oxidation of oxalic acid using platinum catalyst in bubble column reactor. *J. Eng. Sci. Tech. Rev.* **2010**, *3*, 95–107. [CrossRef]
2. Luck, F. Wet air oxidation: Past, present and future. *Catal. Today* **1999**, *53*, 81–91. [CrossRef]
3. Levec, J.; Pintar, A. Catalytic wet-air oxidation processes: A review. *Catal. Today* **2007**, *124*, 172–184. [CrossRef]
4. Cybulski, A. Catalytic wet air oxidation: are monolithic catalysts and reactors feasible? *Ind. Eng. Chem. Res.* **2007**, *46*, 4007–4033. [CrossRef]
5. Bistan, M.; Tišler, T.; Pintar, A. Catalytic and photocatalytic oxidation of aqueous bisphenol a solutions: Removal, toxicity, and estrogenicity. *Ind. Eng. Chem. Res.* **2011**, *51*, 8826–8834. [CrossRef]
6. Pintar, A.; Batista, J.; Tišler, T. Catalytic wet-air oxidation of aqueous solutions of formic acid, acetic acid and phenol in a continuous-flow trickle-bed reactor over Ru/TiO_2 catalysts. *Appl. Catal. B* **2008**, *84*, 30–41. [CrossRef]
7. Erjavec, B.; Kaplan, R.; Djinović, P.; Pintar, A. Catalytic wet air oxidation of bisphenol A model solution in a trickle-bed reactor over titanate nanotube-based catalysts. *Appl. Catal. B* **2013**, *132–133*, 342–352. [CrossRef]
8. Kaplan, R.; Erjavec, B.; Senila, M.; Pintar, A. Catalytic wet air oxidation of bisphenol A solution in a batch-recycle trickle-bed reactor over titanate nanotube-based catalysts. *Environ. Sci. Pollut. Res.* **2014**, *21*, 11313–11319. [CrossRef]
9. Thongprakaisang, S.; Thiantanawat, A.; Rangkadilok, N.; Suriyo, T.; Satayavivad, J. Glyphosate induces human breast cancer cells growth via estrogen receptors. *Food Chem. Toxicol.* **2013**, *59*, 129–136. [CrossRef]
10. Rani, L.; Thapa, K.; Kanojia, N.; Sharma, N.; Singh, S.; Grewal, A.S.; Srivastav, A.L.; Kaushal, J. An extensive review on the consequences of chemical pesticides on human health and environment. *J. Clean. Prod.* **2021**, *283*, 124657. [CrossRef]

11. Muñoz, J.P.; Bleak, T.C.; Calaf, G.M. Glyphosate and the key characteristics of an endocrine disruptor: A review. *Chemosphere* **2021**, *270*, 128619. [CrossRef]
12. Zhang, J.W.; Xu, D.Q.; Feng, X.Z. The toxic effects and possible mechanisms of glyphosate on mouse oocytes. *Chemosphere* **2019**, *237*, 124435. [CrossRef]
13. Kudsk, P.; Mathiassen, S.K. Pesticide regulation in the European Union and the glyphosate controversy. *Weed Sci.* **2020**, *68*, 214–222. [CrossRef]
14. Jacquet, F.; Delame, N.; Vita, J.L.; Huyghe, C.; Reboud, X. The microeconomic impacts of a ban on glyphosate and its replacement with mechanical weeding in French vineyards. *Crop Prot.* **2021**, *150*, 105778. [CrossRef]
15. Matousek, T.; Mitter, H.; Kropf, B.; Schmid, E.; Vogel, S. Farmers' intended weed management after a potential glyphosate Ban in Austria. *Environ. Manag.* **2022**, *69*, 871–886. [CrossRef]
16. Leonelli, G.C. The glyphosate saga continues: 'dissenting'member states and the European way forward. *Transl. Environ. Law* **2023**, *12*, 200–224. [CrossRef]
17. Peruzzo, P.J.; Porta, A.A.; Ronco, A.E. Levels of glyphosate in surface waters, sediments and soils associated with direct sowing soybean cultivation in north pampasic region of Argentina. *Environ. Pollut.* **2008**, *156*, 61–66. [CrossRef]
18. Botta, F.; Lavison, G.; Couturier, G.; Alliot, F.; Moreau-Guigon, E.; Fauchon, N.; Guery, B.; Chevreuil, M.; Blanchoud, H. Transfer of glyphosate and its degradate AMPA to surface waters through urban sewerage systems. *Chemosphere* **2009**, *77*, 133–139. [CrossRef]
19. European Community Council. *L330, 32-5498/83 EEC Directive Concerning the Quality of Water Intended for Human Consumption*; Official Journal of the European Communities: Luxembourg, 1998.
20. Muñoz, P.G.; Dachtler, W.; Altmayer, B.; Schulz, R.; Robert, D.; Seitz, F.; Rosenfeldt, R.; Keller, N. Reaction pathways, kinetics and toxicity assessment during the photocatalytic degradation of glyphosate and myclobutanil pesticides: Influence of the aqueous matrix. *Chem. Eng. J.* **2020**, *384*, 123315. [CrossRef]
21. El Agrebi, N.; Tosi, S.; Wilmart, O.; Scippo, M.L.; de Graaf, D.C.; Saegerman, C. Honeybee and consumer's exposure and risk characterisation to glyphosate-based herbicide (GBH) and its degradation product (AMPA): Residues in beebread, wax, and honey. *Sci. Total Environ.* **2020**, *704*, 135312. [CrossRef]
22. Manassero, A.; Passalia, C.; Negro, A.C.; Cassano, A.E.; Zalazar, C.S. Glyphosate degradation in water employing the H_2O_2/UVC process. *Water Res.* **2010**, *44*, 3875–3882. [CrossRef]
23. Bonansea, R.I.; Filippi, I.; Wunderlin, D.A.; Marino, D.J.G.; Amé, M.V. The fate of glyphosate and AMPA in a freshwater endorheic basin: An ecotoxicological risk assessment. *Toxics* **2017**, *6*, 3. [CrossRef]
24. Van Bruggen, A.H.C.; He, M.M.; Shin, K.; Mai, V.; Jeong, K.C.; Finckh, M.R.; Morris, J.G., Jr. Environmental and health effects of the herbicide glyphosate. *Sci. Total Environ.* **2018**, *616–617*, 255–268. [CrossRef]
25. Gupta, P.; Pandey, K.; Verma, N. Augmented complete mineralization of glyphosate in wastewater via microbial degradation post CWAO over supported Fe-CNF. *J. Chem. Eng.* **2022**, *428*, 132008. [CrossRef]
26. Gupta, P.; Verma, N. Evaluation of degradation and mineralization of glyphosate pollutant in wastewater using catalytic wet air oxidation over Fe-dispersed carbon nanofibrous beads. *J. Chem. Eng.* **2021**, *417*, 128029. [CrossRef]
27. Xing, B.; Chen, H.; Zhang, X. Efficient degradation of organic phosphorus in glyphosate wastewater by catalytic wet oxidation using modified activated carbon as a catalyst. *Environ. Technol.* **2018**, *39*, 749–758. [CrossRef]
28. Žerjav, G.; Kaplan, R.; Pintar, A. Catalytic wet air oxidation of bisphenol A aqueous solution in trickle-bed reactor over single TiO_2 polymorphs and their mixtures. *J. Environ. Chem. Eng.* **2018**, *6*, 2148–2158. [CrossRef]
29. Cojocaru, B.; Andrei, V.; Tudorache, M.; Lin, F.; Cadigan, C.; Richards, R.; Parvulescu, V.I. Enhanced photo-degradation of bisphenol pollutants onto gold-modified photocatalysts. *Catal. Today* **2017**, *284*, 153–159. [CrossRef]
30. Shi, D.; Liu, J.; Ji, S. Preparation of Au/TiO_2 catalyst and the performance of liquid methanol catalytic oxidation to formic acid. *Ind. Eng. Chem. Res.* **2017**, *56*, 11028–11033. [CrossRef]
31. Alvaro, M.; Cojocaru, B.; Ismail, A.A.; Petrea, N.; Ferrer, B.; Harraz, F.A.; Parvulescu, V.I.; Garcia, H. Visible-light photocatalytic activity of gold nanoparticles supported on template-synthesized mesoporous titania for the decontamination of the chemical warfare agent Soman. *Appl. Catal. B* **2010**, *99*, 191–197. [CrossRef]
32. Mrowetz, M.; Villa, A.; Prati, L.; Selli, E. Effects of Au nanoparticles on TiO_2 in the photocatalytic degradation of an azo dye. *Gold Bull.* **2007**, *40*, 154–160. [CrossRef]
33. Žerjav, G.; Say, Z.; Zavašnik, J.; Finšgar, M.; Langhammer, C.; Pintar, A. Photo, thermal and photothermal activity of TiO_2 supported Pt catalysts for plasmon-driven environmental applications. *J. Environ. Chem. Eng.* **2023**, *11*, 110209. [CrossRef]
34. D'Agostino, C.; Brett, G.; Divitini, G.; Ducati, C.; Hutchings, G.J.; Mantle, M.D.; Gladden, L.F. Increased affinity of small gold particles for glycerol oxidation over Au/TiO_2 probed by NMR relaxation methods. *ACS Catal.* **2017**, *7*, 4235–4241. [CrossRef]
35. Žerjav, G.; Roškarič, M.; Zavašnik, J.; Kovač, J.; Pintar, A. Effect of Au loading on Schottky barrier height in TiO_2 + Au plasmonic photocatalysts. *Appl. Surf. Sci.* **2022**, *579*, 152196. [CrossRef]
36. Huang, X.; El-Sayed, M.A. Gold nanoparticles: Optical properties and implementations in cancer diagnosis and photothermal therapy. *J. Adv. Res.* **2010**, *1*, 13–28. [CrossRef]

37. Yu, J.G.; Xiong, J.F.; Cheng, B.; Liu, S.W. Fabrication and characterization of Ag-TiO_2 multiphase nanocomposite thin films with enhanced photocatalytic activity. *Appl. Catal. B* **2005**, *60*, 211–221. [CrossRef]
38. Shu, Y.; Ji, J.; Zhou, M.; Liang, S.; Xie, Q.; Li, S.; Liu, B.; Deng, J.; Cao, J.; Liu, S.; et al. Selective photocatalytic oxidation of gaseous ammonia at ppb level over Pt and F modified TiO_2. *Appl. Catal. B* **2022**, *300*, 120688. [CrossRef]
39. Jovic, V.; Al-Azri, Z.H.N.; Chen, W.T.; Sun-Waterhouse, D.; Idriss, H.; Waterhouse, G.I.N. Photocatalytic H_2 production from ethanol-water mixtures over Pt/TiO_2 and Au/TiO_2 photocatalysts: A comparative study. *Top. Catal.* **2013**, *56*, 1139–1151. [CrossRef]
40. Serpone, N.; Lawless, D.; Khairutdinov, R. Size effects on the photophysical properties of colloidal anatase TiO_2 particles: Size quantization or direct transitions in this indirect semiconductor? *J. Phys. Chem.* **1995**, *99*, 16646–16654. [CrossRef]
41. Nakajima, H.; Mori, T.; Watanabe, M. Influence of platinum loading on photoluminescence of TiO_2 powder. *J. Appl. Phys.* **2004**, *96*, 925–927. [CrossRef]
42. Nakajima, H.; Mori, T. Photoluminescence of Pt-loaded TiO_2 powder. *Phys. B* **2006**, *376–377*, 820–822. [CrossRef]
43. Abazović, N.D.; Čomor, M.I.; Dramićanin, M.D.; Jovanović, D.J.; Ahrenkiel, S.P.; Nedeljković, J.M. Photoluminescence of anatase and rutile TiO_2 particles. *J. Phys. Chem. B* **2006**, *110*, 25366–25370. [CrossRef]
44. Kernazhitsky, L.; Shymanovska, V.; Gavrilko, T.; Naumov, V.; Fedorenko, L.; Kshnyakin, V.; Baran, J. Room temperature photoluminescence of anatase and rutile TiO_2 powders. *J. Lumin.* **2014**, *146*, 199–204. [CrossRef]
45. Dupin, J.-C.; Gonbeau, D.; Vinatier, P.; Levasseur, A. Systematic XPS studies of metal oxides, hydroxides and peroxides. *Phys. Chem. Chem. Phys.* **2000**, *2*, 1319–1324. [CrossRef]
46. Schumacher, B.; Plzak, V.; Cai, J.; Behm, R.J. Reproducibility of highly active Au/TiO_2 catalyst preparation and conditioning. *Catal. Lett.* **2005**, *101*, 215–224. [CrossRef]
47. Liu, H.; Yang, W.; Ma, Y.; Cao, Y.; Yao, J.; Zhang, J.; Hu, T. Synthesis and characterization of titania prepared by using a photoassisted sol−gel method. *Langmuir* **2003**, *19*, 3001–3005. [CrossRef]
48. Chen, H.; Li, P.; Umezawa, N.; Abe, H.; Ye, J.; Shiraishi, K.; Ohta, A.; Miyazaki, S. Bonding and Electron Energy-Level Alignment at Metal/TiO_2 Interfaces: A Density Functional Theory Study. *J. Phys. Chem. C* **2016**, *120*, 5549–5556. [CrossRef]
49. Arshad, M.S.; Trafela, Š.; Rožman, K.Ž.; Kovač, J.; Djinović, P.; Pintar, A. Determination of Schottky barrier height and enhanced photoelectron generation in novel plasmonic immobilized multisegmented (Au/TiO_2) nanorod arrays (NRAs) suitable for solar energy conversion applications. *J. Mater. Chem. C* **2017**, *5*, 10509–10516. [CrossRef]
50. Yang, J.H.; Henao, J.D.; Raphulu, M.C.; Wang, Y.; Caputo, T.; Groszek, A.J.; Kung, M.C.; Scurrell, M.S.; Miller, J.T.; Kung, H.H. Activation of Au/TiO_2 Catalyst for CO Oxidation. *J. Phys. Chem. B* **2005**, *109*, 10319–10326. [CrossRef]
51. Murdoch, M.; Waterhouse, G.I.N.; Nadeem, M.A.; Metson, J.B.; Keane, M.A.; Howe, R.F.; Llorca, J.; Idriss, H. The effect of gold loading and particle size on photocatalytic hydrogen production from ethanol over Au/TiO_2 nanoparticles. *Nat. Chem.* **2011**, *3*, 489–492. [CrossRef]
52. Jovic, V.; Chen, W.-T.; Sun-Waterhouse, D.; Blackford, M.G.; Idriss, H.; Waterhouse, G.I.N. Effect of gold loading and TiO_2 support composition on the activity of Au/TiO_2 photocatalysts for H_2 production from ethanol–water mixtures. *J. Catal.* **2013**, *305*, 307–317. [CrossRef]
53. Zhang, Z.; Yates, J.T. Band bending in semiconductors: Chemical and physical consequences at surfaces and interfaces. *Chem. Rev.* **2012**, *112*, 5520–5551. [CrossRef]
54. Qian, K.; Sweeny, B.C.; Johnston-Peck, A.C.; Niu, W.; Graham, J.O.; DuChene, J.S.; Qiu, J.; Wang, Y.-C.; Engelhard, M.H.; Su, D.; et al. Surface plasmon-driven water reduction: Gold nanoparticle size matters. *J. Am. Chem. Soc.* **2014**, *136*, 9842–9845. [CrossRef]
55. Erjavec, B.; Tišler, T.; Kaplan, R.; Pintar, A. Titanate nanotubes as a novel catalyst for removal of toxicity and estrogenicity of bisphenol A in the CWAO process. *Ind. Eng. Chem. Res.* **2013**, *52*, 12559–12566. [CrossRef]
56. Xu, R.; Gao, Z.; Chen, J.; Yan, W. *From Zeolite to Porous MOF Materials*, 1st ed.; Elsevier: Amsterdam, The Netherlands, 2007.
57. Ibhadon, A.O.; Fitzpatrick, P. Heterogeneous Photocatalysis: Recent Advances and Applications. *Catalysts* **2013**, *3*, 189–218. [CrossRef]
58. Nosaka, Y.; Nosaka, A.Y. Generation and Detection of Reactive Oxygen Species in Photocatalysis. *Chem. Rev.* **2017**, *117*, 11302–11336. [CrossRef]
59. Luxembourg, D.; Flamant, G.; Laplaze, D. Solar synthesis of single-walled carbon nanotubes at medium scale. *Carbon* **2005**, *43*, 2302–2310. [CrossRef]
60. Zhang, L.; Cheng, H.; Zong, R.; Zhu, Y. Photocorrosion suppression of ZnO nanoparticles via hybridization with graphite-like carbon and enhanced photocatalytic activity. *J. Phys. Chem. C* **2009**, *113*, 2368–2374. [CrossRef]
61. Chen, S.; Liu, Y. Study on the photocatalytic degradation of glyphosate by TiO_2 photocatalyst. *Chemosphere* **2007**, *67*, 1010–1017. [CrossRef]
62. Echavia, G.R.M.; Matzusawa, F.; Negishi, N. Photocatalytic degradation of organophosphate and phosphonoglycine pesticides using TiO_2 immobilized on silica gel. *Chemosphere* **2009**, *76*, 595–600. [CrossRef]
63. Barrett, K.A.; McBride, M.B. Oxidative degradation of glyphosate and aminomethylphosphonate by manganese oxide. *Environ. Sci. Technol.* **2005**, *39*, 9223–9228. [CrossRef] [PubMed]

64. Muneer, M.; Boxall, C. Photocatalyzed degradation of a pesticide derivative glyphosate in aqueous suspensions of titanium dioxide. *Int. J. Photoenergy* **2008**, *2008*, 197346. [CrossRef]
65. Žerjav, G.; Arshad, M.S.; Djinović, P.; Zavašnik, J.; Pintar, A. Electron trapping energy states of TiO_2-WO_3 composites and their influence on photocatalytic degradation of bisphenol A. *Appl. Catal. B* **2017**, *209*, 273–284. [CrossRef]

catalysts

MDPI

Article

The Catalytic Wet Oxidation of Excess Activated Sludge from a Coal Chemical Wastewater Treatment Process

Zhongquan Wang [1], Shulin Qin [1], Weicheng Zheng [1], Xiaodan Lou [1], Xu Zeng [2,*] and Taihang Wu [2]

[1] Hangzhou Research Institute of China Coal Technology & Engineering Group, Hangzhou 311201, China; wzhquan@126.com (Z.W.); hzqsl722@163.com (S.Q.); weicheng.zheng@foxmail.com (W.Z.); lxd8982@126.com (X.L.)

[2] State Key Laboratory of Pollution Control and Resource Reuse, College of Environmental Science and Engineering, Tongji University, Shanghai 200092, China; 2232792@tongji.edu.cn

* Correspondence: zengxu@tongji.edu.cn; Tel.: +86-21-55088628

Abstract: An improved catalytic wet oxidation method for the disposal of excess activated sludge from a coal chemical wastewater treatment process by using the prepared Cu-Ce/γ-Al_2O_3 catalyst was reported. The effects of catalyst dosage, reaction temperature and time, and initial oxygen pressure on the degradation of the sludge were investigated. The maximum removal rate of volatile suspended solids, 93.2%, was achieved at 260 °C for 60 min with the catalyst 7.0 g·L^{-1} and initial oxygen pressure 1.0 MPa. The removal rate of chemical oxygen demand was 78.3% under the same conditions. The production of volatile fatty acids, including mainly acetic acid, propanoic acid, and isobutyric acid, increased with the increasing temperature. These acids have the potential to be carbon sources for the biological treatment of wastewater. Scanning electron microscopy images showed that the sludge became a loose porous structure, which is beneficial for dewatering performance. The results of an energy dispersive spectroscopy analysis illustrated that the carbon element in the sludge substantially migrated from solid to liquid phases. Therefore, these results demonstrated that the proposed catalytic wet oxidation method offers a promising pathway for the disposal and utilization of excess activated sludge from the coal chemical wastewater treatment process.

Keywords: catalytic wet oxidation; coal chemical sludge; coal chemical wastewater

1. Introduction

With the fast development of the coal chemical industry, a huge amount of coal chemical wastewater has been produced. Normally, this wastewater is treated with chemical and biological treatment processes. Therefore, a massive excess of activated sludge is generated from the biological treatment plant. The sludge poses an environmental risk unless it is disposed of safely because of the contained hazardous compounds and heavy metals [1,2]. Because of the existence of hazardous materials, the biological method always showed low efficiency [3]. Incineration was chosen for the final method. However, it can produce pollutant problems, such as dioxin and toxic materials. Furthermore, incineration has high cost as a disposal method [4,5]. Nowadays, the treatment of sludge has become a difficult area for the coal chemical industry. A satisfying disposal pathway for coal chemical sludge is strongly desired.

Typically, caprolactam, an important raw material for the production of polymers, was produced massively [6]. And, the increasing trend is still ongoing because of the huge demand of humankind. In this case, a large amount of excess activated sludge was produced simultaneously [7,8]. Caprolactam is normally produced via the toluene process, the phenol process, or the benzene process. Each of them produces hazardous chemical materials, such as benzene and benzene hydrocarbon. Then, it was induced that huge amounts of hazardous materials were dissolved in wastewater [9–11]. In the biological wastewater treatment process, microorganisms will absorb a lot of these hazardous materials. Finally, large amounts of excess activated sludge, containing huge amounts of hazardous materials, were produced. Obviously, the biological treatment process is not suitable for the treatment of wastewater. To date, the disposal of caprolactam sludge is still an urgent problem to be solved.

In recent years, advanced oxidation processes (AOPs) have attracted widespread attention due to their high efficiency in the degradation of hazardous and toxic organic contaminants [12–14]. Among them, wet oxidation (WO) exhibited a high potential for the removal of hazardous contaminants [15–18]. Because of the high oxidation activity of free radical agents, organic compounds could be oxidized to small molecular organic carboxylic acids such as acetic, formic, and oxalic acids, and even CO_2 and H_2O [19,20]. Some commercial wet oxidation plants were established for the treatment of pharmaceuticals, textile effluents, paper mills, and olive effluents [21,22]. However, a high disposal fee is always needed for wet oxidation disposal because of its reaction conditions. Therefore, catalytic wet oxidation technology has been developed in recent years. The addition of a catalyst could enhance the reaction efficiency and decrease the operating temperature and time [23,24]. In the past few decades, transition metals have been proved to be excellent oxidizing candidates [25,26]. Specifically, Cu-based catalysts have attracted considerable attention. Li et al. reported that a Cu-loaded catalyst can accelerate the degradation of acetaldehyde [27–29]. A Cu catalyst was also reported for its efficiency in the oxidation of formic acid and phenol [30]. Furthermore, a ceria catalyst has also been proved to be efficient in wet oxidation [31,32]. Recently, a ceria-based catalyst has obtained increasing attention. However, Cu- and Ce-composite catalysts have been rarely studied.

Volatile fatty acids (VFAs), including oxalic acid, acetic acid, and propanoic acid, could be massively produced in sludge disposal [33]. For example, Gapes reported that carboxylic acids could be realized in the wet oxidation of sludge [34]. Chung reported that the propionic acid concentration increased to 13.5 mg/L at 240 °C in the wet oxidation process [35]. Wu reported on propionic and butyric acid production [36]. A lot of the available literature reported results on maximizing the yield of acetic acid. The effect of the amount of oxygen is the key influence parameter for the production of acetic acid [37]. It should be noted that the VFA solution, produced after the WO process, has the potential to be used as a kind of organic carbon source. Therefore, the production of acetic acid and VFAs through wet oxidation is well worth studying.

In this study, an improved catalytic wet oxidation method for the disposal of excess activated sludge from a coal chemical wastewater treatment process was reported by using the prepared Cu-Ce/γ-Al_2O_3 catalyst. The effects of the catalyst dosage, reaction temperature and time, and the initial oxygen pressure on the reactivity of the system were investigated. In addition, the characteristics of the produced liquid and solid were analyzed.

2. Results

2.1. Effect of Catalyst Dosage

Experiments with the catalyst addition amount from 0 to 7.0 g·L^{-1} were carried out to assess the effects of the catalyst. The reaction conditions included a temperature of 260 °C and time of 60 min, with the initial oxygen pressure 1.0 MPa. The experimental results can be seen in Figure 1. As illustrated in Figure 1, the removal rates of chemical oxygen demand (COD) increased significantly in the presence of the catalyst with the amount increasing. The catalyst used in this study had a considerable catalytic effect on the wet oxidation of the sludge. In contrast, the removal rate of volatile suspended solids (VSSs) was not significant. Comparing the wet oxidation of the sludge and its oxidized intermediates, the VSS removal of the sludge occurred more easily because the removal rate of VSS was above 85% even in the absence of the catalyst, as seen in Figure 1. However, the increase in COD removal influenced the VSS removal with the addition of the prepared catalyst. When the COD removal rate was higher, the hydrolysis of the sludge was enhanced. In our previous study, the catalytic reactivity of a prepared Cu-Ce/γ-Al_2O_3 catalyst was proven to be effective for the wet oxidation of pharmaceutical sludge. It is evident that the catalytic effect is similar for coal chemical sludge. This means that the prepared catalyst has a better capacity with the load of copper and ceria as active elements.

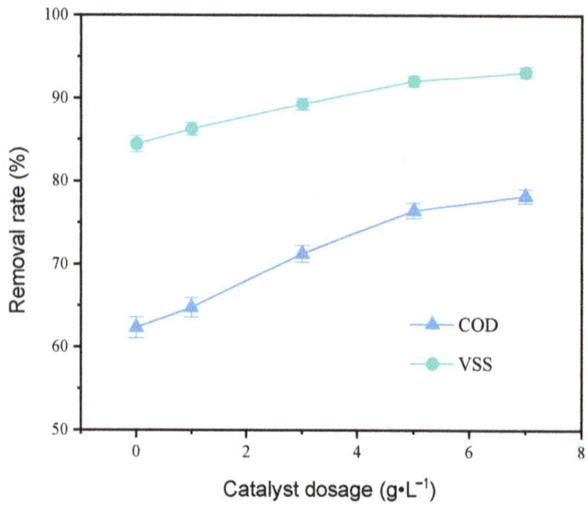

Figure 1. Effect of catalyst dosage (260 °C; 60 min; initial oxygen pressure 1.0 MPa).

2.2. Effect of Reaction Temperature

Reaction temperature always has significant influence in the wet oxidation process. On the one hand, higher temperatures provide higher reaction efficiency due to their higher energy, according to Arrhenius' law. Moreover, the solubility of oxygen gas increased when the temperature increased, which increased the oxidant for the wet oxidation of organic pollutants in the solution. In this study, the influence of temperature (180~260 °C) was investigated with a reaction time of 60 min, a catalyst of 7.0 g·L^{-1}, and an initial oxygen pressure of 1.0 MPa. The experimental results are shown in Figure 2.

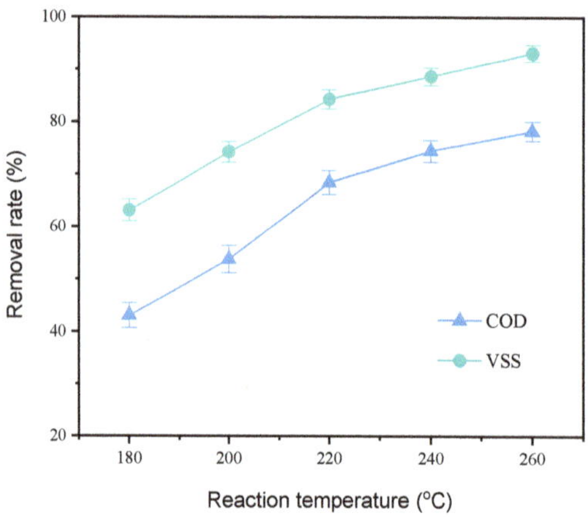

Figure 2. Effect of reaction temperature (60 min; initial oxygen pressure 1.0 MPa; catalyst 7.0 g·L^{-1}).

In Figure 2, it can be seen that the removal rates of the COD and VSSs increased significantly with the temperature increasing. Under the reaction temperature of 180 °C, the removal rates of COD and VSSs are very low. However, the removal rate of COD was almost 40%. Compared to the COD removal in the reaction, the VSS removal rate was much higher. The possible reason may be that the sludge was more easily decomposed due to hydrolysis, which happened easily under hydrothermal conditions. In the wet oxidation process, acetic acid is a common intermediate, which is not easily oxidized. Then, the produced acetic acid could be accumulated, and the COD removal would be inhibited. Therefore, the VSS removal rate was much higher than that of the VSSs, even with surplus oxygen gas in the wet oxidation reaction system. These results were similar to Strong's results, which demonstrated that a VSS removal rate of 93% was achieved at 220 °C for 2 h under a pure oxygen pressure of 2.0 MPa. All these results imply that the reaction temperature is a key reaction parameter in the reaction. And, the reaction temperature should be high enough to acquire better removal results. However, a high temperature will increase the cost for the building and operation of the reaction system. Thus, in the industrial utilization of wet oxidation technology, the reaction temperature should be considered together with the reaction efficiency and the cost for the reaction system.

2.3. *Effect of Reaction Time*

The influence of reaction time on the sludge degradation was investigated with the experiments' reaction times varying from 20 to 60 min.

As shown in Figure 3, the COD and VSS removal rates increased simultaneously with the extension of the reaction time. A linear relationship between the increase in removal rates and the reaction time existed. The possible reason may be that the cell wall was destroyed more completely under higher temperatures during wet oxidation, which induced the solubilization of the sludge, and the decrease in the COD and VSSs when the reaction time was longer than 20 min. With the extension of the reaction time, the wet oxidation of intermediates and the hydrolysis of the sludge took place gradually. More importantly, some intermediates, which were not easily oxidized further, such as acetic acid, accumulated with the process of wet oxidation. Therefore, when the COD removal rate was near 80%, it could not increase further. Therefore, considering the oxidation efficiency and economic factors together, we chose 60 min as the favorable reaction time for the optimum reaction parameter.

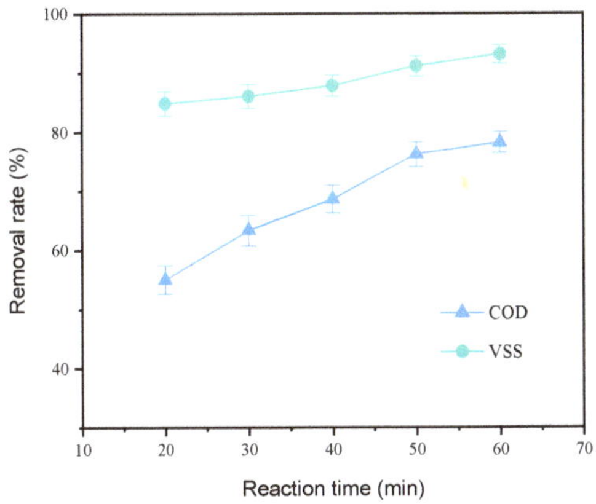

Figure 3. Effect of reaction time (260 °C; initial oxygen pressure 1.0 MPa; catalyst 7.0 g·L^{-1}).

2.4. *Effect of Initial Oxygen Pressure*

To assess the influence on the oxidant in wet oxidation, i.e., the addition of oxygen gas, experiments were carried out with the initial oxygen pressures changing from 0.2 to 1.0 MPa. The experimental results can be found in Figure 4. As seen in Figure 4, the removal of the COD was greatly influenced by the addition of oxygen gas. With the initial oxygen pressure increased, i.e., the increase in the oxidant, wet oxidation took place more easily. In contrast, the VSS removal rate was only slightly influenced by the change in the amount of oxygen. The possible reason for VSS removal could be attributed to the hydrolysis of the sludge; thus, the VSS removal rate was above 80% even with an initial oxygen pressure of 0.2 MPa. As mentioned above, the reaction temperature was the main factor which influenced hydrolysis. Therefore, the influence of the amount of oxygen on VSS removal was small.

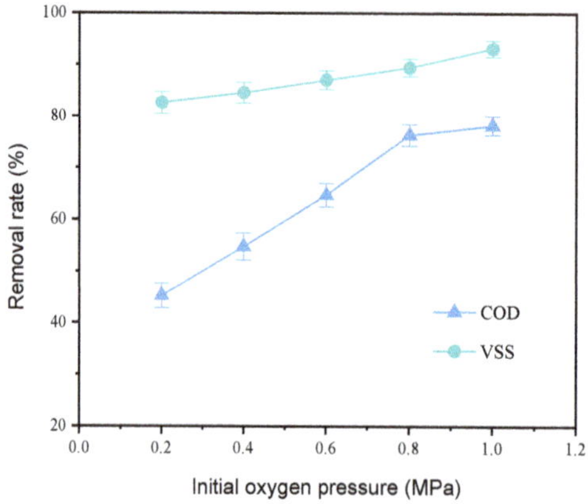

Figure 4. Effect of additional oxygen pressure (260 °C; 60 min; catalyst 7.0 g·L^{-1}).

On the other hand, the production of strong oxidation species was enhanced by the increase in additional amounts of oxygen. The dissolved amount of oxygen in the liquid increased, and the gas–liquid mass transfer speed increased. All these conditions increased the wet oxidation rate. Therefore, the COD removal rate increased significantly with the initial oxygen pressure increasing. However, the surplus addition of oxygen had a small influence on the wet oxidation of the sludge once the wet oxidation was almost stable. As shown in Figure 4, when the initial oxygen pressure increased from 0.8 to 1.0 MPa, the removal rate of the COD increased very little. A possible reason may be that the non-oxidizable intermediates were accumulated in the wet oxidation process, such as acetic acid. Therefore, the additional amount of oxygen gas was enough once it was sufficient for the removal of the COD.

2.5. Removal of COD and SCOD

The COD value can reflect the concentration of most organic pollutants. Therefore, the removal of organic pollutants, i.e., the degradation of the sludge, could be assessed by measuring the COD and the soluble chemical oxygen demand (SCOD). The SCOD means the organic pollutants in the liquid. As shown in Figure 5, it was found that the removal rates of the COD and the SCOD increased significantly with the increasing temperature. The maximum values were obtained at 260 °C. The difference between the COD and the SCOD is mainly due to the different reaction speeds of hydrolysis and oxidation, as the hydrolysis process goes faster than oxidation under subcritical conditions. These results also illustrated the decrease in VSSs under higher reaction temperatures, which implied that the VSSs were transferred to a liquid phase. These results illustrated that the reaction temperature should be high enough to remove the VSSs in the sludge as fast as possible.

2.6. Production of Acetate and VFAs

During the wet oxidation process, organic materials could be oxidized to small molecule organics, i.e., carboxylic acid and even carbon dioxide. VFAs are the dominant component of these intermediates, and they are considered to be reused as value-added materials. The effect of temperature on the concentrations of acetic acid and VFAs is shown in Figure 6.

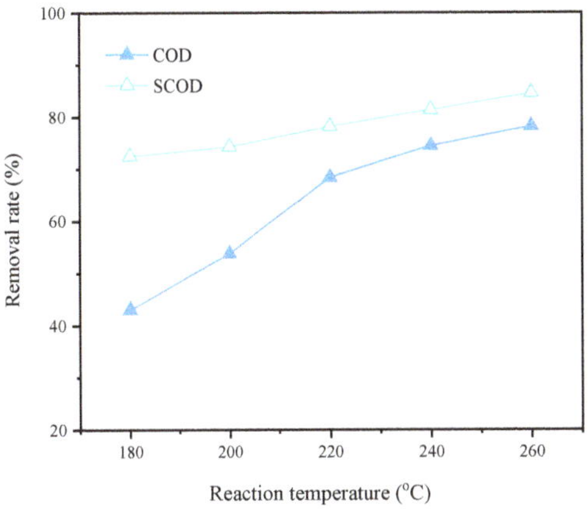

Figure 5. Change in COD and SCOD removal rates with different reaction temperatures.

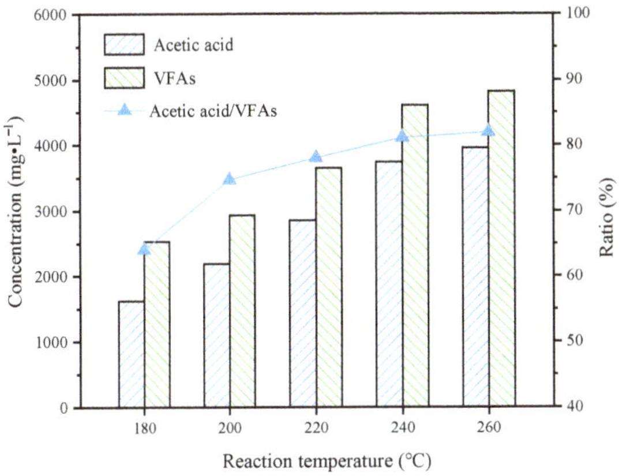

Figure 6. Production of acetic acid and VFAs.

As shown in the figure, the VFA concentration was found to increase dramatically while the reaction temperature rose. For acetic acid, the concentration increased from the minimum (180 °C) to the maximum (260 °C). However, it remained stable from 240 to 260 °C. It has been reported that the maximum value of the VFAs was obtained at 250 °C within 30 min, and the surplus oxidants would induce the oxidation of the VFAs, including acetic acid. Therefore, this steady state might be caused by the slight decomposition of acetic acid as the temperature was quite high and oxidizers were abundant, which reduced the formation of acetic acid from the decomposition of other organic substances. The value of acetic acid/VFAs was also calculated and it was proven that acetic acid was the main component of the VFAs. The value was quite high, almost 80%. Acetic acid dominated the post-reaction VFAs mainly because it was a stable intermediate and was relatively hard to decompose under reaction conditions. Acetate sodium was always selected as an organic carbon source in the biological wastewater treatment process. Therefore, the liquid after wet oxidation has the potential for the utilization as an organic carbon source in the wastewater treatment.

2.7. Characteristics of Sludge before and after WO

To investigate the changes in the microstructure and elemental composition of the sludge before and after the WO reaction, scanning electron microscopy (SEM) and energy dispersive spectroscopy (EDS) analyses were adopted to investigate the solid after wet oxidation. Figure 7 shows the structural changes of the sludge before and after the WO reaction at the scales of 50 and 10 μm. As shown in Figure 7a,b, the sludge was in a cohesive and relatively complete block shape with sparse pores before the WO reaction. As shown in Figure 7c,d, the sludge particles became smaller and the pores increased significantly after the WO reaction. These results illustrated that the solid is conducive to the release of intercellular water and thus improves dehydration performance because of the removal of organic materials in the sludge.

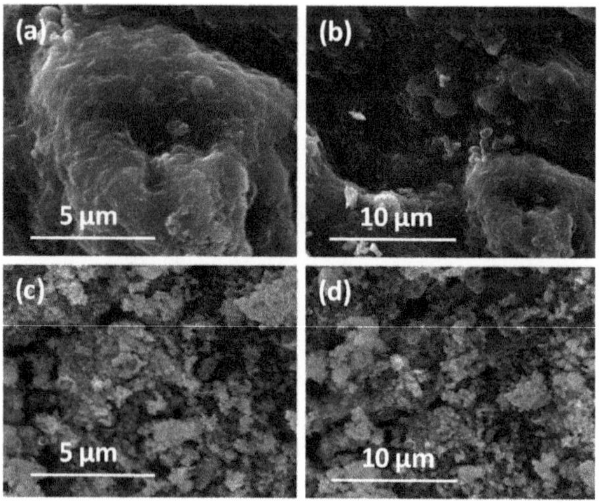

Figure 7. SEM images before (**a**,**b**) and after (**c**,**d**) wet oxidation (260 °C, 1.0 MPa O_2, pH = 8.09; left to right: 100 nm, 500 nm, and 1 μm).

Figure 8 shows the energy spectrum analysis results. As shown in Figure 8, the changes in the content of the carbon and oxygen elements in the sludge after the WO reaction were significantly decreased compared with the results before the WO reaction. The carbon content in the sludge decreased by almost 30% between the results from before and after the reaction, indicating the removal of organic substances. The VSS removal rate was above 90% at 260 °C for 60 min, which means that the amount of solids decreased significantly after the wet oxidation. Based on the calculations, the amount of metals increased. The possible reason may be that the metal ions were oxidized and precipitation happened. Therefore, wet oxidation can not only oxidize the organic pollutants, but it can also remove some heavy metals by changing the pH of the liquid.

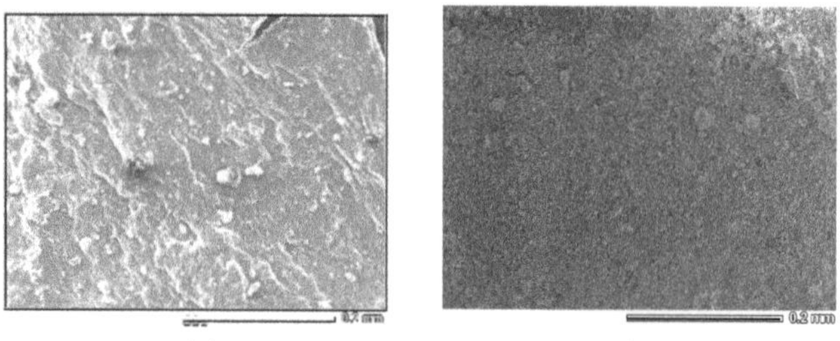

(**a**) (**b**)

Figure 8. EDS images before (**a**) and after (**b**) the sludge WO reaction.

3. Discussion

It is very hard to illustrate the distinct mechanism because the sludge is very complex, and the intermediates were also very complex. The sludge consisted of microorganisms and organic pollutants, but the detailed mechanism of sludge wet oxidation is still unknown [38,39]. However, the wet oxidation reaction is always assumed as a free radical oxidation reaction [40]. Normally, wet oxidation was assumed to be a free radical reaction mechanism. Under hydrothermal conditions, large amounts of free radicals were produced based on the dissolution of oxygen, which possessed strong oxidant activity. Then, the oxidation reaction took place easily, and the pollutants were oxidized and removed.

For the wet oxidation of the sludge, based on the experimental results and discussion above, it is supposed that hydrolysis was the first step. In the hydrolysis process, proteins and lipids decomposed into a liquid, along with the dissolution of inorganic materials. Then, a high VSS removal rate was achieved. Because the reaction temperature was very high, the hydrolysis process took place very quickly. Once the solid materials were transferred to a liquid phase, the compounds were oxidized by the strong oxidant free radicals. Then, a lot of VFAs were produced, including acetic acid, isobutyric acid, and propanoic acid, etc. [41,42]. The BOD_5/COD ratio of the liquid was very high, even above 0.70. In some industries, acetate sodium was always bought as a carbon source for industrial refractory wastewater treatment. Some experimental results proved that the oxidation liquid is useful for nitrogen removal in wastewater treatment. Therefore, the liquid after wet oxidation has the potential to be utilized as an organic carbon source in the biological wastewater treatment process. The advantages of WO also include exothermic reactions, which can be used to amend the energy consumption to maintain the occurrence of the reaction. After the WO reaction, the sludge particles became smaller and the pores significantly increased, which was conducive to the release of intercellular water and thus improved the dehydration performance. The solid could be used for the production of ceramic aggregate or permeable brick. Therefore, WO can be regarded as an ideal method for the volume reduction and resource utilization of pharmaceutical sludge.

In the present study, wet oxidation was conducted in a batch reactor. After the catalyst was used several times, the enhancement of the catalyst for the COD and VSS removal was still above 95%. And, the catalyst became more stable when the catalyst was used for wet oxidation for a long time. Therefore, for the industrial utilization of this method, there is still a need to study it in a continuous reaction system. In addition, the stability of the catalyst still needs to be verified. Under the hydrothermal reaction conditions, the leaching of the active elements, including copper and ceria, should be discussed. Furthermore, the catalyst should be developed further. Once the cost for the wet oxidation process is accepted for commercial utilization, the catalytic wet oxidation method could be broadly industrially utilized.

4. Materials and Methods

4.1. Materials

The excess activated sludge used for the experiments was obtained from a coal chemical industry in China, which produced caprolactam. The initial parameters of the sludge were the following: COD 16,500~17,500 $mg \cdot L^{-1}$, VSS/SS 81.3%, pH, 8.47. The sludge was stored in a refrigerator at 4 °C to avoid possible biological activities. All the reagents used in this study were analytical grade without any treatment.

4.2. WO Reaction System

WO runs were performed in a stainless-steel autoclave reactor with an inner volume of 200 mL. The functions of heating, stirring, and temperature monitoring were integrated and equipped in the reactor. The information on the reaction system can be found elsewhere [43]. A diagram of the reaction system is shown in Figure 9. For a typical experiment run, 100 mL of the sludge solution was added to the reactor initially. Then, oxygen gas at 0.5 MPa was purged and vented three times to remove the air inside. On the fourth purge, oxygen was stored inside the reactor as the oxidant and the pressure was increased to 0.2~1.0 MPa. The reaction temperature was set between 180 and 260 °C and monitored using a thermocouple to maintain a deviation of 2 °C. Once the reactor achieved the set temperature, the reaction time would start to count and was set to 20~60 min. The stirrer speed was kept constant at 300 rpm throughout the process. The experiments for each reaction condition were carried out three times. As the experiments were conducted in batch mode, the inner pressure was self-pressurized. When the reaction finished, the reaction was moved and cooled to room temperature. The liquid and solid were collected and sampled.

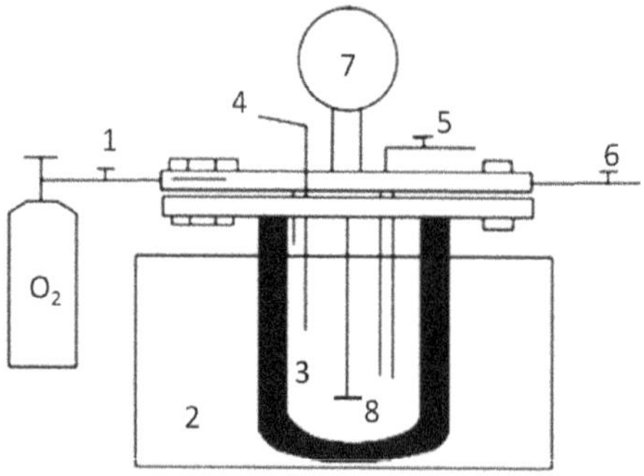

Figure 9. Diagram of wet oxidation reactor; 1—oxygen supply, 2—heating jacket, 3—oxidation reactor, 4—thermocouple, 5—discharge of gas, 6—bleeder valve, 7—pressure meter, and 8—stirrer.

4.3. Preparation of Catalyst

Detailed information about catalyst preparation and characteristics can be found elsewhere [44]. The catalyst was prepared using an impregnation process. The solution of $Cu(NO_3)_2$ and $Ce(NO_3)_3$ with a molar ratio of 1:1 and the concentration 1.0 mol·L^{-1} was prepared. γ-Al_2O_3, as a carrier, was put into the solution. After 24 h of impregnation, the carrier was collected, and then it was baked at 550 °C for 4 h. Finally, the Cu-Ce/γ-Al_2O_3 catalyst was prepared.

4.4. Analysis Method

Various parameters such as the COD, SS, VSSs, and pH were determined according to standard methods [45]. After acidification (pH < 2) and filtration with a 0.22 μm membrane, the VFAs were identified and quantitatively analyzed by using gas chromatography (GC, Persee G5, Beijing, China) with bonded polyethylene glycol capillary columns (DB-FFAP, Agilent, Santa Clara, CA, USA) and a flame ionization detector (FID). Helium gas (Huaxiong Gas Co., Ltd., Jiaxing, China) was selected as the carrier gas. The powdered sample of dried sludge and WO residue were characterized using SEM (Gemini 300, ZEISS, Oberkochen, Germany) and EDS (INCA X-Act, Oxford, United Kingdom) to observe the morphology and element distribution.

5. Conclusions

In the present study, an improved catalytic wet oxidation method for the disposal of excess activated sludge from a coal chemical wastewater treatment process was reported by using the prepared Cu-Ce/γ-Al_2O_3 catalyst. The effects of catalyst dosage, reaction temperature and time, and initial oxygen pressure on the degradation of the sludge were investigated. The highest removal rate of volatile suspended solids, 93.2%, was achieved at 260 °C for 60 min with the catalyst 7.0 g·L^{-1} and an initial oxygen pressure of 1.0 MPa. The removal rate of the chemical oxygen demand was 78.3% under the same conditions. The concentration of volatile fatty acids, including mainly acetic acid, propanoic acid, and isobutyric acid, increased with the temperature increase. The trend of VFAs and acetic acid demonstrated that there was a peak in the promotion of organic acid content by increasing the temperature. These acids have the potential to be carbon sources for the biological treatment of wastewater. Our previous study proved that nitrogen removal could be enhanced with the addition of a wet oxidation liquid in the inlet wastewater in an A/O wastewater treatment system. Scanning electron microscopy images showed that the sludge became a loose porous structure, which is beneficial for the dewatering performance. The results of an energy dispersive spectroscopy analysis illustrated that the carbon element in the sludge substantially migrated from solid to liquid phases. The utilization of the solid could be realized through the production of ceramic aggregate. Future work should be completed for industrial utilization. These results demonstrated that the catalytic wet oxidation process provides a suitable alternative pathway for the disposal and utilization of excess activated sludge from the coal chemical wastewater treatment process.

Author Contributions: Conceptualization, X.Z. and S.Q.; methodology, W.Z.; software, T.W.; validation, Z.W.; formal analysis, T.W.; investigation, X.Z.; writing—original draft preparation, S.Q.; writing—review and editing, X.Z. and X.L.; supervision, X.Z.; project administration, S.Q.; funding acquisition, S.Q. All authors have read and agreed to the published version of the manuscript.

Funding: This work was financially supported by the Science and Technology Innovation Project of the China Coal Technology & Engineering Group (No. 2022-3-KJHZ001).

Data Availability Statement: The data presented in this study are available on request from the corresponding author.

Conflicts of Interest: The authors declare no conflict of interest.

References

1. Arora, P.K.; Srivastava, A.; Singh, V.P. Diversity of 4-chloro-2-nitrophenol-degrading bacteria in a waste water sample. *J. Chem.* **2016**, *2016*, 7589068. [CrossRef]
2. Wei, H.Z.; Yan, X.M.; Li, X.R.; He, S.B.; Sun, C.L. The degradation of Isophorone by catalytic wet air oxidation on Ru/$TiZrO_4$. *J. Hazard. Mater.* **2013**, *244*, 478–488. [CrossRef] [PubMed]
3. Arora, P.K.; Srivastava, A.; Singh, V.P. Bacterial degradation of nitrophenols and their derivatives. *J. Hazard. Mater.* **2014**, *266*, 42–59. [CrossRef] [PubMed]
4. Arora, P.K.; Sasikala, C.; Ramana, C.V. Degradation of chlorinated nitroaromatic compounds. *Appl. Microbiol. Biotechnol.* **2012**, *93*, 2265–2277. [CrossRef] [PubMed]
5. Yin, F.B.; Wang, D.L.; Li, Z.F. Study on anaerobic digestion treatment of hazardous colistin sulphate contained pharmaceutical sludge. *Bioresour. Technol.* **2015**, *177*, 188–193. [CrossRef] [PubMed]
6. Yu, P.; Zhu, Z.; Luo, Y.; Hu, Y.; Lu, S. Purification of caprolactam by means of an electrodeionization technique. *Desalination* **2005**, *174*, 231–235. [CrossRef]
7. Cui, X.M. Supply and demand status and development prospect of caprolactam in China and abroad. *China Fine Spec. Chem.* **2021**, *29*, 12–15.
8. Yi, C. Development of nylon 6 polymerization technology and functional products. *China Syn. Fiber Ind.* **2021**, *44*, 59–65.
9. Zhang, G.H.; Dong, H. Analysis of domestic caprolactam production and market. *Chem. Ind.* **2019**, *37*, 38–41.
10. Liu, J.Q.; Wang, C.Z.; Liu, W.R.; Jiang, B.; Yang, Q.J.; Wu, M. Study on the biodegradability of caprolactam production wastewater. *Technol. Water Treat.* **2009**, *35*, 37–39.
11. Liu, F.; Liu, G.H.; Zhang, M.; Shu, W.; Chen, J.H. Study on Treatment of Nylon-6 Wastewater by Using Hydrolysis. *China Water Wastewater* **2007**, *23*, 104–108.
12. Ramos, M.D.N.; Santana, C.S.; Velloso, C.C.V.; Silva, A.H.M.; Magalhaes, F.; Aguiar, A. A review on the treatment of textile industry effluents through Fenton processes. *Process Saf. Environ. Prot.* **2021**, *155*, 366–386. [CrossRef]
13. Bustillo, C.F.; Mehrvar, M. Cost-effectiveness analysis of TOC removal from slaughterhouse wastewater using combined anaerobiceaerobic and UV/H_2O_2 processes. *J. Environ. Manag.* **2014**, *134*, 145–152. [CrossRef]
14. Yu, H.; Nie, E.; Xu, J. Degradation of diclofenac by advanced oxidation and reduction processes: Kinetic studies, degradation pathways and toxicity assessments. *Water Res.* **2013**, *47*, 1909–1918. [CrossRef]
15. Luck, F. Wet air oxidation, past, present and future. *Catal. Today* **1999**, *53*, 81–91. [CrossRef]
16. Arana, J.; Rendon, E.T.; Rodrıguez, J.D.; Melian, J.H.; Diaz, O.G.; Pena, J.P. Highly concentrated phenolic wastewater treatment by the Photo-Fenton reaction, mechanism study by FTIR-ATR. *Chemosphere* **2001**, *44*, 1017–1023. [CrossRef] [PubMed]
17. Wang, L.; Liu, J.; Li, H. Simultaneous degradation of micro plastics and sludge during wet air oxidation. *Environ. Pollut.* **2023**, *335*, 122348–122357. [CrossRef]
18. Zhu, Y.; Zeng, X.; Fang, K. Enhanced Wet Oxidation of Excess Sludge from Pharmaceutical Wastewater Treatment by NaOH. *Catalysts* **2023**, *13*, 1070. [CrossRef]

19. Zhang, Y. Wet oxidation technology based on organic wastewater treatment. *J. Phys. Conf. Ser.* **2020**, *1549*, 022040–022046. [CrossRef]
20. Dietrich, M.; Randall, T.; Canney, P. Wet air oxidation of hazardous organics in wastewater. *Environ. Prog.* **1985**, *4*, 171–177. [CrossRef]
21. Bertanza, G.; Galessi, R.; Menoni, L.; Salvetti, R.; Slavik, E.; Zanaboni, S. Wet oxidation of sewage sludge: Full-scale experience and process modeling. *Environ. Sci. Pollut. Res.* **2015**, *22*, 7306–7316. [CrossRef] [PubMed]
22. Mantzavinos, D.; Psillakis, E. Enhancement of biodegradability of industrial wastewaters by chemical oxidation pre-treatment. *J. Chem. Technol.* **2004**, *79*, 431–454. [CrossRef]
23. Levec, J.; Pintar, A. Catalytic wet-air oxidation processes: A review. *Catal. Today* **2007**, *124*, 172–184. [CrossRef]
24. Gomesa, H.T.; Figueiredo, J.L.; Faria, J.L.; Serp, P.; Kalck, P. Carbon-supported iridium catalysts in the catalytic wet air oxidation of carboxylic acids: Kinetics and mechanistic interpretation. *J. Mol. Catal. A Chem.* **2002**, *182*, 47–60. [CrossRef]
25. Ribeiro, R.S.; Silva, A.M.T.; Pastrana-Martinez, L.M.; Figueiredo, J.L.; Faria, J.L.; Gomes, H.T. Graphene-based materials for the catalytic wet peroxide oxidation of highly concentrated 4-nitrophenol solutions. *Catal. Today* **2015**, *249*, 204–212. [CrossRef]
26. Huang, K.; Xu, Y.; Wang, L.G.; Wu, D.F. Heterogeneous catalytic wet peroxide oxidation of simulated phenol wastewater by copper metal organic frameworks. *RSC Adv.* **2015**, *5*, 32795–32803. [CrossRef]
27. Li, Y.Z.; Fan, Z.Y.; Shi, J.W.; Liu, Z.Y.; Zhou, J.W.; Shangguan, W.F. Catalytic oxidation of low concentration formaldehyde with the assist of ozone over supported cobalt-manganese composite oxides. *J. Mol. Catal.* **2014**, *1*, 60–66.
28. Shan, H.; Oh, R.; Fan, J. Developing Pt-M/C catalyst (M = Pb, Cu) for efficient catalytic wet air oxidation of phenol wastewater under mild conditions. *J. Environ. Chem. Eng.* **2023**, *11*, 109854–109863. [CrossRef]
29. Chou, B.; Tsai, J.L.; Cheng, S. Cu-substituted molecular sieves as liquid phase oxidation catalysts. *Microporous Mesoporous Mater.* **2001**, *48*, 309–317. [CrossRef]
30. Taran, O.P.; Zagoruiko, A.N.; Ayusheev, A.B.; Yashnik, S.A.; Prihod'ko, R.V.; Ismagilov, Z.R.; Goncharuk, V.V.; Parmon, V.N. Cu and Fe-containing ZSM-5 zeolites as catalysts for wet peroxide oxidation of organic contaminants: Reaction kinetics. *Res. Chem. Intermed.* **2015**, *41*, 9521–9537. [CrossRef]
31. Wang, J.; Zhu, W.; Yang, S.; Wang, W.; Zhou, Y. Catalytic wet air oxidation of phenol with pelletized ruthenium catalysts. *Appl. Catal. B Environ.* **2008**, *78*, 30–37. [CrossRef]
32. Barbier, J.; Oliviero, L.; Renard, B.; Duprez, D. Role of ceria-supported noble metal catalysts (Ru, Pd, Pt) in wet air oxidation of nitrogen and oxygen containing compounds. *Top. Catal.* **2005**, *33*, 77–86. [CrossRef]
33. Kang, K.; Quitain, A.T.; Daimon, H.; Noda, R.; Goto, N.; Hu, H.Y.; Fujie, K. Optimization of amino acids production from waste fish entrails by hydrolysis in sub- and supercritical water. *Can. J. Chem. Eng.* **2001**, *79*, 65–70. [CrossRef]
34. Gapes, D.J.; Stuthridge, T.R.; Strong, P.J.; Lei, R.J.; Aggrey, A. Treatment of Biomass. WO2013128390A1, 6 September 2013.
35. Chung, J.; Lee, M.; Ahn, J.; Bae, W.; Lee, Y.W.; Shim, H. Effects of operational conditions on sludge degradation and organic acids formation in low-critical wet air oxidation. *J. Hazard. Mater.* **2009**, *162*, 10–16. [CrossRef]
36. Wu, Y.C.; Hao, O.J.; Olmstead, D.G.; Hsieh, K.P.; Scholze, R.J. Wet air oxidation of anaerobically digested-sludge. *J. Water Pollut. Control Fed.* **1987**, *59*, 39–46.
37. Baroutian, S.; Smit, A.M.; Andrews, J.; Young, B.; Gapes, D. Hydrothermal degradation of organic matter in municipal sludge using non-catalytic wet oxidation. *Chem. Eng. J.* **2015**, *260*, 846–854. [CrossRef]
38. Shanableh, A. Production of useful organic matter from sludge using hydrothermal treatment. *Water Res.* **2000**, *34*, 945–951. [CrossRef]
39. Mishra, V.S.; Mahajani, V.V.; Joshi, J.B. Wet air oxidation. *Ind. Eng. Chem. Res.* **1995**, *34*, 2–48. [CrossRef]
40. Lin, S.H.; Ho, S.J.; Wu, C.L. Kinetic and performance characteristics of wet air oxidation of high-concentration wastewater. *Ind. Eng. Chem. Res.* **1996**, *35*, 307–314. [CrossRef]
41. Urrea, J.; Collado, S.; Oulego, P.; Díaz, M. Wet oxidation of the structural sludge fractions. *J. Clean. Prod.* **2017**, *168*, 1163–1170. [CrossRef]
42. Jin, F.M.; Zeng, X.; Jing, Z.Z.; Enomoto, H. A potentially useful technology by mimicking nature by rapid conversion of biomass and CO_2 into chemicals and fuels under hydrothermal conditions. *Ind. Eng. Chem. Res.* **2012**, *51*, 9921–9937. [CrossRef]
43. Zeng, X.; Liu, J.; Zhao, J. Wet oxidation and catalytic wet oxidation of pharmaceutical sludge. *Sci. Rep.* **2023**, *13*, 2544–2551. [CrossRef] [PubMed]
44. Zeng, X.; Liu, J.; Zhao, J.F. Highly efficient treatment of pharmaceutical sludge by catalytic wet oxidation using CuO-CeO_2/gamma-Al_2O_3 as a catalyst. *PLoS ONE* **2018**, *13*, e0199520. [CrossRef] [PubMed]
45. Lipps, W.C.; Baxter, T.E.; Braun-Howland, E.B.; Association, A.P.H.; Association, A.W.W. *Standard Methods for the Examination of Water and Wastewater*, 24th ed.; American Public Health Association: Washington, DC, USA, 2023.

catalysts

Article

Rotating Photo-Disc Reactor (RPR) Used in the Photo-Degradation of Pyridine Using Zinc Oxide as a Catalyst Composited with Aluminum Nanoparticles and Irradiated with Natural Light

Carlos Montalvo *, Edith Lemus, Claudia A. Aguilar, Rosa M. Cerón, Julia G. Cerón, Juan C. Robles and Alejandro Ruiz

Faculty of Chemistry, Autonomous University of Carmen, Ciudad del Carmen C.P. 24180, Campeche, Mexico; edith.lemusjauregui@gmail.com (E.L.); caguilar@pampano.unacar.mx (C.A.A.); rceron@pampano.unacar.mx (R.M.C.); jceron@pampano.unacar.mx (J.G.C.); jrobles@pampano.unacar.mx (J.C.R.); amarin@pampano.unacar.mx (A.R.)

* Correspondence: cmontalvo@pampano.unacar.mx or c_montalvo10@hotmail.com

Abstract: Pyridine was degraded in a rotating photo-disc reactor (RPR) using zinc oxide (ZnO) doped with aluminum nanoparticles (ZnO-Al) as a catalyst and natural light lamps. The reactor disks made of clay had a surface area of 329.7209 m^2. The reactor was operated as a semi-batch system, where it handled a volume of 14.8 L and had a hydraulic residence time (HRT) of 72 h at 54 rpm with a constant flow rate. The results indicate an average degradation of 50.6% after an HRT of 72 h, with a maximum degradation of 62%. The characterization results confirm the effectiveness of the doping process, showing an aluminum concentration of 4.11% by mass in the catalyst, as determined by X-ray techniques. Overall, the doping process proved effective for the zinc oxide catalyst, as evidenced by a reduction in the catalyst bandgap from 3.25 eV for undoped ZnO to 3.08 eV for the doped version, making it sufficiently active under artificial visible light.

Keywords: pyridine; rotating photo-disc reactor (RPR); zinc oxide; aluminum

1. Introduction

Industrial development has grown significantly in recent years, leading to a substantial increase in industrial waste discharge into receiving bodies of water. This has resulted in the release of various pollutants into water bodies. Numerous persistent organic compounds have been identified, including phenols, chlorophenols, and pyridine, which are compounds commonly associated with industrial activities [1].

Pyridine is widely used in the synthesis of various compounds, including fertilizers, paints, and pesticides. Advanced oxidation processes are an effective alternative for degrading persistent organic compounds. These processes include non-photocatalytic methods, such as ultrasound, as well as photocatalytic methods, such as heterogeneous photocatalysis.

The increasingly acceptable use of chemical products in industries and homes has led to an increase in organic pollutants in effluents [2].

Various industries release a significant number of aromatic compounds into the environment due to their widespread use in many industries. Among these, heterocyclic aromatic compounds, such as pyridine and its derivatives, are of particular concern as environmental pollutants due to their recalcitrant, toxic, and teratogenic nature [3,4].

Pyridine and substituted pyridines are important intermediates for the synthesis of pharmaceuticals, herbicides, metal corrosion inhibitors, rubber vulcanization accelerators, etc. [5–9].

Pyridine, a chemical compound with the formula C_3H_5N, has caused significant environmental pollution due to its widespread industrial use, especially in wastewater. Even at very low concentrations (0.3 μg/L), pyridine imparts an unpleasant odor to water, and at higher concentrations (0.82 mg/L), it also affects the taste. Consequently, researchers have actively explored effective and economically viable techniques to mitigate pyridine contamination and purify contaminated water [9–13].

ZnO is an important semiconductor material that has a wide range of applications, including transparent conducting oxides, UV light absorbers, and photocatalysis.

Transition metal doping and mixed oxide formation are two widely studied mechanisms to improve the intrinsic properties of binary oxides. Both procedures have been instrumental in the spectacular increase in applications based on zinc oxide (ZnO) and titanium oxide (TiO_2) thin films [14].

Applications derived from metal-doped ZnO in optoelectronic devices include photovoltaic solar cells, flat panel displays, photodetectors, gas sensors, and light-emitting diodes. Most previous works on doped ZnO films focus on doping with group III elements, and in particular, trivalent cations of the elements Al, Ga, and I have frequently been used to enhance the n-type conductivity of ZnO films [15,16].

Preparing doped material is also a competent method for regulating the surface states of ZnO energy levels, which can be further advanced by changing the doping concentration of semiconductor materials.

As one of the most interesting p-type magnetic doping materials, cobalt oxide nanostructures are also recognized as attractive materials with broad applications in various fields, such as doping, catalysts, solid-state sensors, and electrochemical devices [17].

The doping of impurities to create chemical and, in some cases, physical defects in the crystal lattice that would act as the capture and recombination sites of exactions, involves introducing transition metal (M) atoms as impurities in ZnO crystallites to tailor the photocatalytic property [18].

Since the band gap energy of ZnO crystals is about 3.3 eV, ZnO can absorb UV rays with wavelengths under 375 nm. Therefore, ZnO has been regarded as an excellent UV shielding material, with broad UV absorption characteristics and photofastness compared with other organic and inorganic UV shielding materials [19].

The absorbed UV rays excite the valence electrons to the conduction band. When these photo-excited electrons and holes move to the particle surfaces where water and oxygen molecules reside, highly active free radicals, such as superoxide anion ($\bullet O_2$) and $\bullet OH$ radicals, are generated and undergo secondary reactions, such as the decomposition of organic compounds [20].

Photocatalytic reactors come in various configurations, primarily classified by the light source, shape, catalyst, stirring method, bed dynamics, and reactor type. Slurry and immobilized/deposited reactors are two main categories based on the catalyst type. While slurry reactors require secondary separation for catalyst recovery, immobilized reactors offer advantages like increased light penetration and higher efficiency by preventing catalyst agglomeration. To enhance performance, various substrates, such as glass, quartz, alumina, ceramics, stainless steel, activated carbon, and zeolite, are used to support the photocatalyst [21–25].

The deposition of photocatalyst particles on reactor substrates can limit the mass transfer, affecting the overall efficiency. Spinning disc reactors (SDRs) have emerged as a promising solution due to their versatility and ability to enhance the mass transfer and mixing. They have been successfully employed in various processes, including biodiesel production, enzymatic reactions, nanoparticle synthesis, and the photocatalytic degradation of pollutants. Recent studies have explored the potential of SDRs for degrading textile dyes, pharmaceutical compounds, and antibiotics, offering a promising approach to overcome the limitations associated with traditional photocatalytic reactors [26,27].

Spinning disc reactors (SDRs) enhance mixing and mass transfer through centrifugal force. A key limitation in photocatalytic degradation is the reduced efficiency due to insufficient light exposure of the catalyst. The farther the catalyst is from the light source, the lower its activation and degradation efficiency. SDRs offer a solution by creating thin liquid films and fine droplets, maximizing the light penetration to the catalyst surface. This increased light exposure boosts photocatalytic activity, generates more free radicals, and ultimately improves the overall degradation efficiency [28].

Biodisc rotary reactors (BRRs), a well-established technology in secondary wastewater treatment, have been successfully adapted as photo-reactors. Through extensive testing, this system has been optimized, which involved evaluating various catalyst support materials. Clay discs proved to be the most efficient [28,29].

In addition, crucial operating variables have been tested, such as the rotation speed, hydraulic residence time, changes in the type of light source (visible light range and UV radiation), flow rate, initial contaminant concentration, pH, type of disc structure, and flow rate, which has achieved superior performance [30].

These reactors, with batch treatment capacities of between 10 and 30 L, offer the advantage of not requiring external aeration and feature a high contact area/volume ratio, maximizing the process efficiency. Additionally, treatment is performed sequentially thanks to bulkhead separation. Mass transfer limitations, which might otherwise arise, are minimized by reducing the boundary layer to a minimum. This is achieved by means of turbulent flow in the layer formed above the disk by adjusting the number of revolutions applied [31,32].

2. Results

2.1. SEM Results

Figure 1 shows the characterization of the catalysts synthesized by the SEM technique. In Figure 1A, a characteristic shape of undoped ZnO is observed, with a crystal shape with an average size of 100 to 50 μm; a clean surface is observed. In Figure 1B, accumulations of small particles are observed on the surface of the ZnO, which are attributed to the deposited aluminum particles.

The results of the SEM characterization and EDS analysis (Figure 1D) show that the Al^{3+} nanoparticles were indeed deposited on the surface of the zinc oxide. Although this was not a quantitative test but rather a qualitative one, the EDS result was used to give a percentage of 11% by weight of the doped material, and plasmons of this are shown in the microfilms presented (Figure 1C).

2.2. X-Rays

The crystalline structure of zinc oxide (ICDD: 00-036-1451) was identified; see Figure 2A. This sample presented diffraction peaks at −31.8, 34.5, 36.3, 47.6, 56.6, 62.9, 66.4, 67.9, 69.1, 72.6, 76.9, 81.4, 89.6, 92.8, 95.3, 98.6, 102.9, 104.2, 107.5, 110.4, 116.3, 121.6, 125.2, and 133.9, which corresponded to the Miller indices (100), (002), (101), (102), (110), (103), (200), (112), (201), (004), (202), (104), (203), (210), (211), (114), (212), (105), (204), (300), (213), (302), (006), and (205) of zincite [33].

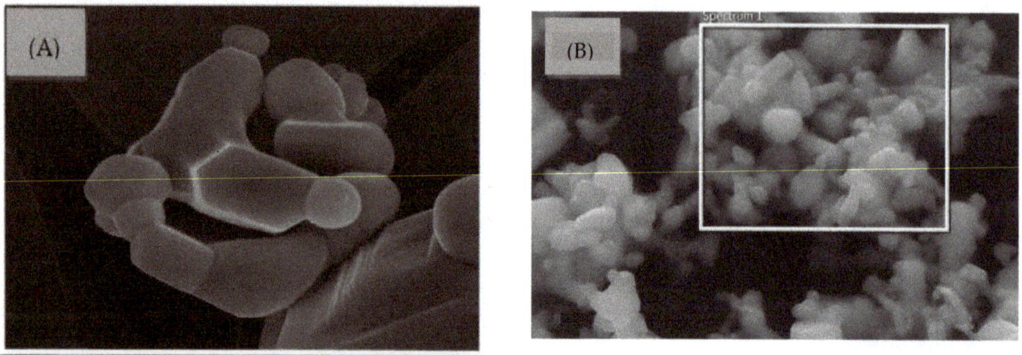

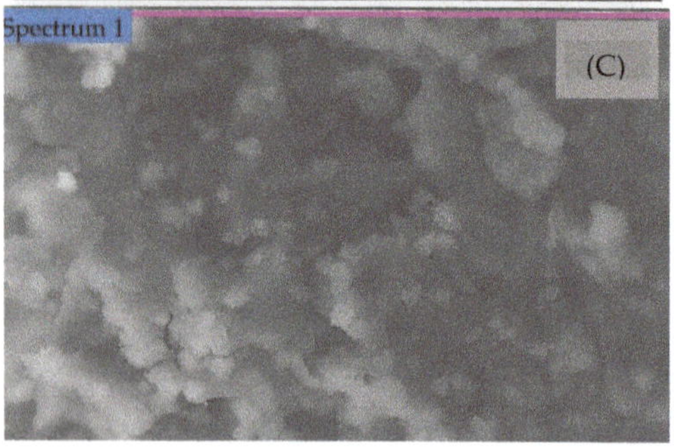

Element	Weight %	Atomic %
O K	35.73	59.70
Al k	11.17	11.06
Zn L	32.50	13.29
Totals	100.00	

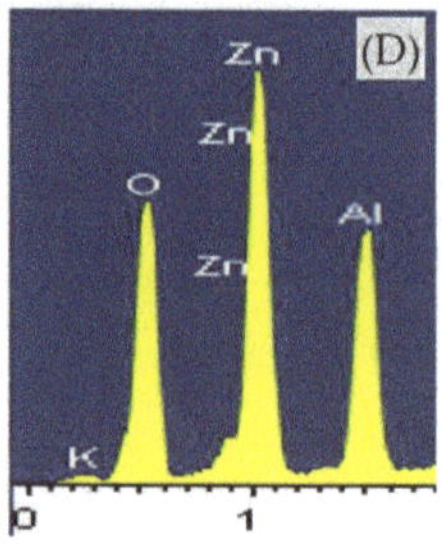

Figure 1. (**A**) Undoped zinc oxide, (**B**) Al^{3+}-doped zinc oxide, (**C**) close-up of the doped area, and (**D**) EDS results of the Al^{3+}-doped zinc oxide.

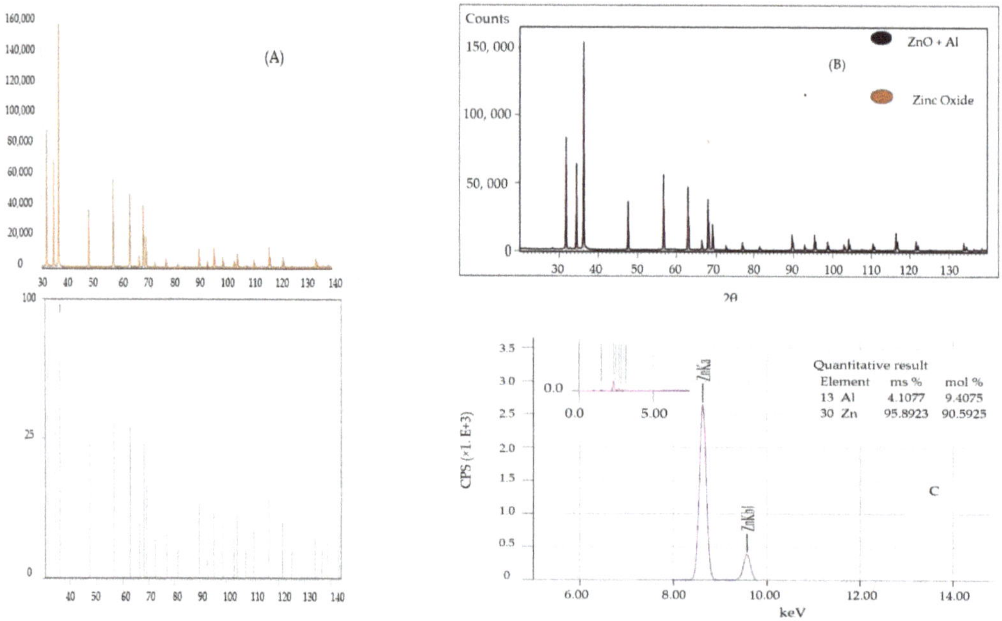

Figure 2. (**A**) X-ray diffraction patterns for ZnO, (**B**) X-ray patterns comparing undoped and doped ZnO, and (**C**) weight percentage of each element observed by EDS analysis (ZnO-Al catalysts).

Figure 2B compares the X-ray spectra of the undoped and aluminum-doped zinc oxide. Only an increase in the intensities of the doped zinc oxide peaks was observed, and small, almost imperceptible, shifts of the peaks, which may have been due to the presence of the dopant in the zinc oxide, which was confirmed by the analysis of the energy-dispersed X-ray studies (Figure 2C).

The elemental study of ZnO-Al was carried out by energy-dispersive X-ray spectroscopy (EDX). Figure 2C represents the EDX results; the peak observed at 8.63 keV represents Zn, while the peaks associated with 1.46 keV that represent Al in the EDS analysis also confirm the presence of aluminum, which was impregnated in the zinc oxide matrix.

Aluminum was present at 4.1%, which is consistent with the mass concentration used in the doping. This study confirmed the presence of aluminum.

2.3. *Diffuse Reflectance*

The optical band gap can be estimated using the following Tauc relationship [34]:

$$\alpha h\nu = B(h\nu - Eg)^{n} \quad (1)$$

where B is a constant; Eg is the forbidden or optical bandwidth of the material; and n is a number characterizing the nature of the electronic transition between the valence band and conduction band, which can have values of 1/2, 2, 3/2, and 3 corresponding to direct allowed, indirect allowed, direct forbidden, and indirect forbidden transitions, respectively.

It is well known that the direct transition through the forbidden band is feasible between the valence band and the conduction band bordering the "k-space". In the transition process, the total energy and momentum of the electron–photon system must be conserved.

It is known that ZnO is a direct band gap semiconductor, so from the above equation, it is clear that the graph of $(\alpha h\nu)^{1/2}$ vs. $h\nu$ will indicate a divergence at an energy value, for example, where the transition occurs. The value of the forbidden band depends on the nature of the transition (i.e., the n value) given.

The transition mode in these ZnO nanoparticles was confirmed to be direct. The absorption coefficient near the band edge was also assumed to show an exponential dependence on the photon energy, and this dependence is given as follows [35–41]:

$$\alpha = \alpha_0 \exp(h\nu/E_u) \quad (2)$$

where α_0 is a constant, and E_u is the Urbach energy interpreted as the width of the tails of the localized states associated with the amorphous state in the forbidden gap.

The estimated band gap from the plot of $(\alpha h\nu)^{1/2}$ versus $h\nu$ for the Al-doped ZnO particles can be seen in Figure 3. The band gap ("Eg") was determined by extrapolating the straight portion to the energy axis at $\alpha = 0$. The linear part shows that the transition mode in these particles was direct. The estimated band gap value of the Al^{+}-doped ZnO was 3.08 eV. The band gap value was smaller than that of undoped ZnO of 3.25 eV [36]; this might have been due to the strain arising from the chemical synthesis of the Al doped ZnO. These microstrains greatly influenced the optical band gap of the material [36].

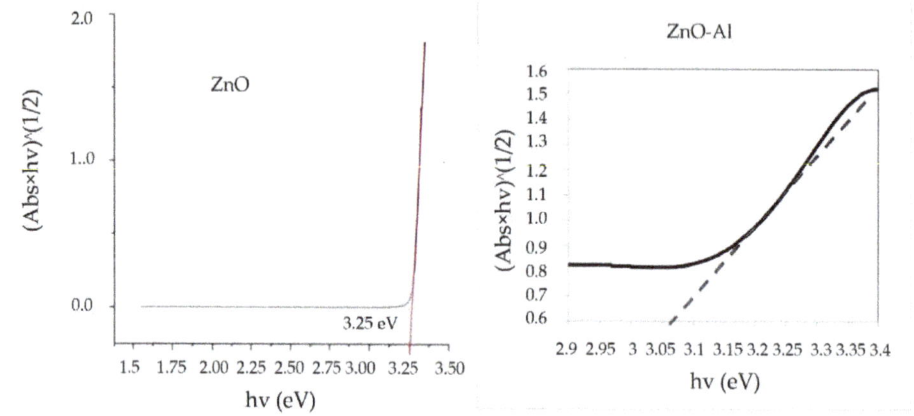

Figure 3. Plots showing $(\alpha h\nu)^{1/2}$ vs. hν for band gap determination of ZnO and ZnO-Al material.

2.4. Raman Spectroscopy

Figure 4 shows the Raman spectrum of undoped (red) and doped (blue line) ZnO. The first shows the characteristic peaks of zinc oxide. In this case, the zinc oxide was obtained from its hexagonal phase (wursite), and its structure belonged to the C_{3V} symmetry group, in which the following vibration modes existed, as determined by group theory:

$$\Gamma = A_1 + 2B_1 + E_1 + 2E_2 \quad (3)$$

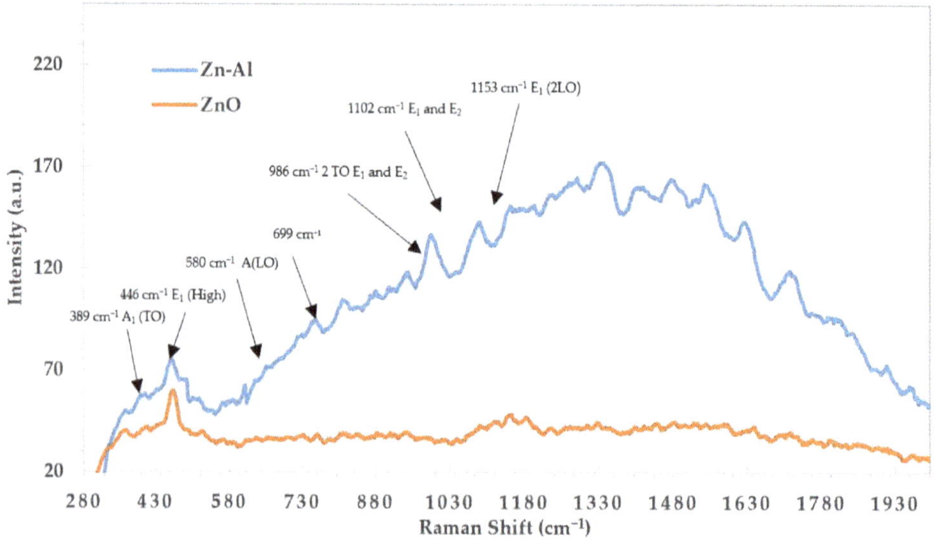

Figure 4. Raman spectrum of undoped ZnO and Al^{3+}-doped ZnO.

The modes A_1, E_1, and E_2 (E_2 (low), E_2 (high)) are Raman active modes. The symmetrical modes A_1 and E_1 are Raman and infrared active modes, E_2 is only Raman active and B1 is a forbidden mode for both Raman and infrared. The polar characteristics of the A1 and E1 vibration modes lead to longitudinal and transverse components designated as A1 (TO), A1 (LO), E_1 (TO), and E_1 (LO).

Zinc oxide (ZnO) exhibits a distinctive peak in the Raman spectrum at 446 cm^{-1}, which disappears upon doping with Al^{3+}. In contrast, the Raman spectrum of Al^{3+}-doped

ZnO exhibited an increase in the intensity of the characteristic peaks at 580 cm^{-1} (LO), 986 cm^{-1} (E_1), and 1102 cm^{-1} (E_2). An average increase of 6 intensity units was observed for these peaks, as well as for the E_1 peak at 1153 cm^{-1} (2LO); suggesting that the presence of aluminum on the zinc oxide surface could be the cause of these variations.

The peak at 437 cm^{-1} is characteristic of the hexagonal phase, while the peak located at 582 cm^{-1} was attributed to the E1 (LO) mode. The latter is generally thought to be related to structural defects (oxygen vacancies, interstitial zinc, free carriers, etc.) in ZnO. The intense presence of the E2 (high) mode and the suppression of the E1 (LO) mode in the Raman spectrum indicate that the ZnO nanostructures obtained in this work were highly crystalline with a hexagonal phase and that the density of the surface defects was minimal.

2.5. Uv–Vis and HPLC Results

The UV–vis results are shown in Figure 5, which shows the characteristic peaks of pyridine at an initial time, with peaks at 256 and 250 nm. Pyridine is a heterocyclic compound that presents several electronic transitions, which are combinations of the π-π* and η-Π* transitions, to which these characteristic peaks are attributed [37–44]. The maximum peak decreased as the degradation time passed; in this case, the total exposure time was 72 h.

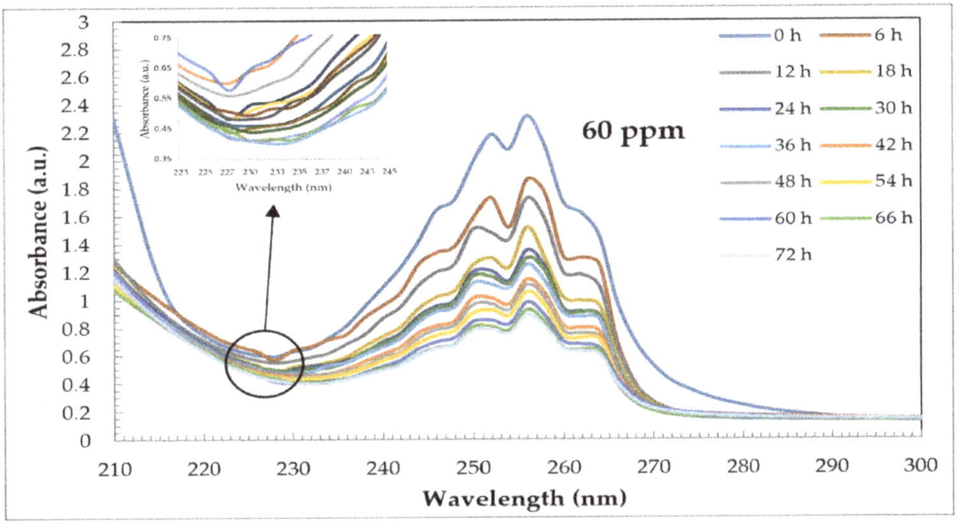

Figure 5. UV–vis spectra of pyridine with an initial concentration of 60 ppm and 72 h of reaction using 15-watt artificial natural light lamps, a rotation speed of 55 rpm, and ZnO-Al as a catalyst.

During the first hours, the percentage of degradation between each 6 h was approximately 39%. For the final times (48 to 72 h), this percentage was drastically reduced to 20%, with a photo-degradation percentage of 62% after 72 h.

Figure 6 shows how the different initial concentrations used decreased over time. In the cases of 60 and 80 ppm, after 24 h, sufficient hydroxyl radicals were produced to attack the molecule. After 18 h in the case of the reaction, a 32% decrease was noted, and as time progressed up to 72 h, this decrease was constant but more minor. This may have been due to the increase in the intermediate compounds that were less susceptible to photo-oxidation.

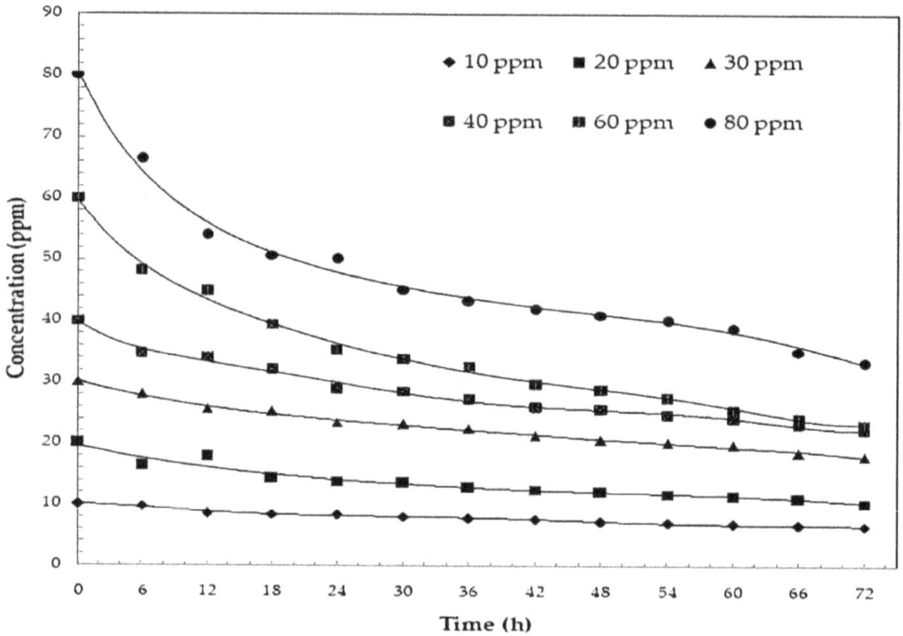

Figure 6. Graph of concentration vs. degradation time of pyridine during 72 h of reaction using 15-watt natural light lamps, a rotation speed of 55 rpm, and ZnO-Al as a catalyst.

Figure 7 shows the percentage degradation values determined using Equation (4):

$$\% \text{ conversion} = (1 - Cf/Ci) \times 100 \quad (4)$$

where Cf—final concentration and Ci—initial concentration

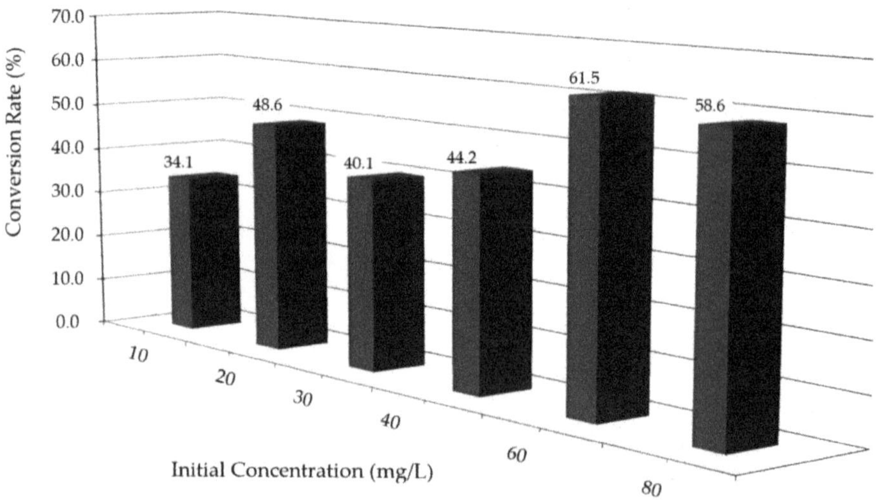

Figure 7. Degradation percentages against initial concentration over a period of 72 h using ZnO-Al with a rotation speed of 55 rpm and natural light lamps.

Figure 7 shows that at low concentrations, the efficiencies were, on average, the same at 40%, which could have been because when the concentration of the organic compound was low, the intermediate compounds were more abundant and more complex and could compete with the main molecule to be oxidized. Meanwhile, at high concentrations (60 and 80 ppm), the pyridine molecule was easily oxidized by the hydroxyl radicals formed, which monopolized most of it considering that natural lights were used.

2.6. Kinetic Analysis

This indicated that the photocatalytic oxidation reactions followed Langmuir–Hinshelwood-type kinetics [9,38–41], as follows:

$$-r_a = -\frac{dC}{dt} = \frac{K_1C}{1 + K_2C + \sum K_iC_i} \tag{5}$$

where K_1C represents the kinetic term of the rate equation, K_2C represents the adsorption term of the reactant, and $\sum(K_i\ C_i)$ represents the adsorption term of all the intermediate products of the degradation reaction of organic compounds.

If the experimental data are analyzed for very short reaction times, the adsorption term of the intermediate products can be neglected.

Based on the above, it can be shown that the following equation can represent the general kinetic form:

$$r_a = -\frac{dC}{dt} = \frac{K_1C^m}{1 + K_2C^n} \tag{6}$$

If the exponents m and n have a value of 1, the constants K_1 and K_2 can be determined directly from the graph of the reaction rate versus the concentration.

Equation (6) can be linearized in the manner recommended by Fogler [42] and Moctezuma [9] using the following initial conditions: $t = 0$, $C = C_0$, and reaction rate, where the following equations are obtained:

$$r_a|_{t=0} = \frac{K_1C_0}{1 + K_2C_0} \tag{7}$$

$$\frac{1}{-r_a|_{t=0}} = \frac{1 + K_2C_0}{K_1C_0} = \frac{1}{K_1C_0} = \frac{K_2C_0}{K_1C_0} \tag{8}$$

$$\frac{1}{-r_a|_{t=0}} = \frac{1}{K_1C_0} + \frac{K_2}{K_1} \tag{9}$$

Equation (9)'s behavior is represented in Figure 8, where the ordinate at the origin is K_2/K_1, and the slope is given by $1/K_1$.

$$-r_{AC} = -\frac{dC_{AC}}{dt} = \frac{K_1C_{AC}}{1 + K_2C_{AC}} \tag{10}$$

From the values obtained in Figure 8, we have the values of the constants, and therefore, the previous equation is expressed as follows:

$$-r_{AC} = \frac{0.0069 \times C_{AC}}{1 + 0.0179 \times C_{AC}} \tag{11}$$

The values of the reaction constants $K_1 = 0.0069\ h^{-1}$ is low if we compare them with the results presented by Elisa [42], but the system was a 300 mL reactor and the catalyst was only zinc oxide, where pyridine was degraded using UV lamps, and the absorption constant $K_2 = 0.0179$ L/mg was slightly lower than the results presented by this author. But the absorption of pyridine is favorable in this system. Therefore, it can be said that the compound had a good affinity for the ZnO surface, which facilitated a closer contact with the hydroxyl radicals formed and allowed for a better photodegradation of the compound.

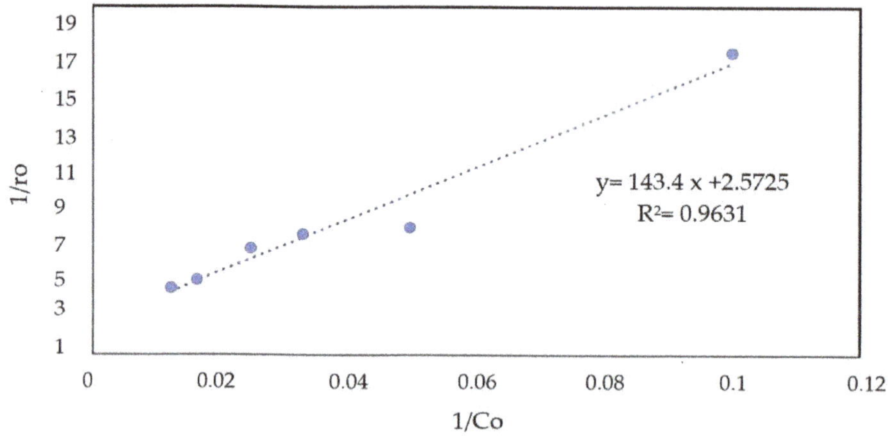

Figure 8. Graph of 1/ro vs. 1/Co to obtain the K_1 and K_2 values of the LH-HW model.

Figure 9 shows the behavior of the LH-HW model of experimental data and the model. The data are observed to have a behavior similar to the model.

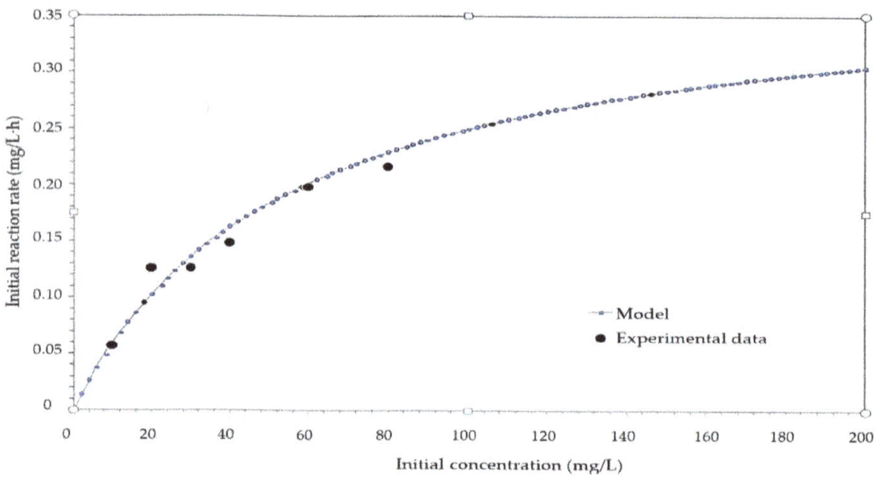

Figure 9. Initial reaction rate as a function of the initial concentration (72 h of reaction using 15-watt natural light lamps, a rotation speed of 55 rpm, and ZnO-Al as a catalyst).

2.7. Proposed Reaction Mechanism

The proposed mechanism for pyridine follows the formation pathway of 2-hydroxy pyridine until the double bonds are weakened and the ring cycle of this compound is broken. It can be assumed that carboxylic acids and other acids are formed.

The experimental results show the formation of 2-hydroxypyridine in the first instance. The inspection of the charged resonance forms suggested that the electron density on the alpha and gamma carbon atoms was especially low; consequently, a beta substitution was expected, also because this position was the only one in which the transition state in the substitution did not have a resonance form with a charge on the trivalent nitrogen.

Once 2-hydroxy pyridine is formed, it can lead to the formation of 2,3-dihydropyridine, but according to the results of the gas mass study, 2,6-dihydropyridine was formed. This weakens the benzene ring, leading to the weakening of the carbon–nitrogen bond, thus forming the monoamine of 3-pentanoic acid or 3-amino pentanoic acid.

Many agree that the next step is the formation of succinic acid or, where appropriate, glutamic acid before it breaks down into carboxylic acid compounds or other derivatives.

Subsequently, the formations of the compound's acetic acid, butyl ester, and 2-butoxy ethanol were observed.

The proposed mechanism is shown in Figure 10, supported by the gas mass analyses.

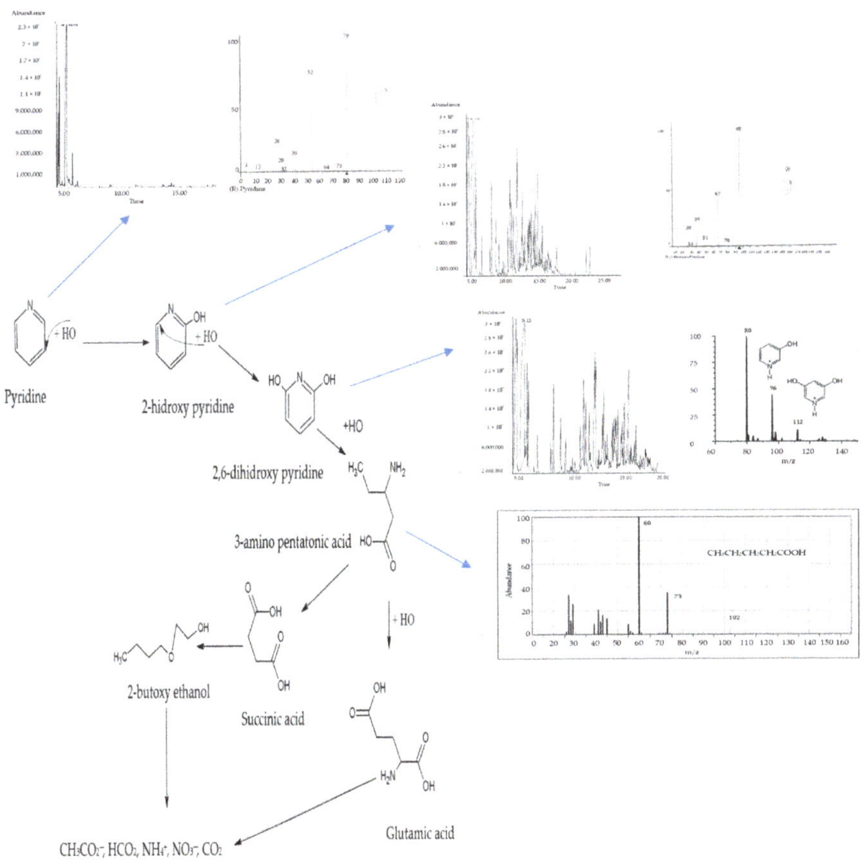

Figure 10. Proposed reaction mechanism for the degradation of pyridine in a rotating photonic reactor in a reaction time of 72 h using 15-watt natural light lamps, a rotation speed of 55 rpm, and ZnO-Al as a catalyst.

3. Methodology

3.1. Reagents and Chemicals

Aldrich brand Al_2O_3 oxide with 99.9% purity and Aldrich brand ZnO were used. For the pH adjustment, Aldrich brand pyridine, sulfuric acid, and sodium hydroxide solutions were used (Aldrich: St. Louis, MO, USA).

3.2. Analytical Methods

3.2.1. Rotating Photo-Disk Reactor (RPR)

For the degradation of pyridine, a rotating photo-disk reactor (RPR) was used, which consisted of 4 stages with 2 disks each. The RPR had a total capacity of 14.8 L and, for each stage, a capacity of 3.7 L; it had a tubular structure support that was 0.58 m wide by 1 m high and 1.35 m long. Each stage had an exit at the bottom of the reactor for sampling.

The RPR was a semi-cylindrical fiberglass tub with a height of 0.31 m and a radius of 0.17 m. Inside the tub, three fiberglass bulkheads divided the reactor into separate stages. The reactor cover was made of stainless steel and measured 0.30 m wide by 0.60 m long. Mounted on the cover were three 15-watt artificial visible light lamps strategically positioned to illuminate each stage, as illustrated in Figure 11.

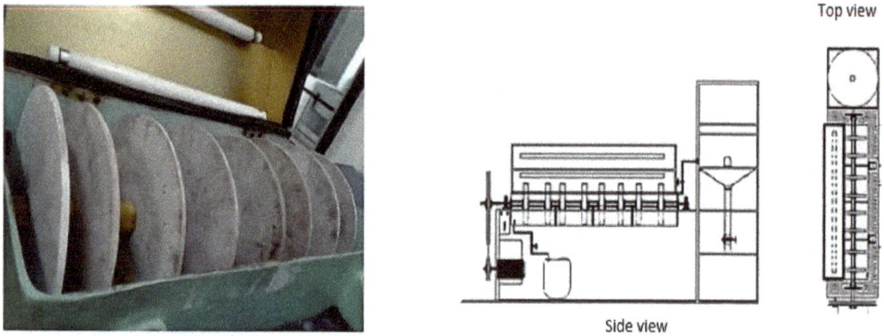

Figure 11. Schematic of the rotating photo-disk reactor (RPR).

The RPR disks were moved by a series of pulleys connected to a stainless-steel shaft, which achieved a speed of 55 rpm. This system worked with a Dayton brand motor of 1/8 hp, 115 W, with a capacity of 1075 rpm and 1.9 amps.

The reactor disks, which had a diameter of 0.23 m and a thickness of 0.008 m, were impregnated with zinc oxide and doped with Al^{3+} metallic nanoparticles. The reactor characteristics are summarized in Table 1.

Table 1. Characteristics of the rotating photo-disk reactor (RPR).

Reactor Characteristic	Value
Number of stages	4
Number of disks per stage	2
Disk diameter	0.23 m
Disk thickness	0.008 m
Total area of undoped disks	0.66476 m^2
Total area of doped disks	329.7209 m^2
Area per stage	41.21512 m^2
Total reactor volume	14.8 L

3.2.2. Doping Process

Clay discs with a diameter of 0.23 m and a thickness of 0.008 m were used. As a first step, the discs were placed in a muffle furnace that had reached 550 °C to eliminate the organic matter residue.

The discs were impregnated with a solution of distilled water and zinc oxide. They were then calcined in a muffle furnace at 550 °C for one hour to improve the catalyst's adhesion to the clay surface.

The photo-deposition technique was used for the ZnO doping using hydrated aluminum sulfate. A 15 L solution was prepared with a concentration of 300 ppm of Al^{3+}. The disks were mounted on the reactor shaft, UV light lamps with a wavelength of 365 nm and

15 watts of power were turned on and allowed to irradiate for 4 h to photo-dope the disks with Al^{3+} particles, and then the disks were placed in the muffle at 550 °C for 1 h [43,44].

3.3. Catalyst Characterization Tests

3.3.1. SEM Tests

Scanning electron microscopy (SEM) is a technique by which it is possible to conclude whether the synthesis of the catalyst and the doping agent has been successful [45]. A JEOL 7600F electron microscope (Akishima, Japan), which was of Japanese origin and operated at 10 kV, was used for this. The samples were analyzed at the Autonomous University of Mexico, Mexico City. The elemental composition was determined utilizing the Oxford INCA X EDX bench, (Oxford Instruments, Abingdon, Oxfordshire, UK) using the JEOL 7600F operated at 10 kV, and the samples were prepared on carbon tape.

This analysis provided information on the sample's composition by providing the percentages of the ZnO catalyst on the support and the presence of aluminum metal ions.

3.3.2. Diffuse Reflectance

To estimate the forbidden band width (Eg), the catalyst was analyzed utilizing a Thermo Scientific Evolution 600 UV spectroscope (Waltham, MA, USA) of USA origin equipped with an ISR-2200 integrating sphere. First, a magnesium oxide blank was analyzed, and then the samples were analyzed in the 200 to 800 nm range. These analyses were developed at the Autonomous University of San Luis Potosi, San Luis Potosi, Mexico.

3.3.3. X-Rays

The photocatalyst was characterized by X-ray diffraction (XDR) to determine the chemical phases and crystallographic properties of the synthesized material [35]; it was carried out using a Bruker D8 advanced X-ray diffractometer made in Karlsruhe, Germany with a 1.54 A copper tube (35 kV, 25 mA) as an X-ray source. The scanning was carried out between 10° and 70° (2 theta) with a step size of 0.03°/s. These analyses were developed at the Metropolitan Autonomous University, Iztapalapa, Mexico City.

To determine the synthesized material's chemical phases and crystallographic properties [46], in most processes, the effective use of nanocatalysts depends on the particle size and ease of manipulation. Therefore, it is of vital importance to characterize them using effective methods at low cost [47].

3.3.4. Raman

Raman microscopy characterized the chemical structure of the undoped and doped zinc oxide. This process was carried out on equipment that used a 785 nm laser and a QE65000 Raman detector from Ocean Optics (Dunedin, FL, USA). This equipment was a homemade assembly in the engineering laboratory of the Autonomous University of Carmen. A 785 nm infrared laser with a laser power of 75 mW was used for undoped ZnO oxide, but for doped ZnO, a laser power of 25 mW was used.

3.4. Preparation of Pyridine Solutions

Pyridine stock solutions with 1000 ppm were prepared, and from these, different dissolutions were made with concentrations of 10, 20, 30, 40, 60, and 80 ppm, where 14.8 L that was the total operating capacity of the photo-reactor for each concentration handled.

The following equation (Equation (12)) was used to find the relationship and prepare the solutions (10, 20, 30, 40, 60, and 80 ppm) of pyridine:

$$C_1V_1 = C_2V_2 \quad (12)$$

where C—concentration and V—volume.

3.5. *Degradation Tests*

The degradation of pyridine was carried out for 72 h, with samples taken continuously every 6 h; an approximately 5 mL aliquot was taken in clean, dry, and labeled amber vials for later analysis using UV–vis Carry 60 spectrometer (Agilent, Santa Clara, CA, USA) and high-performance liquid chromatography (HPLC).

Initially, a pyridine sample was scanned to identify the maximum absorption peak and wavelength. Scans were performed on each sample taken from a wavelength of 200 to 500 nm. Likewise, HPLC was used to analyze each sample.

3.6. *Gas Masses*

Gas chromatography–mass spectroscopy (GC-MS) is a technique that combines the separation capacity of gas chromatography with the sensitivity and selective capacity of a mass detector. This combination allows for the analysis and quantification of trace compounds in complex mixtures with high effectiveness.

This technique is used for the separation of volatile and semi-volatile organic compounds. It consists of vaporizing the sample and injecting it into a capillary chromatographic column.

Elution occurs by the flow of a mobile phase of an inert gas that transports the analytes through the column. The analytes are retained reversibly because of a physical adsorption process. The separated components are eluted from the column and recorded by an MS detector, which obtains a mass spectrum that represents the abundance of different types of ions based on the mass/charge ratio.

The use of GC-MS is restricted to separating compounds with a molecular weight of less than 1000 and a maximum working temperature of about 400 °C.

For this technique, an Agilent 6890N Network GC System gas chromatograph with an Agilent 5973 MSD mass detector (quadrupole) made in the Santa Clara, CA, USA, a DB5-MS column (30 m × 0.25 mm × 0.25 μm), and a split/splitless purge and trap injector (Teledyne Tekmar, Mason, OH, USA) were used. The experiments were carried out at the Faculty of Chemistry of the Autonomous University of Carmen, Mexico.

For this, a temperature ramp that increased by 10 °C every minute until it reached 250 °C was used. This temperature was maintained for 3 min, and then an increase of 10 °C was managed until it reached 355 °C and was maintained for 10 min. The ACQ method CIQ was used, which has already been established.

4. Conclusions

The doping of zinc oxide with aluminum (Al^{3+}) significantly modified its electronic properties, as demonstrated by a decreased band gap from 3.25 eV to 3.08 eV. The EDS and X-ray analyses corroborated the incorporation of the dopant in a proportion of 4.1% by weight. This structural modification, induced by doping, improved the capacity of the material to absorb visible light and, therefore, increased its photocatalytic activity under natural light illumination conditions. On the other hand, this catalyst was suitable for pyridine degradation, which achieved degradations of up to 61.8% for high concentrations and 48.6% for low concentrations, which we concluded was due to the affinity of pyridine for the catalyst surface due to the presence of aluminum.

What was also observed in the values of the determined reaction constants was that the value of the absorption constant (K_2 was 0.0179) was much greater than that of the reaction constant (K_1 of 0.0069), confirming this observed behavior.

According to these results, the degradation kinetic constants followed pseudo-first-order kinetics, which fit perfectly with the LH-HW model for photocatalytic degradation.

The proposed mechanism was based on the formation of hydroxylated compounds, where hydroxyl radicals attack the pyridine ring in the ortho, meta, and para positions for the subsequent rupture of the pyridine ring and the formation of various species until reaching carboxylic acids, carbon monoxide, and water.

The aluminum-doped zinc oxide exhibited photocatalytic activity under natural light and achieved pyridine degradation within 72 h. The experiments were carried out in a rotating photo-disk reactor, where the catalyst was immobilized on clay disks. This system, with a working volume of 14.8 L and a contact area of 329.7 m^2, was adequate for contaminant degradation.

Author Contributions: The contributions of the authors for this manuscript were the following: conceptualization: C.M.; methodology, C.M.; investigation, C.M., C.A.A. and E.L.; resources, J.G.C. and R.M.C.; writing—original draft preparation, C.M.; writing—review and editing, J.G.C., R.M.C., C.M. and E.L.; supervision, J.G.C., R.M.C. and C.M.; project administration, J.G.C., R.M.C. and C.M.; funding acquisition, J.C.R., and A.R. All authors have read and agreed to the published version of the manuscript.

Funding: This work was supported by UNACAR with support from the project registered under number CIPIINOVA/2023/02.

Data Availability Statement: The data presented in this study is available on request from the corresponding author.

Conflicts of Interest: The authors declare that this research was conducted in the absence of any commercial or financial relationships that could be construed as a potential conflict of interest.

References

1. Koe, W.S.; Lee, J.W.; Chong, W.C.; Pang, Y.L.; Sim, L.C. An overview of photocatalytic degradation: Photocatalysts, mechanisms, and development of photocatalytic membrane. *Environ. Sci. Pollut. Res.* **2020**, *27*, 2522–2565. [CrossRef] [PubMed]
2. Careghini, A.; Mastorgio, A.F.; Saponaro, S.; Sezenna, E. Bisphenol A, nonylphenols, benzophenones, and benzotriazoles in soils, groundwater, surface water, sediments, and food: A review. *Environ. Sci. Pollut. Res.* **2015**, *22*, 5711–5741. [CrossRef]
3. Santhi, K.; Manikandan, P.; Rani, C.; Karuppuchamy, S. Synthesis of nanocrystalline titanium dioxide for photodegradation treatment of remazol brown dye. *Appl. Nanosci.* **2015**, *5*, 373–378. [CrossRef]
4. Bello, M.M.; Raman, A. Adsorption and Oxidation Techniques to Remove Organic Pollutants from Water. In *Green Adsorbents for Pollutant Removal*; Springer: Cham, Switzerland, 2018; pp. 249–300. Available online: https://www.researchgate.net/publication/326012580_Adsorption_and_Oxidation_Techniques_to_Remove_Organic_Polltants_from_Water (accessed on 20 March 2025).
5. Grigor'eva Nellya, G.; Filippova Nadezhda, A.; Tselyutina Marina, I.; Kutepov Boris, I. Synthesis of pyridine and methylpyridines over zeolite catalysts. *Appl. Petrochem. Res.* **2015**, *5*, 99–104. [CrossRef]
6. Ashok, B.; Hariram, N.; Siengchin, S.; Rajulu, A.V. Modification of tamarind fruit shell powder with in situ generated copper nanoparticles by single step hydrothermal method. *J. Bioresour. Bioprod.* **2020**, *5*, 180–185. [CrossRef]
7. Frolov, N.A.; Vereshchagin, A.N. Piperidine Derivatives: Recent Advances in Synthesis and Pharmacological Applications. *Int. J. Mol. Sci.* **2023**, *24*, 2937. [CrossRef] [PubMed]
8. Marinescu, M.; Popa, C.-V. Pyridine Compounds with Antimicrobial and Antiviral Activities. *Int. J. Mol. Sci.* **2022**, *23*, 5659. [CrossRef]
9. Moctezuma, E.; López, M.; Zermeño, B. Reaction pathways for the photocatalytic degradation of phenol under different experimental conditions. *Rev. Mex. Ing. Química* **2016**, *15*, 129–137.
10. Elsayed, M. Successive advanced oxidation of pyridine by ultrasonic irradiation: Effect of additives and kinetic study. *Desalination Water Treat.* **2015**, *53*, 57–65. [CrossRef]
11. Aguilar, U.C.; Anguebes, F.F.; Cerón, B.J.; Cerón, B.R.; Córdova-Quiroz, A.V.; Montalvo-Romero, C.; Rangel-Marrón, M.; Robles-Heredia, J.C.; y Zavala-Loría, J.C. *Tópicos Selectos de Ingeniería Química*; UNACAR: Campeche, México, 2012.

12. Song, K.K.; Chen, J.; Song, Y. The degradation of pyridine wastewater by ozonation catalyzed by calcined zinc-magnesium-aluminum hydrotalcites. *J. Beijing Univ. Chem. Technol.* **2021**, *8*, 8–15. [CrossRef]
13. Swarnakar, P.; Kanel, S.R.; Nepal, D.; Jiang, Y.; Jia, H.; Kerr, L.; Goltz, M.N.; Levy, J.; Rakovan, J. Silver deposited titanium dioxide thin film for photocatalysis of organic compounds using natural light. *Sol. Energy* **2013**, *88*, 242–249. [CrossRef]
14. Czech, B.; Rubinowska, K. TiO2-assisted photocatalytic degradation of diclofenac, metoprolol, estrone and chloramphenicol as endocrine disruptors in water. *Adsorption* **2013**, *19*, 619–630. [CrossRef]
15. Kaur, A.; Gupta, G.; Ibhadon, A.O.; Salunke, D.B.; Sinha, A.; Kansal, S.K. A Facile synthesis of silver modified ZnO nanoplates for efficient removal of ofloxacin drug in aqueous phase under solar irradiation. *J. Environ. Chem. Eng.* **2017**, *6*, 3621–3630. [CrossRef]
16. El Nemr, A.; Helmy, E.T.; Gomaa, E.A.; Eldafrawy, S.; Mousa, M. Photocatalytic and Biological Activities of Undoped and Doped TiO_2 Prepared by Green Method for Water Treatment. *J. Environ. Chem. Eng.* **2019**, *7*, 103385. [CrossRef]
17. Rahman, M.M.; Khan, S.B.; Asiri, A.M.; Alamry, K.A.; Khan, A.A.P.; Khan, A.; Rub, M.A.; Azum, N. Acetone sensor based on solvothermally prepared ZnO doped with Co3O4 nanorods. *Microchim. Acta* **2013**, *180*, 675–685. [CrossRef]
18. Islam, M.T.; Jing, H.; Yang, T.; Zubia, E.; Goos, A.G.; Bernal, R.A.; Botez, C.E.; Narayan, M.; Chan, C.K.; Noveron, J.C. Fullerene Stabilized Gold Nanoparticles Supported on Titanium Dioxide for Enhanced Photocatalytic Degradation of Methyl Orange and Catalytic Reduction of 4-Nitrophenol. *J. Environ. Chem. Eng.* **2018**, *6*, 3827–3836. [CrossRef]
19. Sharma, D.K.; Shukla, S.; Sharma, K.K.; Kumar, V. A review on ZnO: Fundamental properties and applications. *Mater. Today Proc.* **2022**, *49*, 3028–3035. [CrossRef]
20. He, R.; Hocking, R.K.; Tsuzuki, T. Local structure and photocatalytic property of sol–gel synthesized ZnO doped with transition metal oxides. *J. Mater. Sci.* **2012**, *47*, 3150–3158. [CrossRef]
21. Binjhade, R.; Mondal, R.; Mondal, S. Continuous photocatalytic reactor: Critical review on the design and performance. *J. Environ. Chem. Eng.* **2022**, *10*, 107746. [CrossRef]
22. Sundar, K.P.; Kanmani, S. Progression of Photocatalytic reactors and it's comparison: A Review. *Chem. Eng. Res. Des.* **2020**, *154*, 135–150. [CrossRef]
23. Mosleh, S.; Ghaedi, M. Chapter 13—Photocatalytic reactors: Technological status, opportunities, and challenges for development and industrial upscaling. *Interface Sci. Technol.* **2021**, *32*, 761–790. [CrossRef]
24. Molinari, R.; Severino, A.; Lavorato, C.; Argurio, P. Which Configuration of Photocatalytic Membrane Reactors Has a Major Potential to Be Used at an Industrial Level in Tertiary Sewage Wastewater Treatment? *Catalysts* **2023**, *13*, 1204. [CrossRef]
25. Sannino, D.; Morante, N.; Sacco, O.; Mancuso, A.; De Guglielmo, L.; Di Capua, G.; Femia, N.; Vaiano, V. Visible light-driven degradation of Acid Orange 7 by light modulation techniques. *Photochem. Photobiol. Sci.* **2023**, *22*, 185–193. [CrossRef]
26. Chakravorty, A.; Roy, S. A review of photocatalysis, basic principles, processes, and materials. *Sustain. Chem. Environ.* **2024**, *8*, 100155. [CrossRef]
27. Hassaan, M.A.; El-Nemr, M.A.; Elkatory, M.R.; Ragab, S.; Niculescu, V.-C.; El Nemr, A. Principles of Photocatalysts and Their Different Applications: A Review. *Top. Curr. Chem.* **2023**, *381*, 31. [CrossRef]
28. Hassard, F.; Biddle, J.; Cartmell, E.; Jefferson, B.; Tyrrel, S.; Stephenson, T. Rotating biological contactors for wastewater treatment—A review. *Process. Saf. Environ. Prot.* **2015**, *94*, 285–306. [CrossRef]
29. Cortez, S.; Teixeira, P.; Oliveira, R.; Mota, M. Rotating biological contactors: A review on main factors affecting performance. *Rev. Environ. Sci. Bio/Technol.* **2008**, *7*, 155–172. [CrossRef]
30. Boiarkina, I.; Norris, S.; Patterson, D.A. Investigation into the effect of flow structure on the photocatalytic degradation of methylene blue and dehydroabietic acid in a spinning disc reactor. *Chem. Eng. J.* **2013**, *222*, 159–171. [CrossRef]
31. Imayama, S. Studies of the Rotating-Disk Boundary-Layer Flow. Technical Reports from Royal Institute of Technology KTH Mechanics. December 2014, Stockholm, Sweden. Available online: https://www.diva-portal.org/smash/get/diva2:781517/SUMMARY01.pdf (accessed on 6 March 2025).
32. Imayama, S.; Alfredsson, P.H.; Lingwood, R.J. On the laminar–turbulent transition of the rotating-disk flow: The role of absolute instability. *J. Fluid Mech.* **2014**, *745*, 132–163. [CrossRef]
33. Wang, L.; Liu, S.; Wang, Z.; Zhou, Y.; Qin, Y.; Wang, Z.L. Piezotronic Effect Enhanced Photocatalysis in Strained Anisotropic ZnO/TiO_2 Nanoplatelets *via* Thermal Stress. *ACS Nano* **2016**, *10*, 2636–2643. [CrossRef]
34. Porrawatkul, P.; Pimsen, R.; Kuyyogsuy, A.; Rattanaburi, P.; Nuengmatcha, P. Morphology-dependent photocatalytic performance of ZnO nanostructures in organic dye and antibiotic degradation. *Int. J. Environ. Sci. Technol.* **2024**, *21*, 7397–7414. [CrossRef]
35. Salah, N.; Hameed, A.; Aslam, M.; Babkair, S.S.; Bahabri, F. Photocatalytic activity of V doped ZnO nanoparticles thin films for the removal of 2- chlorophenol from the aquatic environment under natural sunlight exposure. *J. Environ. Manag.* **2016**, *177*, 53–64. [CrossRef]
36. Raliya, R.; Avery, C.; Chakrabarti, S.; Biswas, P. Photocatalytic degradation of methyl orange dye by pristine titanium dioxide, zinc oxide, and graphene oxide nanostructures and their composites under visible light irradiation. *Appl. Nanosci.* **2017**, *7*, 253–259. [CrossRef]

37. Medellin, C.N.; Ocampo, R.; Leyva, R.R.; Sanchez-Polo, M.; Rivera-Utrilla, J.; Méndez-Díaz, J.D. Removal of diethyl phthalate from water solution by adsorption, photo-oxidation, ozonation and advanced oxidation process (UV/H_2O_2, O_3/H_2O_2 and O_3/activated carbon). *Sci. Total Environ.* **2013**, *1*, 26–35. [CrossRef]
38. Montalvo, C.; Aguilar, C.; Alcoser, R.; Ramirez-Elias, M.; Cordova-Quiroz, V. Semi-Pilot Photocatalytic Rotating Reactor (RFR) with Supported TiO_2/Ag Catalysts for Water Treatment. *Molecules* **2018**, *23*, 224. [CrossRef]
39. Aguilar, C.A.; Montalvo, C.; Ceron, J.G.; Moctezuma, E. Photocatalytic Degradation of Acetaminophen. *Int. J. Environ. Res.* **2011**, *5*, 1071–1078.
40. Fogler, H.S. *Elements of Chemical Reaction Engineering*, 3rd ed.; Pearson Education Limited: Hongkong, China, 1999; ISBN 9780135317167.
41. Matthews, R.W.; McEvoy, S.R. Photocatalytic degradation of phenol in the presence of near-UV illuminated titanium dioxide. *J. Photochem. Photobiol. A Chem.* **1992**, *64*, 231–246. [CrossRef]
42. Elisa, L.; Carlos, M.; Edgar, M.; Socorro, L. Photocatalytic degradation of pyridine in water solution using ZnO as an alternative catalyst to TiO_2. *J. Ceram. Process. Res.* **2008**, *9*, 455–462.
43. He, B.; Zhao, Q.; Zeng, Z.; Wang, X.; Han, S. Effect of hydrothermal reaction time and calcination temperature on properties of Au@CeO_2 core–shell catalyst for CO oxidation at low temperature. *J. Mater. Sci.* **2015**, *50*, 6339–6348. [CrossRef]
44. Thaha, S.K.S.M.; Nalandhiran, P.; Kaliyamoorthy, S.; Mizota, I.; Mangalaraja, R.V.; Sathishkumar, P. Critical Review of Photocatalytic Reactor Designs for Environmental Applications. In *Photocatalysis for Energy and Environmental Applications. Green Energy and Technology*; Sathishkumar, P., Ed.; Springer: Singapore, 2024. [CrossRef]
45. Wang, Z.; Luo, C.; Zhang, Y.; Gong, Y.; Wu, J.; Fu, Q.; Pan, C. Construction of hierarchical TiO_2 nanorod array/graphene/ZnO nanocomposites for high-performance photocatalysis. *J. Mater. Sci.* **2018**, *53*, 15376–15389. [CrossRef]
46. Zyoud, A.; Zu'bi, A.; Helal, M.H.S.; Park, D.; Campet, G.; Hilal, H.S. Optimizing photo-mineralization of aqueous methyl orange by nano-ZnO catalyst under simulated natural conditions. *J. Environ. Health Sci. Eng.* **2015**, *13*, 46. [CrossRef]
47. Gupta, N.; O'Loughlin, E.J.; Sims, G. Microbial degradation of pyridine and pyridine Derivatives. In *Microbial Metabolism of Xenobiotic Compounds*; Microorganisms for, sustainability, Arora, P., Eds.; Springer: Singapore, 2019.

catalysts

Article

3D-Printed Reactor for Coupling Photoelectrochemical (Sea)Water Splitting with Solid-State H_2 Storage

Paweł Wyżga [1,2,*], Joanna Macyk [1], Yuan-Chih Lin [3], Emil Høj Jensen [3], Matylda N. Guzik [3], Krzysztof Bieńkowski [4], Renata Solarska [4] and Wojciech Macyk [1,2,*]

[1] InPhoCat—Innovative Photocatalytic Solutions Sp. z o.o., ul. Brzask 49, 30-381 Krakow, Poland; jmacyk@inphocat.pl
[2] Faculty of Chemistry, Jagiellonian University, ul. Gronostajowa 2, 30-387 Krakow, Poland
[3] Department of Technology Systems, University of Oslo, NO-2027 Kjeller, Norway; yuali611@gmail.com (Y.-C.L.); e.h.jensen@its.uio.no (E.H.J.); matylda.guzik@its.uio.no (M.N.G.)
[4] Centre of New Technologies, Laboratory of Molecular Research for Solar Energy Innovations, University of Warsaw, ul. Banacha 2C, 02-097 Warsaw, Poland; k.bienkowski@cent.uw.edu.pl (K.B.); r.solarska@cent.uw.edu.pl (R.S.)
* Correspondence: pwyzga@inphocat.pl (P.W.); macyk@chemia.uj.edu.pl (W.M.)

Abstract: The modular photoelectrochemical (PEC) reactor accommodating eight photoelectrodes with a total active area of up to 46 cm^2 has been designed and manufactured using the fused deposition modeling method. The device was equipped with an electrolyte flow system, a relay module for the photoelectrode connection, and a feedback-loop module for switching between counter electrodes. The performance and durability of the system were tested within three case study experiments. The water splitting process was successfully combined with an in situ hydrogen storage in the form of metal hydride phases (confirmed by powder X-ray diffraction) using Fe_2O_3- or WO_3-based photoanodes and $LaNi_5$-based cathodes. The PEC water oxidation at the anodes was realized either in a strongly alkaline electrolyte (pH > 13.5) or in acidified synthetic seawater (pH < 2) for Fe_2O_3 and WO_3 electrodes, respectively. In the latter case, the photoresponse of the anodes decreased the cell charging voltage by 1.7 V at the current density of 60 $mA \cdot g^{-1}$. When the seawater was used as an anolyte, the oxygen evolution reaction was accompanied by the chlorine evolution reaction. The manufactured PEC-metal hydride reactor revealed mechanical and chemical stability during a prolonged operation over 300 h and in the broad range of pH values.

Keywords: PEC reactor; 3D-printing; photoelectrochemical water splitting; Fe_2O_3; WO_3; metal hydride; hydrogen storage; solar battery; chlorine evolution

1. Introduction

Solar radiation is a virtually inexhaustible source of energy with 120 PW of light power constantly crossing the Earth's atmosphere [1]. However, the amount of available solar energy is governed by seasonal and hourly sunlight fluctuations, often incompatible with the highest energy demand periods. Consequently, temporal storage of the harvested energy, e.g., in the form of chemical bonds, and its subsequent release on-demand are required to fully exploit the potential of solar energy. The storage can be based on the formation of stable fuel, for instance, hydrogen, that can be later reused, e.g., in fuel cells for electricity generation. One of the promising solar-to-chemical energy conversion routes is based on photoelectrochemical (PEC) water splitting, in which a hydrogen evolution reaction (HER) and an oxygen evolution reaction (OER) take place at the cathode and anode, respectively, and at least one of these electrodes is photoactive. Although the state-of-the-art efficiency of the photoelectrolysis of water is below 10% (without additional bias from photovoltaic/dye-sensitized cells), it could replace the more complex PV-electrolyzer unit, which often requires the use of noble metal HER and OER catalysts, as well as other power control devices like inverters [2,3]. Nevertheless, to store and reuse produced hydrogen, both systems must be coupled with external units, e.g., gas cylinders with compressors, and fuel cells. These additional components can, in principle, be avoided when a standard cathode in a PEC cell is replaced by a hydride-forming material that functions as a hydrogen-evolving/storing electrode [4]. This can be a single element like Pd or Zr [5–7] or an intermetallic compound like $LaNi_5$ [8,9]. Hydrogen is reversibly introduced into the crystal structure of cathode material in an electrochemical manner, and it can be released in the same way (in the PEC cell) or through thermal desorption (outside the PEC cell) [4]. The working principle of such a PEC-metal hydride (MH) hybrid system, when equipped with an additional counter electrode (CE) for the discharge process, can be considered a photo-assisted rechargeable battery [10,11].

In the first approach, the PEC-MH device can be designed as a standard PEC cell operating with photoanodes, in which the cathode/counter electrode is replaced with a hydride-forming electrode [4]. For the operation of photoelectrodes, at least one face of the cell must be transparent, or a photoelectrode port should be provided. The reactor should consist of individual anodic and cathodic compartments separated by a porous diaphragm or an ion-exchange membrane (IEM) to minimize the crossover of evolved gases, an additional counter electrode (for electrochemical discharge) or a straightforward system for an introduction/removal of the cathode (for thermal discharge), a degassing option in at least the anodic compartment, and an electrolyte feeding system. Moreover, an efficient charging rate of the MH-based cathode requires a minimum electric current provided by the photoactive anode. Taking into account typical values of current densities needed for charging electrodes based on different hydride-forming compounds [12], photoactive areas larger (i.e., $\gg$1 cm^2) than those usually used in lab-scale investigations may be required. Consequently, scaling up the whole reactor may be considered. The number of reported large PEC reactors is, however, rather low. A. Mendes et al. provided important contributions to the field by designing and investigating PEC reactors called *Porto cell* and *CoolPEC* [13–15]. The devices were characterized by a modular design and equipped with photoelectrodes with an active area of 20–100 cm^2. The *CoolPEC* reactor revealed a stable operation over 42 days with a single photoelectrode having an active area of 50 cm^2. The authors stressed two detrimental effects in the operation of such large photoelectrodes: (i) notable ohmic losses caused by the resistance of the transparent conducting oxide (TCO) substrate layer, and (ii) an increased recombination rate within the photoactive material (in that case, Fe_2O_3). To address these challenges, they proposed a modified version of the reactor, in which a single large photoelectrode was replaced by eight smaller ones with a total active area of about 28 cm^2. As prompted by the simulation of electric potential and current vectors, they introduced internal separators that minimize overpotential losses and parasitic ionic paths. The multi-electrode approach allowed not only for the minimization of problems reported for a previous setup version but also increased the efficiency of the whole cell due to the higher homogeneity of the smaller electrodes. A similar-sized reactor (the $BiVO_4$ photoanode with an active area of 64 cm^2) coupled with a Si PV cell was proposed in the ArtipHyction project [16,17]. The authors recognized the problem of efficiency reduction (down to 33% of the initial value) during the long-term operation (about 1000 h) due to mass transfer limitations and bubble formation at the photoanode. To minimize the latter, the cell was equipped with separators that formed parallel channels. The follow-up simulation study discussed the pros and cons of this modified design [18]. Another example of a larger reactor (active area of 50 cm^2) was reported by Becker et al. [19]. The cell body was made of poly(methyl methacrylate) (PMMA) and accommodated circular electrodes/windows as well as a compartment separator. The results of the study supported the conclusions from [18] that modular design provides a straightforward assembly and flexibility necessary for system optimization (e.g., an interelectrode distance, a type of intercompartment separator, etc.).

The manufacturing of compartments/bodies in the standard electrochemical and PEC reactors made of polymers (including already presented examples) usually follows two steps: (i) the injection molding or extrusion of a simple polymer block, and (ii) the fabrication of details (windows, ports, threads, and grooves) using subtractive manufacturing methods (e.g., milling, turning, and grinding). The commonly utilized materials include PMMA, polytetrafluoroethylene (PTFE), polycarbonate (PC), or polyether ether ketone (PEEK). Such a fabrication route is highly robust and used routinely in industry; however, it might not be suitable for fabricating complex details (e.g., internal flow channels). In the previous decade, additive manufacturing methods, also known as 3D printing, emerged as an interesting alternative for producing not only early-stage prototypes but also specialized end products [20]. Among several polymer-based 3D printing techniques, fused deposition modeling (FDM) is the most common. During operation, a thermoplastic filament is extruded through the heated nozzle onto a build plate. The pre-programmed movement of the nozzle and build plate in X/Y− and Z−directions, respectively, results in the fabrication of the object in a layer-by-layer manner. Although the printing resolution of FDM is limited to ~250 μm, there is a broad selection of compatible polymer filaments (including electrically conductive ones), and FDM desktop printers are nowadays readily available. Reported examples of 3D-printed PEC cells or their components include simple open cells with electrodes fully immersed in the electrolyte [21–23] as well as small modular reactors [24,25]. Their designs were adjusted to the investigated photoelectrodes and sometimes equipped with additional ports for electric contacts, electrolyte circulation, or gas collection. Nevertheless, none of the reported reactors has reached the size and complexity of demonstrators fabricated with traditional subtractive manufacturing.

In this work, we designed a photoelectrochemical reactor accommodating eight photoelectrodes with up to 46 cm^2 in total active area and operating under the flow of water-based electrolytes. The reactor was manufactured by the FDM technique, followed by necessary post-processing. The constructed system was used in several (photo)electrochemical experiments (presented as three case studies). This report demonstrates that additive manufacturing can be successfully utilized for producing fully functional and flexible scaled-up reactors. The presented results provide a proof-of-concept example of the PEC-MH hybrid device with $LaNi_5$-based hydrogen-storage cathodes, supplemented by simultaneous seawater splitting. The manufactured system can be utilized for various photocatalytic and photoelectrochemical reactions and processes.

2. Results and Discussion

2.1. Case Study 1—PEC Water Splitting Coupled with MH-Based Hydrogen Storage in Alkaline Electrolyte

The concept of combining PEC water splitting with the in situ storage of generated hydrogen by hydride-forming intermetallic compound was thoroughly discussed by Lin et al. [4]. When such a device operates in an aqueous solution, water oxidation takes place at the anode, whereas water reduction at the cathode is completed by the formation of a hydride phase. The hydrogen absorption is reversible and thus allows for the recovery of stored hydrogen, in either thermal or electrochemical way. The combination of a PEC system with an MH-forming cathode extends the functionality of the former so that it follows the concept of a solar battery. The photocurrent generated by photoelectrode contributes to the charging current and effectively decreases the absolute value of the applied potential required for charging. For these experiments, Fe_2O_3 photoanodes and the $LaNi_5$-based cathodes were chosen and operated in strongly alkaline electrolytes.

The performance of the Fe_2O_3 photoelectrodes was studied by linear sweep voltammetry (LSV). The scans were measured for each electrode separately and for all of them connected electrically in parallel (bold line in Figure 1b–d and Figure S2 in Supplementary Materials). Photocurrents generated by individual photoanodes (during constant illumination) did not exceed 1.3 mA in the analyzed potential range. Recombination spikes were visible below 1.4 V vs. RHE during chopped illumination. The observed values of photocurrent densities were much lower [at least three times at 1.5 V vs. RHE for the C3 sample (Figure S2), taking into account lower light power used in this study] than those reported for analogous but smaller Ti-doped Fe_2O_3 electrodes [26], which indicates the problem with the scaling-up process and homogeneity of the samples. Interestingly, the photocurrent generated by the combination of eight electrodes was around 0.97 mA at 1.23 V vs. RHE, approximately 33% of the sum of photocurrents measured individually for each photoanode (2.92 mA). These discrepancies were decreasing along with the applied potential, i.e., 2.8 mA was observed at 1.5 V, which was 95% of the sum of individual photocurrents. Large losses at low potentials were observed previously in the multi-electrode system with one electrode shaded [15], and they were analogous to mismatch losses between different solar cells connected in parallel or series [27]. In such cases, the worst-performing electrode (i.e., the one generating the lowest photocurrent) limits the overall photocurrent of the combination of electrodes. The presented variation in photocurrents for the individual electrodes (Figure 1b) comes not only from differences in the PEC efficiency but also from the high spatial inhomogeneity of incoming light. These two factors must be improved to generate a higher overall photocurrent. To exclude the effect of light variation, a comparison of photocurrent density normalized by light power density can be made (Figure 1c). There are clearly two groups of electrodes—the one (electrodes C2–C4) with higher and the second (electrodes C1, C5, C6, and C8) with lower photoresponse. A similar trend was observed during measurements in a standard lab cell with constant power of light, and, thus, the influence of light power on electrode photoactivity plays a minor role in this case. The above-discussed phenomena and a high onset potential for the Fe_2O_3 electrodes (>900 mV) resulted in negligibly low applied bias photon-to-current efficiency (ABPE, three-electrode system).

The photo-assisted charging of hydride-forming cathodes ($LaNi_{4.7}Al_{0.3}$ and $LaNi_{4.3}Co_{0.4}Al_{0.3}$) vs. eight Fe_2O_3 photoanodes and dark discharging vs. Ni-helix electrode are presented in Figure 2. The average light power illuminating the reactor was in the range of 50–54.5 $mW{\cdot}cm^{-2}$. To highlight the effect of generated photocurrents on the charging potential, the modulated illumination (ON/OFF every 15 min) was applied to photoelectrodes. The protocol of charging/discharging (C/D) experiments comprised (i) a single C/D cycle at the current density of 60 $mA{\cdot}g^{-1}$ up to a 60% state of charge (SoC), (ii) a single C/D cycle at 60 $mA{\cdot}g^{-1}$ up to a 90% SoC, and (iii) ten C/D cycles at 200 $mA{\cdot}g^{-1}$ up to a 90% SoC. Such a sequence was found to be optimal for reaching a high and stable specific capacity of the cathodes. Values of specific current densities were chosen based on the peak current value determined for the hydride formation by chronovoltammetry scans (CV, Figure S3) and previous literature reports. Steps (i) and (ii) of the experimental protocol are magnified in Figure 2c,d. The dark cell voltage required to charge cathodes at the current density of 60 $mA{\cdot}g^{-1}$ varied between −1.82 to −1.77 V and −1.85 to −1.78 V for $LaNi_{4.7}Al_{0.3}$ and $LaNi_{4.3}Co_{0.4}Al_{0.3}$, respectively. The higher absolute values, as expected from CV measurements, were required for charging at 200 $mA{\cdot}g^{-1}$: −2.08 to −2.01 V and −2.08 to 2.00 V, respectively. These numbers are comparable for both electrodes. The cathode potential was between −0.05 and −0.01 V vs. RHE for 60 $mA{\cdot}g^{-1}$ and between −0.09 and −0.16 V vs. RHE for 200 $mA{\cdot}g^{-1}$. When photoanodes were irradiated, the charging potential increased (its absolute value decreased) by 70–80 mV at 60 $mA{\cdot}g^{-1}$ for both samples. After increasing the current density to 200 $mA{\cdot}g^{-1}$, the changes in potential dropped to 20–35 mV. The dark potential of CE vs. RHE (calculated as the difference between $E_{WE\text{-}CE}$ and $E_{WE\text{-}RHE}$) was approx. 1.7–1.8 V and 1.9 V for 60 $mA{\cdot}g^{-1}$ and 200 $mA{\cdot}g^{-1}$, respectively, which was higher than the dark current onset potential for Fe_2O_3 photoanodes (Figure S4). The small changes in the cell voltage upon illumination indicated that the water oxidation reaction at the photoanodes was driven mainly by the dark current. Increasing the specific current density (i.e., magnitude of current flowing in the system) forces photoelectrodes to operate under conditions in which their photoresponse plays a minor role. Photo-induced changes in cell potential for particular current density values remained unchanged in the subsequent charging steps, indicating a stable photoanode operation over a prolonged period.

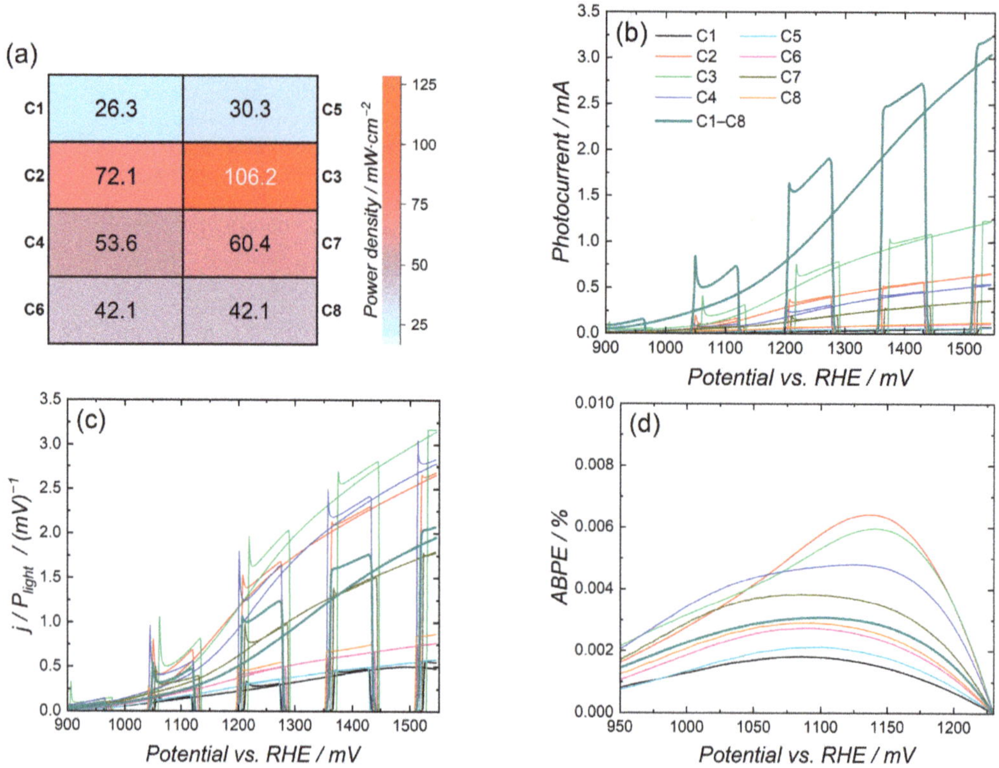

Figure 1. Linear sweep voltammetry (LSV, scan rate: 10 mV·s^{-1}) experiments with Fe_2O_3 photoanodes, conducted in 1M NaOH under the illumination of a solar simulator: (**a**) Power density of simulated solar light for each photoelectrode. Electrodes are highlighted with bold lines and labels. The power density was measured at three points per electrode, and, then, the average value was taken for further analysis. (**b**) The photocurrent for individual Fe2O3 electrodes and all together connected in parallel (bold line). Measurements were collected under constant and modulated illumination. Reference electrode (RE): Hg/HgO; Counter electrode (CE): Ni-helix. (**c**) Photocurrent density normalized by light power density. (**d**) Applied bias photon-to-current efficiency (ABPE) derived from data in (**b**).

The specific discharge capacity derived from discharge time is presented in Figure 3. Capacity for both samples reached saturation after 2–3 cycles at 200 mA·g^{-1} with the values of 160 and 190 mAh·g^{-1} for $LaNi_{4.7}Al_{0.3}$ and $LaNi_{4.3}Co_{0.4}Al_{0.3}$, respectively. In general, determined capacities fall into the range of values reported in the literature for analogous active materials [28–31]. However, it is difficult to directly compare measurements performed in small laboratory cells (usually with Ar purging and the more concentrated KOH electrolyte) with our experiments conducted in a scaled-up reactor due to different compositions of the cathode mixture, a larger distance between the WE and the CE or the two-electrode mode of operation, to mention just a few reasons.

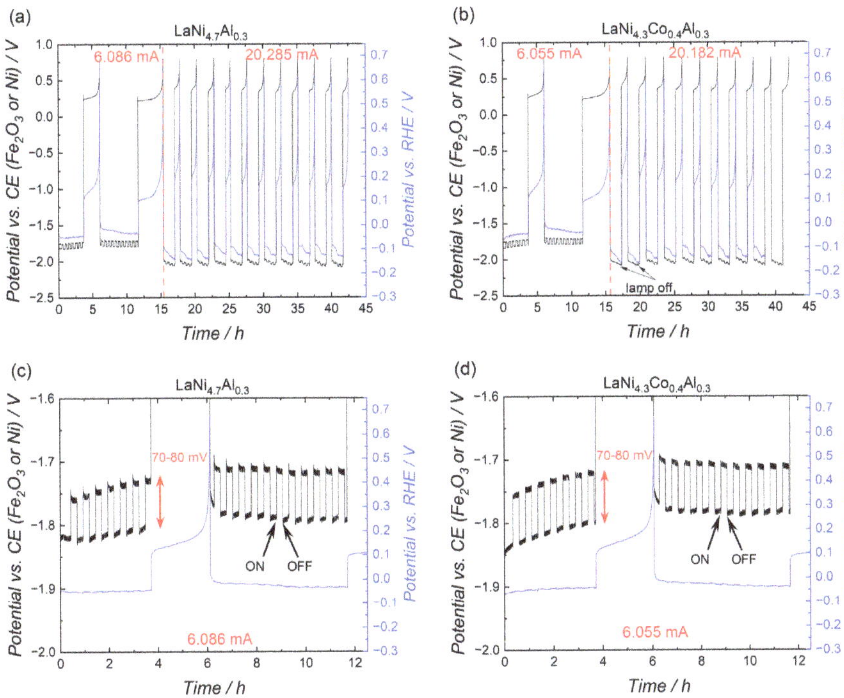

Figure 2. Photo-assisted charging/discharging (C/D) experiments of (**a**,**c**) $LaNi_{4.7}Al_{0.3}$ and (**b**,**d**) $LaNi_{4.3}Co_{0.4}Al_{0.3}$ cathodes. The absolute values of charging/discharging currents and potential changes upon illumination are given in red.

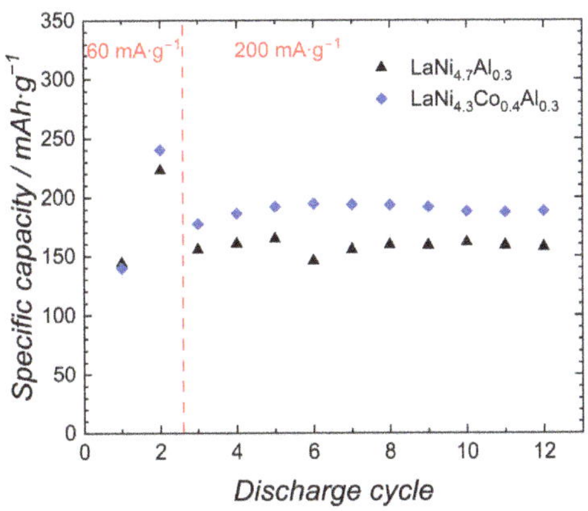

Figure 3. Specific discharge capacities for $LaNi_5$-based cathodes derived from C/D measurements. The discharge current densities are given in red.

To directly prove the formation of metal hydrides and, thus, hydrogen sorption by the studied cathodes, C/D experiments with an analogous protocol were performed in a small laboratory cell (with the three-electrode setup) using cathodes with a larger powder mixture (300 mg). After the last charging cycle, the powders were extracted from the cathodes and analyzed by powder X-ray diffraction (PXRD). The results of the Rietveld refinements for the collected diffraction patterns are presented in Figure 4 with calculated values of unit-cell parameters given in Table S1. The collected data clearly indicated the presence of characteristic Bragg reflections associated with the formation of the pristine $LaNi_5$-based intermetallic phases (space group P6/mmm). However, the occurrence of the second set of diffraction peaks, shifted towards lower 2θ angles, was also apparent and indicated the formation of the second phase with the same hexagonal symmetry but increased unit-cell volume. The sample phase analysis confirmed that the latter was associated with the formation of the hydrogenated $LaNi_5$-based compound. After electrochemical charging (hydrogenation), the unit-cell volumes increased to 13.59(1) Å^3 and 10.82(1) Å^3 for $LaNi_{4.7}Al_{0.3}$ and $LaNi_{4.3}Co_{0.4}Al_{0.3}$, respectively, which correspond to the following chemical compositions of hydride phases: $LaNi_{4.7}Al_{0.3}H_{4.218(3)}$ and $LaNi_{4.3}Co_{0.4}Al_{0.3}H_{3.371(3)}$. Hydrogen concentrations were derived from the estimated volume expansion of 3.21 Å per hydrogen atom in the $LaNi_5$-based hydrides [32].

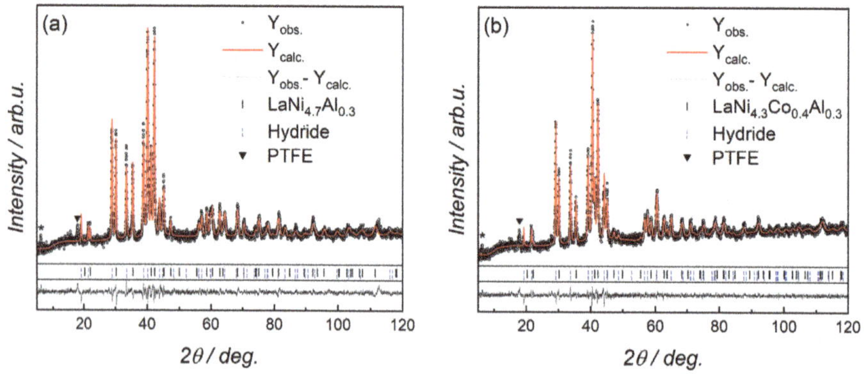

Figure 4. Experimental (dot), calculated (red line), and differential (black line) profiles from Rietveld refinement of electrochemically charged $LaNi_{4.7}Al_{0.3}$ (**a**) and $LaNi_{4.3}Co_{0.4}Al_{0.3}$ (**b**) samples. Bragg peaks indicated with * appeared after C/D experiment (see Figure S5) in both samples but could not be indexed with any well-known secondary phase.

2.2. Case Study 2—PEC (Sea)Water Splitting

Performing water splitting in seawater attracts more and more attention because of the naturally large abundance and low cost of this electrolyte [33]. From the (photo)anode perspective, the utilization of seawater may introduce a chlorine evolution reaction (ClER, $E°$ = 1.36 V [34]) taking place alongside the oxygen evolution reaction ($E°$ = 1.23 V). Depending on the current density range, either the ClER or OER dominates the electrode reaction [35,36]. Moreover, dissolved chlorine undergoes disproportionation ($E°$ = 1.59 V [34]), or chloride ions may hydrolyze ($E°$ = 1.49 V [34]), leading to the formation of hypochlorite HClO acid and further acidification of the solution. Although OER reveals the lowest standard potential among the other mentioned reactions, the chlorine-involving ones require lower overpotentials because of their two-electron character (OER is the four-electron process). As suggested by Jadwiszczak et al. [5], such a chlorine-containing solution could be reused as a disinfecting agent, which would be a supplementary benefit of water splitting in addition to hydrogen generation. Case study 2 comprises short LSV (three-electrode setup) and prolonged chronoamperometry (CA, two-electrode setup) measurements using eight WO_3 photoanodes as the working electrode, steel foil as the counter electrode (CE), and Ag/AgCl (sat. KCl) as the reference electrode (RE). The experiments were performed in acidified synthetic seawater (pH < 2) in both compartments separated by Nafion 117 membrane. A pH below four is required for the stable operation of the WO_3 electrodes [37].

WO_3 electrodes revealed a relatively high photocurrent, ranging from 0.4 to 3.1 mA at 1230 mV vs. RHE, which almost saturated at 1500 mV vs. RHE (Figure 5). The photocurrent onset potential was observed approx. at 500 mV vs. RHE. Recombination peaks were not observed, which indicated the long lifetime of charge carriers in the studied electrodes. Interestingly, the photocurrent generated by all electrodes connected equaled 62% of the sum of individual photocurrents at 500 mV vs. RHE (onset potential) and exceeded 99% at 1125 mV vs. RHE. Smaller differences between both values for WO_3 in comparison to Fe_2O_3 could be associated with the higher electrical conductivity of the former [38] and the larger uniformity within a single film and between different WO_3 electrodes. Nevertheless, the variation in normalized photocurrent densities among studied electrodes (Figure 5c) suggests that the reproducibility of films could be improved, and their activity may depend on the power of incident light. The individual WO_3 electrodes revealed maximum ABPE efficiency in the range of 780–880 mV. The highest value of 0.23 % at 835 mV was determined for the W6 electrode, which was also illuminated with the strongest light. In general, higher photocurrent density and lower (by approx. 400 mV) onset potential for WO_3 anodes enable reaching higher by two orders of magnitude ABPE efficiency (three electrode setup) in comparison to Fe_2O_3.

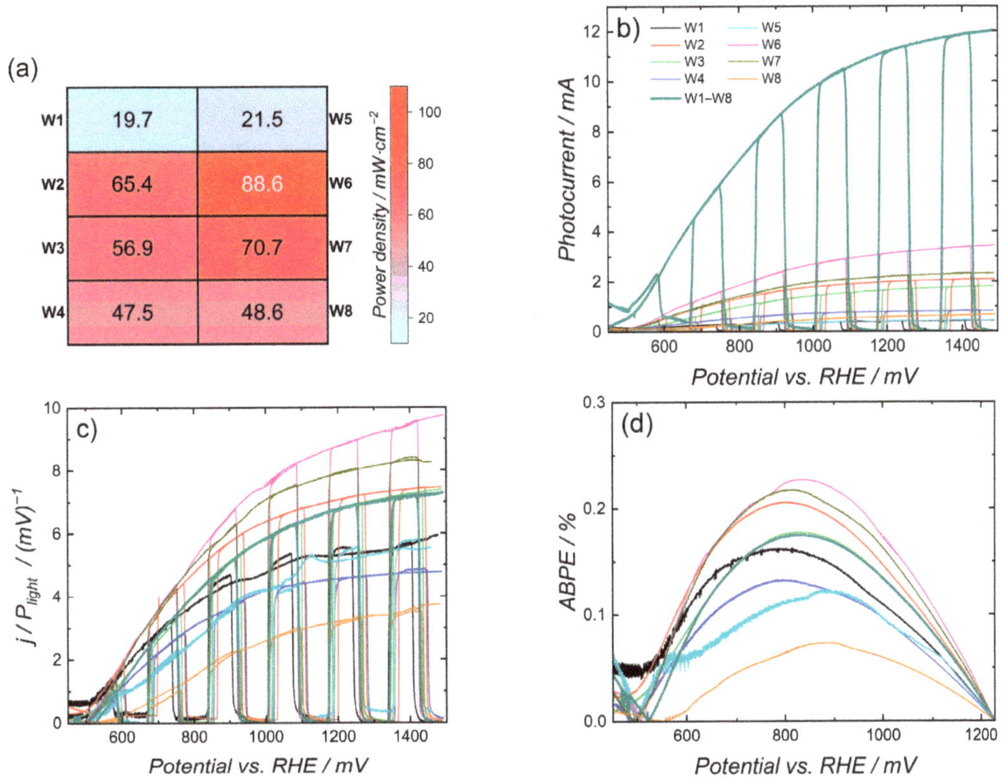

Figure 5. LSV (scan rate: 20 $mV \cdot s^{-1}$) experiments with WO_3 photoanodes, conducted in acidified synthetic seawater (pH = 1.55) under the illumination of the solar simulator: (**a**) Light power density per electrode (average of three points). The average power density equals 52.4 $mW \cdot cm^{-2}$. (**b**) The photocurrent for individual electrodes and all-together connected electrically in parallel (bold line). (**c**) Photocurrent density normalized by light power density. (**d**) ABPE efficiency derived from (**b**). The noisy signal for W1 and W5 samples (top row of electrodes) is probably connected to poor contact with the current collector. The color code is the same in subfigures (**b–d**). RE: Ag/AgCl (sat. KCl), CE: Pt-coil.

The PEC activity of WO_3 photoanodes was additionally evaluated in the acidified Na_2SO_4 solution (pH = 1.55), in which only the OER takes place. Although the onset potential measured in Na_2SO_4 was almost the same as in seawater (approx. 500 mV vs. RHE), the increase in the photocurrent below 1000 mV is much slower in Na_2SO_4 electrolyte. For some samples (e.g., W2, W3, W4), values of the photocurrent at potentials > 800 mV vs. RHE overcame those measured in seawater (Figures 6a and S6). Consequently, maximum ABPE values were shifted towards higher potentials > 900 mV, and, thus, they were slightly lower. Similar behavior was also found for nanorod-TiO_2 electrodes (measurements performed in 0.5M of NaCl and 0.5M Na_2SO_4) [39], suggesting that the ClER (and possibly other reactions involving Cl-species) is more efficient than the OER at low overpotentials (<1000 mV), independent of the type of photoanode.

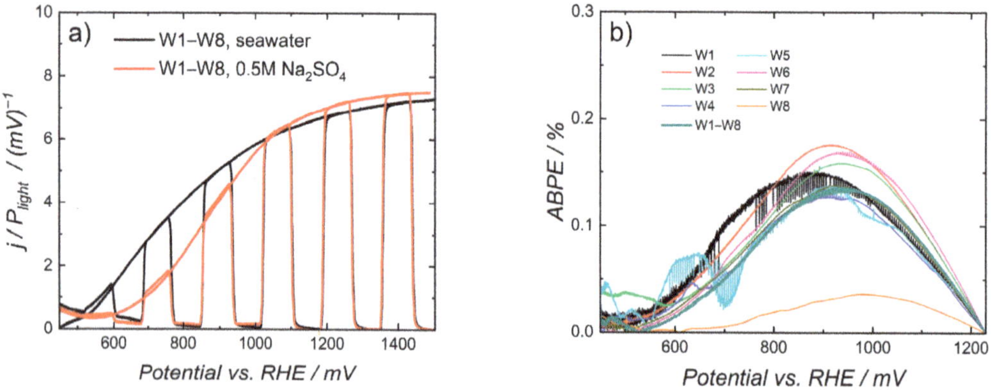

Figure 6. LSV (scan rate: 20 $mV{\cdot}s^{-1}$) experiments with WO_3 photoanodes, conducted in acidified 0.5M Na_2SO_4 (pH = 1.55) under the illumination of the solar simulator: (**a**) Photocurrent density normalized by light power density. (**b**) Applied bias photon-to-current efficiency derived from the photocurrent. The noisy signal for W1 and W5 samples (top row of electrodes) is probably connected with poor contact with the current collector. RE: Ag/AgCl (sat. KCl), CE: Pt-coil.

The stability of the system in seawater splitting was evaluated during a prolonged CA experiment (two-electrode configuration) with eight photoanodes connected as the WE and stainless steel foil acting as the CE (Figure 7a). The measurement was performed at 1230 mV (500 mV in the first 60 s) vs. CE with anolyte circulating between the anodic half-cell and an external reservoir. Additionally, the potential of the WE vs. the Ag/AgCl electrode (WE-RE potential) was independently measured (Figure 7b). Imposing the bias of 1230 mV between the WE and the CE resulted in WE-RHE potential of 1168 mV (3. min), which indicated that the overpotential connected with resistances of electrolyte, membrane, etc., equaled 62 mV at these conditions (inset in Figure 7b). The dark current density during the first 3 min was below 0.01 $mA{\cdot}cm^{-2}$ and rapidly increased to 0.3 $mA{\cdot}cm^{-2}$ after switching on the light, which was associated with the drop of WE-RHE potential by 141 mV. On the other hand, the start of electrolyte circulation caused the increase in both the WE-RHE potential by 20 mV and current density by 7 $\mu A{\cdot}cm^{-2}$. During 17.7 h of experiment, current density decreased to 0.177 $mA{\cdot}cm^{-2}$ (drop of 33%), whereas the WE-RHE potential decreased to 829 mV (drop of 339 mV). The plausible reasons for such a behavior are (i) a decrease in light power during the measurement, (ii) a degradation of photoelectrodes, or (iii) an increase of overpotentials. The two first hallmarks seem to play only a minor role in the observed changes (see discussion in Supplementary Materials, Figures S7 and S8). On the other hand, the color of the stainless steel electrode was changed after the CA experiment and likely indicates a corrosion process occurring at the surface and, thus, the possible increase in the cell overpotential. This type of electrode is clearly not suitable for prolonged operation in acidified seawater, even under cathodic bias.

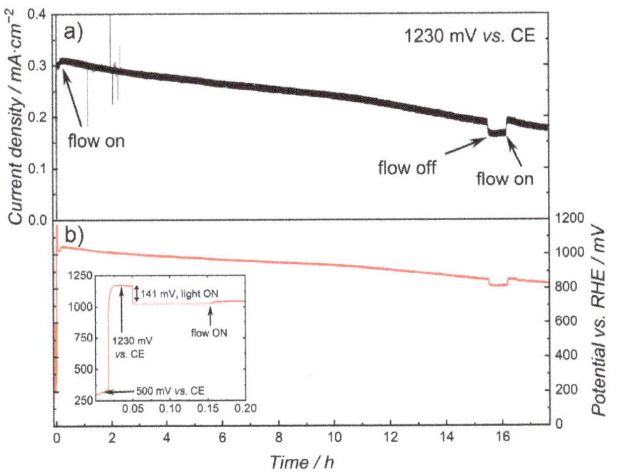

Figure 7. (**a**) The two-electrode chronoamperometric (CA) measurement in acidified synthetic seawater (pH = 1.55) under approx. 0.5 Sun illumination at the potential of 1.23 V vs. CE. WE: 8 WO_3 photoanodes connected electrically in parallel, CE: stainless steel foil. (**b**) The potential of WE vs. RE (Ag/AgCl electrode). It was measured independently from the CA setup using auxiliary potentiostat.

A characteristic odor of chlorine was readily noticeable, and the anolyte was turning more and more yellow during the CA experiment. The pH of anolyte and catholyte increased from 1.55 to 1.61 and from 1.52 to 1.69, respectively. Such an observation can be expected in the cathodic half-cell because of the hydrogen evolution reaction (HER). Concerning anodic reactions, the ClER does not influence the pH, while the OER and other reactions involving Cl-species result in acidification of the solution. An increase in pH seems to result only from the transfer of H_3O^+ ions across the membrane to minimize the pH gradient between both compartments. Such behavior suggests that the efficiency of the OER was lower than the HER because of competition with the ClER. The upper limit of the total amount of produced chlorine (calculated from the total charge, ~486 C) reached almost 450 mg·L^{-1}, which translates into the concentration of approx. 440 ppm after the conclusion of the CA experiment (assuming 100% Faradaic efficiency of the ClER). This value is rather highly overestimated due to the preference for the OER process at current densities below 1 mA·cm^{-2} [35]. The simple estimation of chlorine concentration in the anolyte was conducted using test strips. The intensive purple color of the upper pad (Figure S9) indicated the presence of at least 10 ppm of chlorine in the solution after the CA experiment. Moreover, no chlorine was detected in the catholyte, suggesting no crossover of Cl-containing ions between compartments. Further analysis of the chlorine formation was performed within case study 3 (see text below).

2.3. Case Study 3—PEC Seawater Splitting Coupled with MH-Based Hydrogen Storage

The last case study combines two former, i.e., photo-assisted seawater splitting in the anodic half-cell was coupled with in situ hydrogen storage in the cathodic compartment. The use of WO_3 photoanodes (in an acidic electrolyte) and the $LaNi_{4.3}Co_{0.4}Al_{0.3}$ cathode (in an alkaline electrolyte) implicated a significant overpotential (~0.7 V according to the Nernst equation) due to the pH gradient between both compartments. Although the Na-substituted Nafion membrane was again used as the separator of compartments, the proton transfer seemed to be the dominating conduction mechanism through the membrane, in agreement with a general order of cation permeability in cation-exchange membranes [40]. This process led to a decrease in the inter-compartment pH gradient. Consequently, to maintain pH < 4 in the anolyte solution (required for WO_3 photoanodes), the larger volume of electrolyte (2 L) was circulating between the anodic half-cell and the external reservoir.

The photo-assisted galvanostatic C/D sequence, analogous to the one from case study 1 (CS-1), is presented in Figure 8. The experiment was performed in a two-electrode mode with charging of the cathode vs. eight WO_3 photoanodes and discharging vs. the Ni-helix electrode. Moreover, cathode potential was independently measured vs. the Hg/HgO electrode placed in the cathodic half-cell. Due to a different setup (two electrolytes, membrane separation, different photoanodes) compared with CS-1, cell potentials during charging in the dark at current densities of 60 $mA \cdot g^{-1}$ and 200 $mA \cdot g^{-1}$ were much lower (i.e., higher absolute values), reaching −3.4 V and −3.8 V, respectively. Moreover, WE potentials also decreased to −0.16 V and −0.29 V vs. RHE. Upon irradiation (~50 $mW \cdot cm^{-2}$) the cell potential increased by 1.65–1.75 V at 60 $mA \cdot g^{-1}$ (Figure 8b), which was more than 20 times higher than in the case of Fe_2O_3 photoelectrodes (Figure 2). Such a voltage drop at the given current density (absolute current of 6.06 mA) decreased the power withdrawn from the galvanostat by 10–11 mW. Taking into account the total active area of the studied photoelectrodes (28.5 cm^2), the light-to-electricity efficiency of the system (calculated as the ratio of electrical and light power) reached 0.7%. Moreover, the absolute cell potential under illumination was also lower than in the CS-1, even with a large pH gradient. The CE potential at 60 $mA \cdot g^{-1}$ under illumination was equal to 0.8 V vs. RHE, at which the maximum ABPE efficiency of the WO_3 electrodes was found (Figure 5d). The latter observation explains a high cell voltage reduction due to the photoresponse of the anodes. WE-RE potential remained unaffected during light modulation. After increasing the current density to 200 $mA \cdot g^{-1}$, the photoanodes operated at a potential >2.7 V vs. RHE, which was much higher than the onset potential of the dark current. Consequently, the dark current dominated the reaction at the anodes; thus, applied illumination had a minor impact on the cell potential (reduction of around 40 mV), similar to the observations in CS-1 (Figure 2). The changes in cell potential remained constant for over 17 h (for 200 $mA \cdot g^{-1}$), indicating operational stability of the system in dual electrolyte setup.

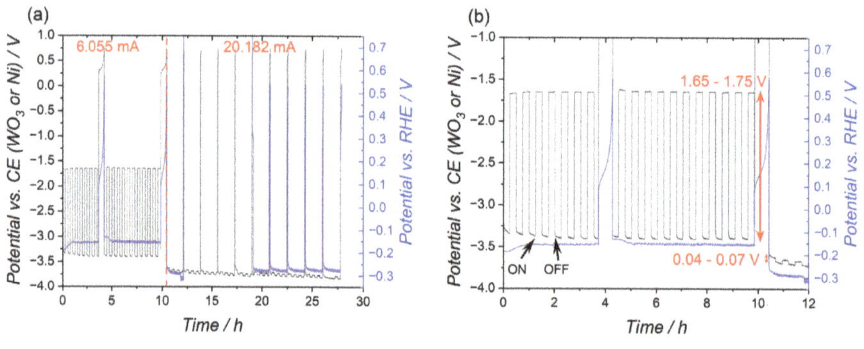

Figure 8. (**a**) Photo-assisted C/D experiments of the $LaNi_{4.3}Co_{0.4}Al_{0.3}$ cathode. Data of potential vs. RHE during 4–7. cycles are missing because the RE was temporarily above the electrolyte. (**b**) The first twelve hours of the experiment. The absolute values of charging/discharging currents and potential changes upon illumination are given in red.

The influence of the specific current density on the magnitude of the cell voltage change was verified during short charging experiments (Figure 9), in which the light was switched ON/OFF every 60 s (15 min was used for 200 $mA \cdot g^{-1}$). The dark charging potential decreased (absolute value increased) along with the growth of the charging current, as expected from the cyclic voltammetry curves (Figure S3). On the other hand, cell voltage change was gradually suppressed for larger currents, which is coherent with the fact that photovoltage decreases for higher (photo)currents. The highest voltage gain was approx. 1.8 V at 20 $mA \cdot g^{-1}$ (2.02 mA), which is almost 57% of the dark charging voltage. This suggests that using a more intense illumination or multiplying photoelectrodes could bring the device closer to bias-free charging for the given current density. Nevertheless, to facilitate hydride formation at the cathode (i.e., to yield hydrogen storage), a minimum charging current of approx. 5 mA for $LaNi_5$-based cathodes (Figure S3) is required, which limits the maximum available voltage gain from photoelectrodes operating in the galvanostatic mode.

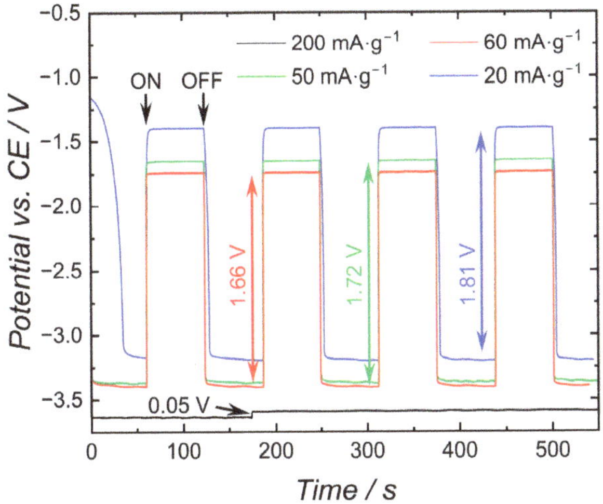

Figure 9. Short photo-assisted C/D experiments with different charging rates. The specific current densities of 20, 50, 60, and 200 mA·g^{-1} correspond to absolute current values of 2.02, 5.05, 6.06, and 20.18 mA, respectively. Modulated light was applied to photoanodes during experiments.

The specific discharge capacity was significantly lower (7–35 mAh·g^{-1}, Figure S10) than values determined in CS-1 (140–240 mAh·g^{-1}, Figure 3), which may be connected with the lower ionic strength of the catholyte (0.5M NaOH in CS-3 vs. 1M KOH in CS-1), differences in the experimental setup, and/or the possible degradation/oxidation of cathode active materials between CS-1 and CS-3 (a period of four months).

The chlorine evolution in the anodic compartment was again confirmed by test strips (Figure S11) and by the formation of the intense characteristic odor and a color change in the electrolyte. No chlorine was detected in the catholyte, as in experiments within CS-2. Two electrolytes were compared: (i) the first one (Figure S12a) after shorter experiments partly presented in Figure 9, and (ii) the second (Figure S12b) after the complete C/D experiment shown in Figure 8. Interestingly, the yellowish color was significantly more intense in the former case despite the fact that a much smaller electric charge transferred during the process (288 C vs. 1420 C). This suggests a higher concentration of the colorless ionic chlorine-based species in the second case. The analysis of a free chlorine (Cl_2)-based species with the spectrophotometric DPD (N,N-diethyl-p-phenylenediamine) method [41] revealed the concentration between 20 and 30 mg·L^{-1} (approx. 19–29 ppm) for the second electrolyte. These values indicate that no more than 11% of the total charge transferred during the C/D experiment was involved in the ClER, which is in line with the preference towards the OER for currents in the mA-range [35]. The observed chlorine concentration was much lower than 5%, the concentration commonly found in commercial antifungal/antimold products. Thus, such solutions might be used as a disinfection agent or oxidating reagent with a lowered pH (final value of around 2).

2.4. Stability of the Reactor

As discussed in the introduction, the 3D-printed reactors reported in the literature are typically early-stage prototypes operating for up to 30 h in a specified electrolyte [21–23,25]. In contrast, the PEC cell manufactured in this study was used for several prolonged experiments, each lasting between 17 and 45 h (e.g., four experiments analogous to those in CS-1 were not presented), giving more than 300 h in total. The designed and tested setup revealed mechanical and chemical stability, as well as tightness in various chemical environments (with the pH ranging from 1 to 14) and under the flow of diverse electrolytes. After the C/D experiments in seawater, the initially transparent impregnation layer inside the anodic half-cell, the outer layer of electrolyte tubing (only inside the external reservoir, where tubes were immersed in the electrolyte), and the multi-electrode gasket changed color (Figures S13 and S14) due to the contact with chlorine species (in the form of Cl_2, HClO, ClO^-, etc.), which are known to be strong oxidative agents. Nevertheless, these changes did not alter the durability of these components. The tailor-made gaskets provided sufficient water tightness; however, they were single-use because of the substantial deformation, which remained after the disassembly of the reactor (Figure S14). Moreover, the noisy signal observed for the top-row electrodes suggests the need for improving the design of the current collectors and the uniformity of pressure applied to the set of electrodes (the type of gasket also affects this factor).

3. Materials and Methods

3.1. Manufacturing of the Reactor

The reactor and additional components, including compartment bodies, top lids, the cathode holder, the diaphragm, closing plates, stoppers, adapters, caps for electrolyte bottles, and additional boxes were designed in the Freecad v. 0.19 parametric-modeler [42]. The slicing of the 3D models, setting the printing conditions, and the generation of .gcode files were performed in *Ultimaker Cura* v. 4.9.1 software [43]. Fused deposition modeling was conducted with the Ender 5 Plus printer (Creality, Shenzhen, China) equipped with an enclosure, which minimizes temperature gradients between the printer's build plate and the environment and reduces the cooling rate of the printout. Several printing parameters were optimized to obtain objects with the best properties, e.g., wall thickness, single-layer height, infill type and density (the fraction of the object's internal volume filled with filament), nozzle and build-plate temperatures, printing speed, and print orientation. They affect the quality of the printouts, the water tightness of certain faces, the printing time, and the amount of building material. The exemplary parameters for different objects were compared in Table 1. The ASA filament (acrylonitrile-styrene-acrylate filament, Polar White, Spectrum Filaments, Pecice, Poland) was used to print all objects. This polymer was chosen due to its chemical stability in acidic and alkaline aqueous solutions, and its resistance to UV radiation. To increase the adhesion of the printout to the build plate, the latter was manually coated with a thin ASA layer. After printing, all supporting elements were removed, and rough faces were smoothed with acetone (if applicable and needed). The interior of cell compartments was additionally impregnated with transparent epoxy resin 049 (AG TermoPasty, Sokoly, Poland) to ensure water tightness. Gaskets for electrode/lid ports (methyl vinyl silicone, RubberPro, Liszki, Poland) were glued to the cell body/lid using a cyanoacrylate adhesive (Technicqll, Trzebinia, Poland). Softer gaskets in contact with photoanodes and a membrane/diaphragm were cast in-house from mold-forming silicone (SiliForm 25, AgBet, Dobrzelin, Poland) using 3D-printed tailor-made molds.

Table 1. Examples of parameters used for printing different objects.

<table>
<tr><th>Parameter</th><th>Compartment Body</th><th>Cathode Holder</th><th>Box for Pumps</th><th>Gasket Mold</th></tr>
<tr><td>Filament material</td><td colspan="4">ASA</td></tr>
<tr><td>Nozzle diameter/mm</td><td>0.4</td><td>0.1</td><td>0.4</td><td>0.4</td></tr>
<tr><td>Nozzle/bed temperature/°C</td><td colspan="4">245/60</td></tr>
<tr><td>Layer height/mm</td><td>0.1</td><td>0.08</td><td>0.2</td><td>0.2</td></tr>
<tr><td>Infill/%</td><td colspan="3">20</td><td>15</td></tr>
<tr><td>Wall thickness/mm</td><td>4</td><td>4</td><td>1.2</td><td>1.2</td></tr>
<tr><td>Top/bottom thickness/mm</td><td>4</td><td>4</td><td>0.6</td><td>0.6</td></tr>
<tr><td>Top/bottom printing speed/mm·s^{-1}</td><td colspan="4">20</td></tr>
</table>

3.2. Design of the Reactor and Peripheral Modules

The manufactured reactor was based on a modular design, which allows for various configurations of electrodes and processes. The multi-electrode half-cell (no. 1 in Figure 10) accommodated up to eight (photo)electrodes with dimensions of 50 mm × 22mm (an active area up to 5.8 cm^2 and 46 cm^2 for a single electrode and the whole set, respectively). On the contrary, the single-electrode compartment (no. 2 in Figure 10) was suitable for a larger electrode of size 100 mm × 100 mm (an active area up to 71 cm^2, not presented in this report). In both variants, electrode(s) were pressed against the half-cell equipped with an intermediate gasket and Cu foil acting as an electrical connection (no. 6 in Figure 10). Depending on the type of experiment/process studied, the electrode window in each half-cell could be alternatively closed by the PMMA plate (no. 11 in Figure 10). Compartments were separated by an IEM (e.g., Nafion 117, Sigma Aldrich, Burlington, MA, USA) or a porous diaphragm, which provided an ionic connection between both parts. Top lids enabled the introduction of additional electrodes or tubing to the cell. The removal of evolved gases out of the compartments was assumed to take place along with a separated circulation of anolyte and catholyte between the cell and external reservoirs. For this reason, each compartment contained at least ten standard 1/4″- hydraulic ports for the connection of the electrolyte flow system. Figure 10 illustrates the PEC-MH configuration of the reactor, with eight photoanodes and the cathode introduced through the lid, whereas the disassembled reactor is presented in Figure S15. The multi- and single-electrode half-cells are characterized by internal volumes of approx. 170 and 240 mL, respectively. The electrolyte circulation system was based on a set of membrane pumps with a maximum flow of up to 900 mL·min^{-1}, inert tubing (Versilon SE-200, Saint-Gobain, Courbevoie, France), and plastic reservoirs. The individual switching of each electrode was realized by a computer-controlled relay module.

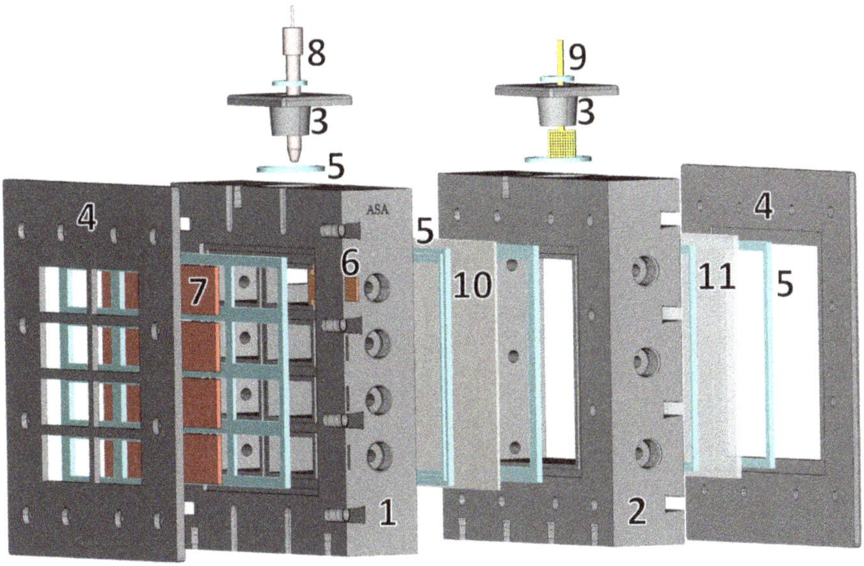

Figure 10. A schematic assembly of the PEC-MH reactor: (1) an anodic compartment, (2) a cathodic compartment, (3) lids with ports for electrodes, (4) fixing plates with windows, (5) silicone gaskets, (6) an electric contact (the other seven are not shown), (7) a photoanode, (8) a reference electrode (optional), (9) cathode, (10) an ion-exchange membrane/diaphragm, and (11) the PMMA/glass plate. For clarity, some elements of the assembly are not shown, e.g., the cathode holder, additional counter electrode, hydraulic fittings, electrolyte tubing, screws, and nuts.

3.3. Preparation of Cathode Assembly

Hydride-forming intermetallic compounds ($LaNi_{4.7}Al_{0.3}$ and $LaNi_{4.3}Co_{0.4}Al_{0.3}$) were synthesized by arc melting. Stoichiometric amounts of La (99.9%, Alfa Aesar, Ward Hill, MA, USA), Ni (99.98%, Alfa Aesar), Co (99.95%, Alfa Aesar), and Al (99.99%, Alfa Aesar) were melted together in a pure argon (oxygen-free) atmosphere, in an arc furnace (MAM-1, Edmund Bühler GmbH, Bodelshausen, Germany) utilizing an electric arc formed between tungsten and copper electrodes. The arc temperature was adjusted for each composition (up to max. 3500 °C) by varying the applied electric current. To reach a high chemical and structural homogeneity, the obtained ingots were repeatedly melted and flipped upside down multiple times. In the final stage of the synthesis process, the as-cast button-shaped samples were once again melted and cast into rod-shaped samples. To further increase the homogeneity of the synthesized materials, the rod-shaped ingots were subsequently enclosed in stainless steel tubes under the Ar atmosphere and annealed at 1200 K for 96 h. In the next step, the annealed samples were pulverized and then mixed with binding and conducting materials to form a cathode. The studied cathodes comprised a powder mixture of an annealed hydrogen-storing intermetallic compound (the active material, 90 wt.%), the binder (PTFE, 5 wt.%), and carbon black (5 wt.%, to increase electrical conductivity) of approx. 120 mg total mass, which was cold-pressed between two sheets of Ni-foam. To minimize powder losses and to increase the mechanical stability of such a sandwich-type electrode, the cathode was enclosed in a 3D-printed cathode holder supplemented with two silicone gaskets and a Ni-frame acting as an electrical connection (Figure S16).

3.4. Preparation of Photoanodes

Thin films of Ti-doped Fe_2O_3 were deposited on FTO-glass substrates using iron(III) acetylacetonate (FeAcac, 99.9 %, Sigma Aldrich) as the precursor following the procedure reported before [26]. Shortly, an appropriate amount of $TiCl_4$ (99.9 %, Sigma Aldrich) and FeAcac were sequentially dissolved in ethanol and homogenized using sonication. FTO substrates were cleaned (detergent solution, de-ionized water, and isopropyl alcohol), masked at one edge (for electrical contact), and placed onto a hot plate (450 °C). The precursor solution was then sprayed onto FTO glass with the Ar carrier gas pressure of 100 psi. After 10 min, the samples were calcinated at 650 °C for 1 h.

Approximately 2.5 µm thick WO_3 films were obtained by using a sol–gel method based on a tungstic acid/polyethylene glycol (PEG) 300 precursor following a previously described procedure [37]. FTO glass (resistance 8–10 ohm·sq^{-1}, Sigma Aldrich) was used as substrate. The WO_3 films were formed during six consecutive applications of the colloidal precursor onto the substrates with a surface area of 5.8 cm^2 using the doctor-blade technique. Each application was followed by heating in flowing oxygen at 500 °C for 30 min. The annealing provides highly crystalline WO_3 films, which crystallize in a monoclinic structure.

3.5. PEC Characterization

Linear sweep voltammetry, chronoamperometry, and chronopotentiometric charge/discharge measurements were controlled by potentiostat/galvanostat OrigaStat 100 (OrigaLys, Lyon, France), whereas additional voltage measurements were performed with the potentiostat 2.0 (Instytut Fotonowy, Krakow, Poland). The source of solar radiation was the solar simulator consisting of the xenon arc lamp (XBO, 150 W, Instytut Fotonowy) and the AM1.5G filter, coupled with a water filter. The light spectrum (Figure S17) was measured using a Black-Comet-SR spectrometer (StellarNet Inc., Tampa, FL, USA) equipped with a cosine receptor and fiber optic cable. The used solar simulator was characterized by a small light power variation over time (less than 5 %). However, it produced a highly focused beam (spot smaller than 1 cm^2) because of the installed elliptic reflector. Consequently, light power over 100 cm^2 area (size of the photoanode port) was highly inhomogeneous, and each photoanode was illuminated differently. To account for this effect, the spatial distribution of light power was measured before each experiment at three points per electrode using a light calibrator (Instytut Fotonowy), fitted with a reflective neutral density filter (required to avoid saturation of sensor, optical density 2, Edmund Optics, Barrington, NJ, USA), and a tailor-made holder resembling the geometry of the multi-electrode side (three points per electrode, Figure S18). The average power for each electrode was used to calculate ABPE efficiency. The photoanode active areas in multi-electrode compartments were determined graphically (Figure S19) using ImageJ v. 2 software [44]. Electrolytes were freshly prepared before experiments. The pH values of electrolytes were measured with the CX-461 multifunction meter (Elmetron, Zabrze, Poland) coupled with EPS-1/EPX-4 electrodes and a temperature probe.

3.6. Case Study 1 (CS-1)

The photo-assisted charging of $LaNi_{4.7}Al_{0.3}$- and $LaNi_{4.3}Co_{0.4}Al_{0.3}$-based cathodes (working electrode WE) was performed under simulated solar radiation with eight Fe_2O_3 photoanodes connected electrically in parallel (CE). The discharging step was performed in dark vs. the Ni-helix electrode (CE) with a cut-off potential of 0.7 and 0.79 V vs. CE (0.52 V and 0.58 V vs. RHE) for $LaNi_{4.7}Al_{0.3}$ and $LaNi_{4.3}Co_{0.4}Al_{0.3}$, respectively. Although C/D experiments were performed in two-electrode mode (to simulate real operation), the potential of WE vs. Hg/HgO was additionally measured by an independent potentiostat (Figure S20a). The automatic switching of two types of counter electrodes (photoanodes or Ni-helix electrodes) between charging and discharging was realized by a feedback-loop module based on a microcontroller (Seeeduino Nano, Seeed Studio, Shenzhen, China) coupled with a relay. Both compartments of the reactor were separated by a porous membrane made of an ASA polymer, allowing the mixing of electrolytes (1M KOH, pH = 13.5–13.7, prepared from KOH pellets, 88%, VWR, Radnor, PA, USA) but limiting the crossover of evolved gases. The scheme with distances between electrodes is given in Figure S21a, and pictures of anodic and cathodic compartments are presented in Figure S1. The total volume of the electrolyte in the system amounted to 1.2 L. Powder X- ray diffraction experiments were performed with the Rigaku Miniflex diffractometer (Cu $K_{\alpha 1}$ radiation, 2θ-range = 5–120°, $\Delta 2\theta = 0.02°$).

3.7. Case Study 2 (CS-2)

Three-electrode LSV and two-electrode CA measurements were performed with eight WO_3 photoanodes connected electrically in parallel (WE) and Pt-coil or stainless steel foil (CE), respectively. Ag/AgCl (sat. KCl) was used as the RE. The experiments were performed in synthetic seawater (AF Perfect Water, Aquaforest, Brzesko, Poland) and 0.5M Na_2SO_4 (from powder, 99%, Alfa Aesar), acidified with HCl (35–38%, Chempur, Piekary Slaskie, Poland) and H_2SO_4 (95%, POCH, Gliwice, Poland) to pH = 1.55 and 1.7, respectively. The measurements were performed in a two-compartment configuration with a Na-substituted Nafion 117 membrane (ion exchange in 0.5–1M NaCl lasted at least 5 days). Both compartments were initially filled with the same starting electrolyte. The set of membrane pumps maintained the electrolyte flow. Total volumes of circulating anolyte and catholyte were approx. 0.4 L each. The presence of chlorine in each compartment was verified qualitatively using test strips (Aqua Star).

3.8. Case Study 3 (CS-3)

The photo-assisted C/D experiment for $LaNi_{4.3}Co_{0.4}Al_{0.3}$-based cathode coupled with simultaneous seawater oxidation was performed in a two-electrode configuration. The setup was analogous to CS-1, except for the Na-substituted Nafion 117 membrane separating both compartments (Figures S20b and S21b). For the anolyte and catholyte, acidified synthetic seawater (pH = 1.7, 2 L) and 0.5M NaOH (pH = 13, 0.8 L, from NaOH pellets, 98 % Honeywell, Charlotte, NC, USA) were used, respectively.

4. Conclusions

The objective of this work was the design and production of a photoelectrochemical reactor for coupling PEC (sea)water splitting with in situ hydrogen storage in hydride-forming intermetallic compounds. The large-scale modular PEC-MH cell was manufactured using the 3D-printing method supplemented by appropriate impregnation. The reactor size was larger than a standard laboratory cell and comparable with the upscaled devices manufactured by traditional subtractive methods. It can operate with up to eight photoactive electrodes with a total active area of up to 46 cm^2. The device was equipped with an electrolyte flow system, a relay module for the photoelectrode connection, and a feedback-loop module for switching between counter electrodes.

The fabricated system was tested in three case-study experiments. Photo-assisted galvanostatic charging/discharging of $LaNi_5$-based cathodes with the Fe_2O_3 photoanodes resulted in the formation of the metal hydride phases (confirmed by powder X-ray diffraction) and proved a successful operation of the upscaled PEC-MH device. Owing to the low efficiency of the tested photoanodes (the total photocurrent density for eight electrodes of 0.03 $mA \cdot cm^{-2}$) and a potential–current mismatch between anodes and the cathode, the reduction in the cell charging voltage due to the photoresponse from the photoanodes was only ~4% (charging at 6 mA). Significantly higher photoactivity was observed for WO_3 photoelectrodes (the total photocurrent density of 0.34 $mA \cdot cm^{-2}$) towards oxygen evolution reaction (performed in Na_2SO_4 solution) and simultaneous chlorine and oxygen evolution reactions (in synthetic seawater). The higher PEC efficiency of WO_3 electrodes and the optimal potential–current conditions for both the anodes and the cathode were substantiated by the significant reduction in the cell charging voltage by 50% (charging at 6 mA). The experiment was performed in a dual electrolyte system (the catholyte: 0.5M NaOH, the anolyte: acidified seawater), with the intrinsic overpotential of approx. 0.7 V caused by the pH gradient. It was proposed that increasing the power density of light would bring the system closer to bias-free operation. The experiments involving seawater as an electrolyte were associated with the simultaneous presence of the ClER and the OER. The estimated concentrations of dissolved Cl_2 were in the range of 20–30 ppm. These values indicate that the OER is the dominating process when operating the device at the mA-current range.

The performed experiments confirmed the versatility of the manufactured device, which can be operated in a broad range of pH and the presence of aggressive chemical species. The reactor revealed stability over several measurements, lasting between 17 and 43 h each. It was recognized that a single set of tailor-made gaskets is suitable only for experiment(s) that do not require cell disassembly. Moreover, the quality of electric contacts and the uniformity of the pressure applied on photoelectrodes by the fixing plate should be improved. The suitability of the constructed cell is not limited to the few examples of water splitting presented in this work as it could serve as a flow reactor for (photo)electrochemical synthesis or water purification.

5. Patents

The manufactured two-compartment photoelectrochemical flow cell and cathode holder were registered as industrial designs in the European EUIPO database with IPR numbers: 015055914-0001 and 015055830-0001, respectively.

Supplementary Materials: The following supporting information can be downloaded at https://www.mdpi.com/article/10.3390/catal14120941/s1, Figure S1: The PEC-MH setup used in charge/discharge experiments. (a) Multi-electrode anodic compartment, (b) cathodic compartment with transparent PMMA window. 1—Fe_2O_3 photoanode, 2—Cathode holder with the cathode, 3—Ni-helix electrode, and 4—ASA porous diaphragm; Figure S2: Photocurrent density derived from linear sweep voltammetry (LSV) measurements (presented in Figure 2 in the main text); Figure S3: Ten chronovoltammetry cycles for $LaNi_{4.3}Co_{0.4}Al_{0.3}$ cathode measured in ~1M of KOH solution (pH = 13.5). RE: Hg/HgO, CE: Ni-helix, scan rate: 10 $mV \cdot s^{-1}$. Peaks corresponding to H_2-absorption/desorption are highlighted; Figure S4: Dark LSV (scan rate: 10 $mV \cdot s^{-1}$) experiment with Fe_2O_3 photoanodes connected in parallel, conducted in 1M of NaOH; Table S1: Unit-cell parameters for the pristine and hydrogenated $LaNi_5$-based phases derived from the Rietveld refinements; Figure S5: Comparison of the PXRD patterns for active material and powder mixture embedded in $LaNi_{4.7}Al_{0.3}$-based cathode. Bragg peak indicated with * could not be indexed with any well-known secondary phase; Figure S6: LSV (scan rate: 20 $mV{\cdot}s^{-1}$) experiments with WO_3 photoanodes, conducted in acidified 0.5M of Na_2SO_4 (pH = 1.55) under the illumination of the solar simulator. (a) Light power density per electrode (average of three points). (b) The photocurrent for each electrode and all-together connected in parallel (bold line). (c) Photocurrent density. The noisy signal for W1 and W5 samples (top row of electrodes) is probably connected to poor contact with the current collector. The color code is the same in subfigures (b,c). RE: Ag/AgCl (sat. KCl), CE: Pt-coil; Figure S7: Changes in light power density during experiments of case study 2 for individual electrodes. The change in average total power was smaller than 1.5%. The first light measurement was performed 3.5 h before the start, while the second was 2.5 h after the end of CA experiments (period of approx. 24 h); Figure S8: LSV curves in the form of current density normalized by light power density measured before and after the CA experiment (case study 2). Data were collected for eight WO_3 electrodes connected in parallel. The largest deviation at 1500 mV is lower than 5.7 %. RE: Ag/AgCl (sat. KCl); Figure S9: Semi-qualitative chlorine detection using test strips. Although the Cl_2-pad turns yellow in acidic electrolytes, the intensive purple color clearly indicates the presence of free chlorine in the solution after the CA experiment; Figure S10: Discharge specific capacity of the $LaNi_{4.3}Co_{0.4}Al_{0.3}$-based cathode derived from charging/discharging measurements in CS-3; Figure S11: Semi-qualitative chlorine detection using test strips. Although the Cl_2-pad turns yellow in acidic electrolytes, the intensive purple color clearly indicates the presence of free chlorine in the anolyte after the C/D experiment in CS-3. No trace of free chlorine was detected in the catholyte; Figure S12: (a) Anolyte after short charging experiments with different C-rates (Figure 9 in the main text) and two charging steps at 20 $mA{\cdot}g^{-1}$ (total charge transferred through photoanodes ~288 C). (b) Anolyte after a prolonged C/D experiment (Figure 8 in the main text, total charge transferred ~1420 C); Figure S13: Elements of PEC-MH device after experiments with seawater oxidation. (a) Anodic half-cell. The internal impregnation layer was colored yellow. (b) Catholyte tubing. (c) Anolyte tubing—only the external layer of tubing was oxidized; Figure S14: Multi-electrode gasket after experiment with seawater in CS-2. Deformations and discoloration are clearly visible; Figure S15: Components of the PEC-MH device. 1—Multi-electrode half-cell accommodating eight transparent photoanodes with electric contacts to photoanodes (Cu-foil) and hydraulic ports for electrolyte circulation. 2—Cathodic half-cell. 3—Backplate for fixing PMMA window. 4—Front plate for fixing photoelectrodes. 5, 6—Back and front thermal shields. 7—PMMA back window. 8—Fe_2O_3 photoanodes. 9—Ni-helix electrode. 10—Top lids with electrode ports. 11, 12—Photoelectrode gaskets. 13—Fixing bolts. The compartment separator (membrane or diaphragm) and its gaskets are not shown; Figure S16: The subsequent steps (a–d) of assembly of the cathode holder. The holder was 3D-printed with the ASA polymer, with screws and nuts made of polypropylene. The Ni-contact is combined with an adapter fitting the electrode port in the lid of the reactor; Figure S17: Spectral irradiance of the solar simulator (xenon arc lamp + AM1.5G filter) coupled with water filter. The spectrum was measured at an arbitrary distance from the lamp; Figure S18: A holder with a photodiode for calibration of light power over 100 cm^2. The holder resembles the geometry of a multi-electrode compartment (three points per electrode); Figure S19: Active areas of electrodes in cm^2: (a) case study 1, (b) experiments in seawater in case study 2, (c) experiments in 0.5M of Na_2SO_4 in case study 2 and case study 3; Figure S20: Schemes of electrical circuits during (a) charging and (b) discharging experiments. The potential of WE vs. RE was measured with an auxiliary potentiostat (labeled as a voltmeter in the schemes); Figure S21: Schemes of experimental setups used in case study 1 (a) and 3 (b). Horizontal distances are given with an accuracy of $\pm$3 mm due to the irreproducible shrinkage of gaskets.

Author Contributions: Conceptualization, P.W., J.M., Y.-C.L., M.N.G., K.B. and W.M.; methodology, P.W., Y.-C.L., E.H.J. and K.B.; validation, P.W., J.M., Y.-C.L. and E.H.J.; investigation, P.W., E.H.J. and K.B.; formal analysis, P.W. and E.H.J.; data curation, P.W. and E.H.J.; writing—original draft preparation, P.W.; writing—review and editing, P.W., J.M, Y.-C.L., E.H.J., M.N.G., R.S. and W.M.; visualization, P.W.; supervision, M.N.G., R.S. and W.M.; project administration, J.M. and M.N.G.; funding acquisition, M.N.G., R.S. and W.M. All authors have read and agreed to the published version of the manuscript.

Funding: This research was funded by the National Center for Research and Development (NCBR) within the HERA (Hydrogen Energy Rechargeable Architectures) project (NOR\POLNOR\HERA\0043\2019).

Data Availability Statement: The original contributions presented in this study are included in this article/Supplementary Materials, and further inquiries can be directed to the corresponding authors.

Acknowledgments: The authors gratefully acknowledge Alicja Macyk for chlorine analysis and Marcin Kobielusz for the fruitful discussions.

Conflicts of Interest: Authors Paweł Wyżga, Joanna Macyk and Wojciech Macyk are employed by the InPhoCat – Innovative Photocatalytic Solutions Sp. z o.o. The remaining authors declare that the research was conducted in the absence of any commercial or financial relationships that could be construed as a potential conflict of interest.

References

1. Chen, Z.; Dinh, H.N.; Miller, E. Photoelectrochemical Water Splitting. In *Standards, Experimental Methods, and Protocols*; Springer: Berlin/Heidelberg, Germany, 2013; ISBN 9781461482987.
2. Kim, J.H.; Hansora, D.; Sharma, P.; Jang, J.W.; Lee, J.S. Toward Practical Solar Hydrogen Production-an Artificial Photosynthetic Leaf-to-Farm Challenge. *Chem. Soc. Rev.* **2019**, *48*, 1908–1971. [CrossRef]
3. Hamdani, I.R.; Bhaskarwar, A.N. Recent Progress in Material Selection and Device Designs for Photoelectrochemical Water-Splitting. *Renew. Sustain. Energy Rev.* **2021**, *138*, 110503. [CrossRef]
4. Lin, Y.-C.; Wyżga, P.; Macyk, J.; Macyk, W.; Guzik, M.N. Solar-Driven (Photo)Electrochemical Devices for Green Hydrogen Production and Storage: Working Principles and Design. *J. Energy Storage* **2024**, *82*, 110484. [CrossRef]
5. Jadwiszczak, M.; Jakubow-Piotrowska, K.; Kedzierzawski, P.; Bienkowski, K.; Augustynski, J. Highly Efficient Sunlight-Driven Seawater Splitting in a Photoelectrochemical Cell with Chlorine Evolved at Nanostructured WO3 Photoanode and Hydrogen Stored as Hydride within Metallic Cathode. *Adv. Energy Mater.* **2020**, *10*, 1903213. [CrossRef]
6. Schoonman, J.; Kiliaan, H.S. In-Situ Storage of Solar Hydrogen in a PEC Cell. *Solid. State Ion.* **1983**, *9–10*, 1087–1092. [CrossRef]
7. Schoonman, J. Electrochemical Conversion and Storage of Solar Energy. *Berichte Bunsenges. Phys. Chem. Chem. Phys.* **1982**, *86*, 660–664. [CrossRef]
8. Danko, D.B.; Sylenko, P.M.; Shlapak, A.M.; Khyzhun, O.Y.; Shcherbakova, L.G.; Ershova, O.G.; Solonin, Y.M. Photoelectrochemical Cell for Water Decomposition with a Hybrid Photoanode and a Metal-Hydride Cathode. *Sol. Energy Mater. Sol. Cells* **2013**, *114*, 172–178. [CrossRef]
9. Kongsawatvoragul, K.; Chomkhuntod, P.; Sawangphruk, M. Solar-Driven Energy Storage Enhancement of Nickel Hydroxide Nanomaterials. *Electrochim. Acta* **2021**, *388*, 138654. [CrossRef]
10. Akuto, K.; Sakurai, Y. A Photorechargeable Metal Hydride/Air Battery. *J. Electrochem. Soc.* **2001**, *148*, A121. [CrossRef]
11. Lei, B.; Li, G.R.; Chen, P.; Gao, X.P. A Solar Rechargeable Battery Based on Hydrogen Storage Mechanism in Dual-Phase Electrolyte. *Nano Energy* **2017**, *38*, 257–262. [CrossRef]
12. Gholami, T.; Pirsaheb, M. Review on Effective Parameters in Electrochemical Hydrogen Storage. *Int. J. Hydrogen Energy* **2021**, *46*, 783–795. [CrossRef]
13. Lopes, T.; Dias, P.; Andrade, L.; Mendes, A. An Innovative Photoelectrochemical Lab Device for Solar Water Splitting. *Sol. Energy Mater. Sol. Cells* **2014**, *128*, 399–410. [CrossRef]
14. Vilanova, A.; Lopes, T.; Spenke, C.; Wullenkord, M.; Mendes, A. Optimized Photoelectrochemical Tandem Cell for Solar Water Splitting. *Energy Storage Mater.* **2018**, *13*, 175–188. [CrossRef]
15. Vilanova, A.; Lopes, T.; Mendes, A. Large-Area Photoelectrochemical Water Splitting Using a Multi-Photoelectrode Approach. *J. Power Sources* **2018**, *398*, 224–232. [CrossRef]
16. Tolod, K.R.; Hernández, S.; Russo, N. Recent Advances in the BiVO4 Photocatalyst for Sun-Driven Water Oxidation: Top-Performing Photoanodes and Scale-up Challenges. *Catalysts* **2017**, *7*, 13. [CrossRef]
17. ArtipHyction Project: "Fully Artificial Photo-Electrochemical Device for Low Temperature Hydrogen Production". Available online: https://www.artiphyction.org/ (accessed on 22 November 2023).
18. Torras, C.; Lorente, E.; Hernández, S.; Russo, N.; Salvadó, J. Hydrodynamics and Oxygen Bubble Characterization of Catalytic Cells Used in Artificial Photosynthesis by Means of CFD. *Fluids* **2017**, *2*, 25. [CrossRef]
19. Becker, J.P.; Turan, B.; Smirnov, V.; Welter, K.; Urbain, F.; Wolff, J.; Haas, S.; Finger, F. A Modular Device for Large Area Integrated Photoelectrochemical Water-Splitting as a Versatile Tool to Evaluate Photoabsorbers and Catalysts. *J. Mater. Chem. A Mater.* **2017**, *5*, 4818–4826. [CrossRef]
20. Ambrosi, A.; Pumera, M. 3D-Printing Technologies for Electrochemical Applications. *Chem. Soc. Rev.* **2016**, *45*, 2740–2755. [CrossRef]
21. Andrei, V.; Hoye, R.L.Z.; Crespo-Quesada, M.; Bajada, M.; Ahmad, S.; De Volder, M.; Friend, R.; Reisner, E. Scalable Triple Cation Mixed Halide Perovskite–$BiVO_4$ Tandems for Bias-Free Water Splitting. *Adv. Energy Mater.* **2018**, *8*, 1801403. [CrossRef]

22. Taştan, Ü.; Ziegenbalg, D. Getting the Most out of Solar Irradiation: Efficient Use of Polychromatic Light for Water Splitting. *Chem. A Eur. J.* **2016**, *22*, 18824–18832. [CrossRef]
23. Mousset, E.; Huang Weiqi, V.; Foong Yang Kai, B.; Koh, J.S.; Tng, J.W.; Wang, Z.; Lefebvre, O. A New 3D-Printed Photoelectrocatalytic Reactor Combining the Benefits of a Transparent Electrode and the Fenton Reaction for Advanced Wastewater Treatment. *J. Mater. Chem. A Mater.* **2017**, *5*, 24951–24964. [CrossRef]
24. Pornrungroj, C.; Andrei, V.; Reisner, E. Thermoelectric–Photoelectrochemical Water Splitting under Concentrated Solar Irradiation. *J. Am. Chem. Soc.* **2023**, *145*, 13709–13714. [CrossRef]
25. Dinesh, G.K.; Dessì, P.; Tong, W.; González-Gómez, R.; Farràs, P. 3D-Printed Photoelectrochemical Cell and Its Application in Evaluation of Bismuth Vanadate Photoanodes: Synthesis, Cell Design and Testing. *Johns. Matthey Technol. Rev.* **2022**, *66*, 386–392. [CrossRef]
26. Ramalingam, K.; Bieńkowski, K.; Nawaz, N.; Solarska, R. Enhanced PEC Generation of Hydrogen from Seawater Driven by Efficient and Stable Ti-Fe2O3 Photoanode. *Appl. Surf. Sci.* **2025**, *679*, 161246. [CrossRef]
27. Tobías, I.; del Cañizo, C.; Alonso, J. Crystalline Silicon Solar Cells and Modules. In *Handbook of Photovoltaic Science and Engineering*; Wiley: Hoboken, NJ, USA, 2003; pp. 255–306.
28. Todorova, S.; Abrashev, B.; Rangelova, V.; Mihaylov, L.; Vassileva, E.; Petrov, K.; Spassov, T. Hydrogen Gas Phase and Electrochemical Hydriding of LaNi5−xMx (M = Sn, Co, Al) Alloys. *Materials* **2020**, *14*, 14. [CrossRef] [PubMed]
29. Giza, K.; Bala, H.; Drulis, H.; Hackemer, A.; Folcik, L. Gaseous and Electrochemical Hydrogenation Properties of LaNi4.3(Co, Al)0.7 -XInx Alloys. *Int. J. Electrochem. Sci.* **2012**, *7*, 9881–9891. [CrossRef]
30. Dymek, M.; Bala, H.; Hackemer, A.; Drulis, H. Hydrogenation and Electrochemical Corrosion Properties of LaNi4.5Co0.5 Alloy Doped with Aluminum. *Solid. State Ion.* **2015**, *271*, 116–120. [CrossRef]
31. Feng, F. Electrochemical Behaviour of Intermetallic-Based Metal Hydrides Used in Ni/Metal Hydride (MH) Batteries: A Review. *Int. J. Hydrogen Energy* **2001**, *26*, 725–734. [CrossRef]
32. Fukai, Y. *The Metal-Hydrogen System*; Springer: Berlin/Heidelberg, Germany, 2005; Volume 21, ISBN 978-3-540-00494-3.
33. Dresp, S.; Dionigi, F.; Klingenhof, M.; Strasser, P. Direct Electrolytic Splitting of Seawater: Opportunities and Challenges. *ACS Energy Lett.* **2019**, *4*, 933–942. [CrossRef]
34. Pourbaix, M. *Atlas of Electrochemical Equilibria in Aqueous Solutions*; National Association of Corrosion Engineers: Houston, TX, USA, 1974.
35. Bennett, J.E. Electrodes for Generation of Hydrogen and Oxygen from Seawater. *Int. J. Hydrogen Energy* **1980**, *5*, 401–408. [CrossRef]
36. Dionigi, F.; Reier, T.; Pawolek, Z.; Gliech, M.; Strasser, P. Design Criteria, Operating Conditions, and Nickel-Iron Hydroxide Catalyst Materials for Selective Seawater Electrolysis. *ChemSusChem* **2016**, *9*, 962–972. [CrossRef] [PubMed]
37. Santato, C.; Ulmann, M.; Augustynski, J. Photoelectrochemical Properties of Nanostructured Tungsten Trioxide Films. *J. Phys. Chem. B* **2001**, *105*, 936–940. [CrossRef]
38. Kafizas, A.; Godin, R.; Durrant, J.R. Charge Carrier Dynamics in Metal Oxide Photoelectrodes for Water Oxidation. In *Semiconductors and Semimetals*; Elsevier Inc.: Amsterdam, The Netherlands, 2017; Volume 97, pp. 3–46.
39. Wyżga, P.; Tabari, T.; Trochowski, M.; Macyk, W. Optimizing the Morphology of Titania Nanorods for Enhanced Solar Seawater Splitting. *Results Eng.* **2023**, *17*, 100921. [CrossRef]
40. Luo, T. *Porous Ion Exchange Membranes for Vanadium Redox Flow Battery*; Rheinisch-Westfälischen Technischen Hochschule Aachen: Aachen, Germany, 2017.
41. Harp, D.L. *Current Technology of Chlorine Analysis for Water and Wastewater*; HACH Company: Loveland, CO, USA, 2002.
42. Riegel, J.; Mayer, W.; van Havre, Y. FreeCad 0.19 [Software]. Available online: https://www.freecad.org (accessed on 1 June 2021).
43. Ultimaker B.V. *Ultimaker Cura 4.9.1 [Software]*; Ultimaker B.V.: Utrecht, The Netherlands, 2021.
44. Schneider, C.A.; Rasband, W.S.; Eliceiri, K.W. NIH Image to ImageJ: 25 Years of Image Analysis. *Nat. Methods* **2012**, *9*, 671–675. [CrossRef] [PubMed]

catalysts

Review

The Synthesis and Photocatalytic Efficacy of Distinct Nano-Oxides in the Breakdown of Organic Contaminants

Jelena Pavlović [1], Nataša Novak Tušar [2,3] and Nevenka Rajić [4,*]

1 Institute of Soil Science, Teodora Drajzera 7, 11000 Belgrade, Serbia; soils.pavlovic@gmail.com
2 National Institute of Chemistry, Hajdrihova 19, SI-1001 Ljubljana, Slovenia; natasa.novak.tusar@ki.si
3 Graduate School, University of Nova Gorica, Vipavska 13, SI-5000 Nova Gorica, Slovenia
4 Faculty of Ecology and Environmental Protection, University "Union, Nikola Tesla", Cara Dušana 62–64, 11158 Belgrade, Serbia
* Correspondence: nrajic@unionnikolatesla.edu.rs

Abstract: Nano metal oxides (NMOs), with their unique physico-chemical properties and low toxicity, have become a focus of research in heterogeneous catalysis. Their distinct characteristics, which can be tailored based on size and structure, make them highly efficient catalysts. NMOs have the potential to significantly contribute to the degradation of numerous environmental pollutants through photolytic decomposition. This work comprehensively analyzes the synthesis, catalytic performance, and applications of photocatalytically active metal oxides, specifically titanium, zinc, copper, iron, silver, tin, and tungsten oxides. The primary objective is to demonstrate how the effectiveness of photocatalytic processes can be enhanced and optimized by incorporating metals, non-metals, and metalloids into their structure and forming heterostructures. Furthermore, the aim is to understand the underlying process of photocatalytic oxidation thoroughly. Photocatalysis, a promising approach in advanced oxidation processes, has garnered significant interest in these fields.

Keywords: nano oxides; photocatalysis; advanced oxidation processes; synthesis; organic pollutants

1. Introduction

Nanoparticles (NPs) exhibit distinct physical and chemical properties due to their large surface area relative to their volume and the impact of their small size at the quantum level. NPs have diverse applications in electronics, catalysis, energy storage, environmental remediation, biology, and other fields [1,2]. Nano metal oxides (NMOs) exhibit exceptional catalytic activity, specific adsorption properties, high electrical conductivity, and magnetic traits. They are frequently utilized as catalysts in chemical reactions or as sensors in electronic systems [3–5]. Extensive research into NMOs is leading to innovative methods for their production [6], new methodologies for their characterization, and uncovering new uses for these adaptable NPs. With the ongoing progress of technology, NMOs will probably have a growing significance in influencing the future of many areas.

Advanced oxidation processes (AOPs) are chemical treatment methods for eliminating organic and inorganic contaminants from water and wastewater. This is achieved by generating hydroxyl radicals (•OH), which possess strong reactivity and can effectively degrade complex pollutants such as organic dyes, pesticides, or pharmaceuticals [7–10]. AOPs are highly efficient in breaking down diverse organic contaminants resistant to removal through traditional treatment methods.

Photolytic advanced oxidation processes (PAOPs) employ ultraviolet (UV) radiation to produce •OH through the oxidation of molecules like hydrogen peroxide (H_2O_2) or ozone (O_3). Hydroxyl radicals are generated by activating photocatalysts, such as the widely used titanium dioxide (TiO_2), by UV radiation and initiating the generation of electron–hole pairs in the photocatalyst, which react with water and oxygen, creating •OH.

Due to their unique features and high efficiency, semiconductor NMOs have been extensively studied and utilized in photocatalysis. In Figure 1, band gap positions of different semiconductors in contact with an aqueous electrolyte are shown.

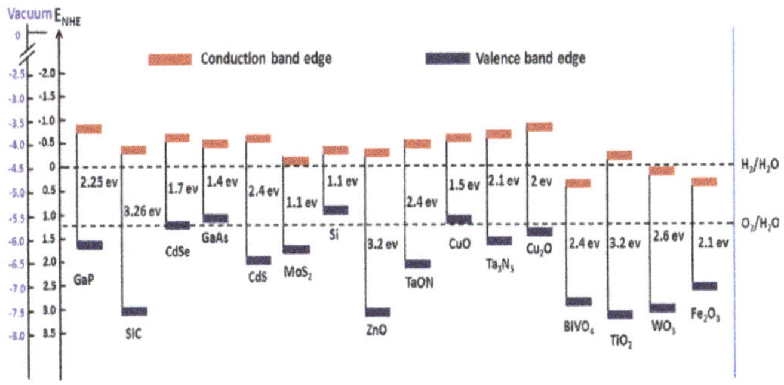

Figure 1. A comparative view of band edge locations of several semiconductors in contact with an aqueous electrolyte (pH = 0) is conducted on the conventional hydrogen electrode, the vacuum level, and the redox potentials of water splitting [11].

The most frequently employed semiconductor NMOs in photocatalysis are titanium dioxide (TiO_2), zinc oxide (ZnO), and iron oxide (Fe_2O_3). Several more candidates show promise for utilization in photocatalysis. They exhibit photocatalytic behavior upon exposure to light, typically in the ultraviolet (UV) spectrum, as they generate electron–hole pairs upon photon absorption. On the catalyst's surface, the energized electrons and vacancies can participate in oxidation–reduction reactions, leading to the decomposition of organic pollutants, the splitting of water into hydrogen, or other specific chemical transformations. NMOs have several advantages in photocatalysis, such as a large surface area to volume ratio, customizable band gap energy, chemical stability, and cost-effectiveness. Recently, various approaches have been explored to enhance the light to energy conversion efficiency of NMOs. These include modifying their surface properties, incorporating additional elements, and fabricating nanostructures to optimize light absorption and charge separation. In general, NMOs exhibit significant promise for applications in environmental remediation, energy conversion, and other disciplines that require efficient and sustainable photocatalytic processes. It should be noted that using suspended photocatalysts for wastewater treatment is common, but it presents numerous drawbacks, such as photocatalyst saturation or the necessity for recovery and post-treatment following photocatalytic procedures, which are expensive and challenging to execute. One approach is to immobilize photocatalysts on supports to address these issues. Using NMO photocatalysts in the membrane diminishes fouling and improves its effectiveness. Improving membrane applications by incorporating photocatalysts and membrane technology is linked to the enhanced hydrophilicity and antibacterial properties of photocatalysts. The decreased contact angle of the treated membrane indicates the hydrophilic properties of the composite membrane. The composite membrane's anti-fouling characteristics promote the development of a hydrophilic thin layer via hydrogen bonding, thus preventing the attachment of fouling materials to the membrane surface [12].

The use of NMO NPs in photocatalysis has attracted considerable attention recently, leading to increased published literature, including review papers. A bibliometric analysis has demonstrated the evolution of the importance of this research topic in the literature over the last decade [13]. This increase is due to the necessity for developing water purification solutions in response to rising global environmental degradation. Review articles concentrate on specific components, such as contaminants, particular metal oxide particles, or innovative materials like BiOBr nanocomposites. Srinisava performed a comprehensive review highlighting current progress in applying diverse photocatalysts to remove emerging pharmaceutical contaminants in wastewater [14]. Krishnan et al.

conducted a study of current research trends in transition metal oxide-based photocatalysis for the elimination of organic dyes from water [15]. The review included synthesizing BiOBr-based nanocomposites, the mechanisms that boost photocatalytic activity, and their present research state, and possible applications in destroying organic pollutants was also published [16].

This paper thoroughly examines synthetic methods and the catalytic efficacy and uses of photocatalytically active metal oxides, namely titanium, zinc, copper, iron, silver, tin, and tungsten oxides. The main goal is to show how the efficacy of photocatalytic processes can be improved and maximized by integrating metals, non-metals, and metalloids into their framework, forming heterostructures. Importance is also given to the methods of synthesis of metal oxide particles, with particular emphasis on the advantages and disadvantages of individual methods.

2. Synthesis

2.1. Wet Chemical Methods

2.1.1. Sol–Gel Methods

Due to its compatibility with industrial use, the sol–gel method is a widely studied wet chemical technique for manufacturing NMOs. One of its benefits is the ability to manage the size and uniformity of NMOs effectively. Nevertheless, this method is constrained by an organic solvent, which can bring environmental risks, restricted availability, costly starting materials, and a protracted reaction duration.

Usually, metal alkoxide, the molecular precursor, is dissolved in water or alcohol and converted to gel by heating and stirring by hydrolysis/alcoholysis. Producing a homogenous sol from the molecular precursor and its subsequent transformation into a gel is the most crucial step in the sol–gel process. Most often, the citrate sol–gel approach is used. Citric acid is a chelating agent that bonds metal cations and lowers reaction temperatures [17]. The citric acid sol–gel method effectively synthesizes nano mesoporous La_2O_3 [18]. The procedure applies to synthesizing binary, ternary, and quaternary metal oxides in crystalline and amorphous forms. The essential advantage of this method, as with the more traditional sol–gel process, is the homogeneity of the starting material.

The sol–gel technique provides numerous benefits for the creation of NMOs. It allows for manipulating the size and shape of nanoparticles, producing uniform and consistent particles [19]. By using proper solvents and pH in the synthesis of TiO_2, a high photocatalytic efficiency was achieved by band gap engineering [20]. However, it is essential to note that the sol–gel method may not be appropriate for synthesizing specific types of nanoparticles or materials [21]. Furthermore, this method has several drawbacks. The technique can be laborious and require several sequential stages [22].

2.1.2. Hydrothermal Method

The hydrothermal process occurs at high pressures and temperatures above the boiling point in water-based solutions. NMOs usually form in solution from metal precursors at 80–200 °C. This method allows for tuning the morphology of NPs, improving specific properties for specific applications. It has been traditionally used to create different types of ZnO nanoparticles, such as nanorods and nanowires [23]. Hexagon-shaped nano SnO_2 was successfully prepared by a simple hydrothermal route [24], as well as Ga_2O_3 nano cuboids [25], NiO in the form of nano petals [26], nano-MnO [27], and nano-V_4O_9 plates [28]. The main disadvantages of this approach are the synthesis at high pressure and temperatures, the complicated experimental setup, and, most importantly, the limited potential for scale-up manufacturing.

2.1.3. Coprecipitation

This is an economically efficient and uncomplicated technique that enables the large-scale manufacturing of NMOs. The process entails manipulating variables such as temperature, pH, solvent selection, and precipitating agent to customize the characteristics of the nanoparticles. The process is effective for the synthesis of ZnO and CuO. However, it may lead to the formation of clumps, necessitating adjustments. CuO nanoparticles with a high purity level and a crystalline structure were synthesized using $CuCl_2$ and NaOH as precursors and CH_3COOH as a capping agent. The particle size was narrowly from 21.88 nm to 22.14 nm [29]. The synthesis of nano-ZnO powder was achieved via oxalate coprecipitation methodology. Zinc oxalate dehydrate was converted into zinc oxalate on calcination at 500 °C and into ZnO at 700 °C [30].

The cocurrent coprecipitation method was developed as an enhanced version of the traditional coprecipitation method. This method involves the simultaneous addition of a metal salt solution and a precipitant while maintaining the constant pH value of the reaction system. The approach ensures the simultaneous precipitation of multiple cations. Nano-powder ZrO_2, garnet, and perovskite were obtained by cocurrent coprecipitation [31]. Recently, the method was applied to synthesize complex nano-sized Dy_2O_3–Sc_2O_3 co-stabilized ZrO_2 powders [32].

The hypergravity coprecipitation technique is used in the industrial production of nano calcium carbonate, nano iron oxide, and nano cerium dioxide powder. This technique uses a hypergravity rotating bed to create a chemical reaction between two liquid phases, resulting in minor films, filaments, and droplets. This process enhances micro-mass transfer and separation between the metal salt solution and precipitant, forming a protective layer on the particles. The hypergravity method addresses uneven particle size, inadequate powder dispersion, and particle clumping, making it suitable for industrial manufacturing [33].

Chemical precipitation is a cost-effective and scalable method for producing NMOs with controlled structures and morphologies. It effectively synthesizes ZnO and CuO nanoparticles but may cause agglomeration issues, requiring modifications.

2.1.4. Electrochemical Methods

The increasing interest in electrochemical synthesis for preparing NMOs is motivated by its simplicity, low-temperature operation, low energy consumption, greater product purity, and ecological friendliness. Recent advancements in this method have shown the ability to manipulate the composition and morphology of these structures without the need for adsorbing capping agents [34]. Electrochemical deposition (cathodic and anodic), cathodic corrosion, and galvanic exchange reactions are usually applied [35]. Applying a pulsed potential waveform during Cu_2O electrodeposition in alkaline electrolytes enables the formation of Cu_2O films with a controllable morphology without altering the electrolyte composition or temperature [36]. Cathodic corrosion produces NMOs and mixed NMOs with an excellent particle size and shape homogeneity, significantly improving photocatalyst nanoparticle synthesis procedures. Amorphous TiO_2 and crystalline H_2WO_4 and $BiVO_4$ nanoparticles were prepared (Figure 2) with a preferential crystallographic orientation starting from the base metal as the reactant [34]. This method is ripe for industrial scale-up, as it avoids large volumes of organic solvents and significant investments in heating, cleaning, safety, and disposal issues.

Nanotubular TiO_2, possessing an average inner pore width of roughly 110 nm, was synthesized in a meticulously regulated process using electrochemical anodization in an aqueous phosphoric acid solution, including hydrofluoric acid [37]. The anodization cell voltage affects the pore diameter of TiO_2 nanotubes, with a voltage of 20 V yielding the biggest pore diameter and the maximum conversion efficiency under light. Moreover, hydrofluoric acid has a substantial role in pore formation and dissolution. The produced materials were assessed as photoanodes for dye-sensitized solar cells utilizing metal phthalocyanine dyes.

Galvanic replacement reaction also yields metal nanostructures with controlled shapes, morphologies, and compositions [38]. In this process, one metal (often known as a sacrificial template) is oxidized by the ions of another metal having a higher reduction potential. The template undergoes oxidation and dissolution by coming into touch with a solution. Simultaneously, the ions of the second metal undergo reduction and are deposited onto the external surface of the template. Galvanic replacement reactions provide a simple and versatile route for producing hollow nanostructures with controllable pore structures. Nanocrystals of Mn_3O_4 react with $Fe(ClO_4)_2$, forming hollow box-shaped nanocrystals of Mn_3O_4/γ-Fe_2O_3 ("nanoboxes"), which are ultimately transformed into hollow cage-like nanocrystals of γ-Fe_2O_3 ("nanocages") [39]. It is important to note that this approach can only be used for a limited number of highly reactive metal oxides. Using somewhat raised temperatures is necessary due to the relatively sluggish kinetics, which might be considered a disadvantage compared to standard electrochemical approaches.

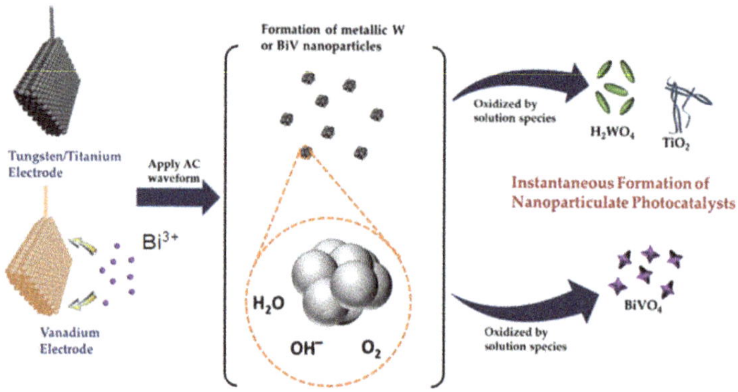

Figure 2. Schematic illustration of the cathodic corrosion processes for preparing H_2WO_4, TiO_2, and $BiVO_4$ [34].

2.1.5. Chemical Vapor Deposition

Chemical vapor deposition (CVD) is a flexible method for creating NMOs, offering benefits like homogeneity, scalability, and control over flakes. However, it has drawbacks like cost and hazardous gas safety issues. Recently, a graphene coating on nano-TiO_2 was applied using CVD, resulting in a composite material with superior photocatalytic activity and anti-fouling properties. Graphene improved pollutant adsorption and increased the TiO_2's charge separation and transportation, resulting in a positive anti-fouling effect [40]. A highly active photocatalyst based on Fe-doped nano-sized TiO_2 was successfully synthesized by chemical vapor deposition (CVD) using $FeCl_3$ as an iron source [41]. The growth of Mo-doped ZnO thin films on glass substrates was achieved by aerosol-assisted CVD, operating at ambient pressure [42]. Mo is present in the 4+ oxidation state, contributing two electrons to electrical conduction for every Zn^{2+} ion replaced in the lattice. SnO_2 nanowires and nanoribbons (doped and pure) have been synthesized using the CVD method and Sn and SnO_2 powders as precursors [43].

2.1.6. Spray Pyrolysis

This method of the production of NMO powder involves consistently blending chemical components in solution to obtain a complex solution, which is then sprayed into a high-temperature reaction furnace (Figure 3). An ultra-fine MO powder is created in the furnace through an instantaneous spray pyrolysis reaction. The following are some of this method's benefits: (1) the whole process can be made relatively simple by eliminating steps like the mixing, calcining, and milling of solid powder; and (2) mixing with impurities can be minimized, and the various pyrolysis reaction conditions can adjust the characteristics of the produced particles. Many manufacturers—including Merck, Scimarec, and the American company SSC—create beneficial ceramic powders by spray pyrolysis [44]. The method produced nano-NiO powder using $NiCl_2$ solution as a raw material. The concentration of the solution and atmospheric pressure influence the powder's properties. The average particle size at 700 °C is about 20 nm, increasing by temperature [44]. Similarly, spray pyrolysis can prepare nano-CoO powder with an average particle size below 50 nm from

$CoCl_2$. The authors reported that as the reaction temperature increases, the crystallinity gradually increases, yet the specific surface area gradually decreases [45].

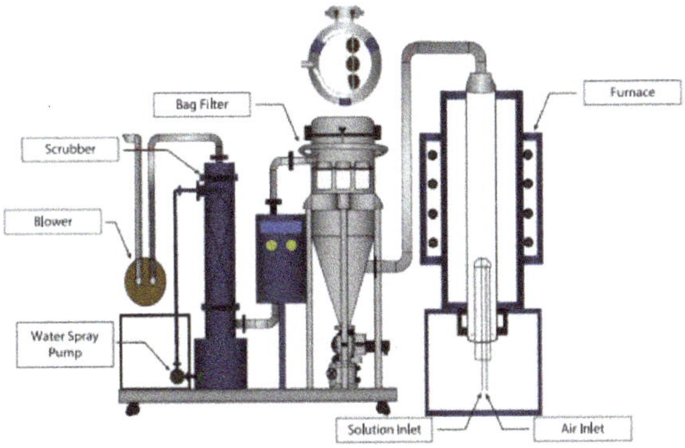

Figure 3. Schematic diagram of a spray pyrolysis system [45].

Although spray pyrolysis is a versatile, simple technique, it has certain limitations, such as requiring the precise control of deposition parameters, the complex reaction mechanism that disables the prediction of the final product, and several environmental concerns due to solvent emissions. Moreover, equipment, solvents, and energy consumption costs can be significant, especially for large-scale production.

2.1.7. Microwave Method

Microwave-assisted synthesis (MW) is frequently used to synthesize NPs, ensuring a uniform heat distribution. This results in regulated physicochemical properties and a restricted size distribution. Recent developments in MW synthesis have mainly been due to the introduction of high-tech microwave reactors, leading to rapid advancements in nanomaterial synthesis [46]. The technique helps produce ZnO-NPs in various sizes and morphologies. This hydrothermal technique has been successfully used to manufacture flower-type, needle-type, and spherical-type NPs [47]. By MW heating for 20 min at 180 °C, fine anatase TiO_2-NPs with a crystal size of around 7 nm and a specific surface area of 220 $m^2\ g^{-1}$ can be produced [48]. A high level of photocatalytic activity has been reported. Furthermore, the crystallization of nano-CeO_2 photocatalysts doped with F and Fe is possible by an MW–hydrothermal treatment, which enhances photoelectric characteristics, electron and hole separation, and visible light catalytic activity due to homogeneous crystal structures and large specific surface areas [49].

The MW technique was recently used to prepare heterostructures like ZnSe/ZnO (Figure 4), which have a high visible area adsorption efficiency [50]. A facile one-pot MW-assisted synthesis also obtained metal-free In_2S_3/In_2O_3 nanosheets (Figure 5) [51] effective in photocatalytic water splitting under blue LED light irradiation.

While MW-assisted synthesis is a valuable technique for producing nanomaterials, it also has disadvantages, including the necessity for real-time reaction monitoring in MW reactors, which makes reaction parameter optimization difficult, and scaling up for commercial production.

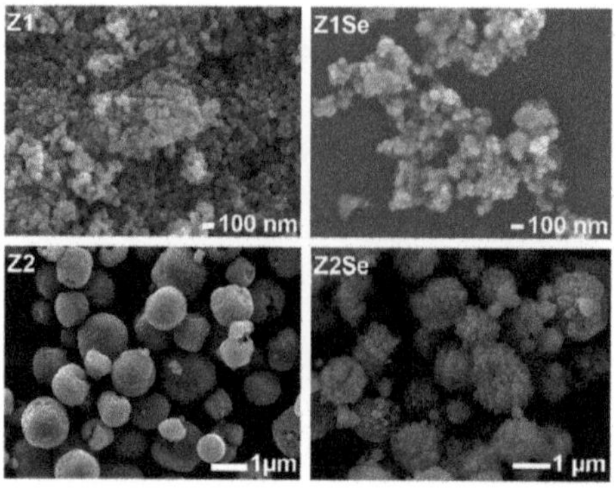

Figure 4. SEM images of ZnO heterostructures Z1 and Z2, synthesized with the microwave-assisted hydrothermal method with zinc acetate and zinc nitrate as a source of Zn^{2+} ions, respectively, along with the corresponding Se-containing ZnO heterostructures on Z1 and Z2 [50].

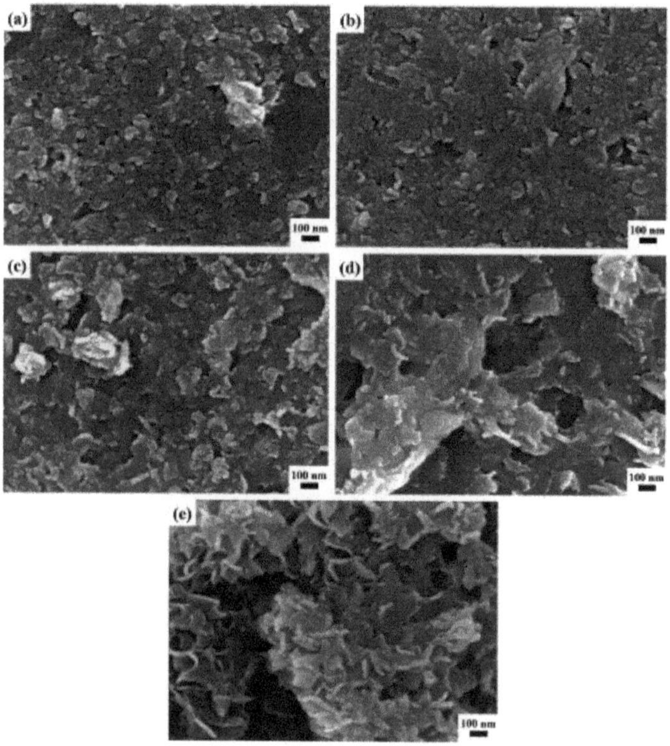

Figure 5. FESEM images of In_2S_3/In_2O_3 nanosheets synthesized at different concentrations of $InCl_3$ at 150 °C for 30 min. The concentrations of $InCl_3$ are (**a**) 2.5, (**b**) 5, (**c**) 10, (**d**) 20, and (**e**) 30 mM, respectively [51].

2.2. Solid-Phase Synthesis

Solid-phase synthesis obviates the need for a liquid medium to produce NPs generated from solid precursors, addressing a significant constraint of liquid-phase synthesis. This solution-phase approach allows for the dependable and reproducible synthesis of nanomaterials with precise dimensions and configurations. However, the extensive use of solvents, capping agents, and additional separation procedures required for this method leads to the generation of dangerous waste. It limits the scalability of the synthesis due to its high costs.

Solid-phase synthesis has garnered attention due to its potential to produce NPs with distinct features. The process often entails mechanical milling and solid-state processes conducted in a controlled environment, such as an inert gas, to avoid oxidation or contamination. Cooperative coprecipitation in the solid state can occur when precursor solids are combined and heated to enhance reactivity, crystallization, and sol–gel processes using solid precursors.

Notable disadvantages include restricted particle size and form control, high energy demands, challenges in NPs aggregation, and a lengthy synthesis duration. However, although solid-phase synthesis has several disadvantages, it is nonetheless a valuable, easy, and cost-effective method for synthesizing NPs. This is especially true when combined with other techniques or tailored for specific purposes.

Solid-state synthesis yielded a visible light-driven photocatalytic system verified for model methylene blue decomposition based on ZnO nanoparticles, natural clay, and TiO_2 nanoparticles [52]. The synthesis included mechanically grinding the components and calcinating the mixture at 700 °C. $Ni(OH)_2$-TiO_2 nanocomposites (with a particle size of about 50 nm) with an acceptable degree of dispersion were designed and successfully synthesized by facile, simple, and eco-friendly solid-state synthesis at room temperature [53].

2.3. Green Methods

Green synthesis processes use plant extrication, plant parts, bacteria, yeast, fungi, and algae to create nanostructures [54]. These methods are cost-effective, pollution-free, and easier to process than traditional methods. The green approach reduces environmental damage and allows for the development of impurity-free nanomaterials. The synthesis of ZnO-NPs using natural materials, such as microbes and plant parts, is advantageous due to their phytochemical constituents acting as capping and reducing agents. Natural materials reduce zinc to the zero-valance state, adding oxide through calcination. Moreover, zinc ions in natural extract solutions form complexes with phytochemicals, and then through hydrolysis, ZnO-NPs are formed. Thus, ZnO nanoparticles were effectively manufactured via a green, straightforward, and environmentally acceptable method, wherein ethanolic extract of *O. europaea* fruit waste served as a capping and reducing agent [55].

Microbes are another biological approach for preparing ZnO-NPs. However, this technique has certain limitations in contrast to plant extract synthesis. It is essential to screen the microbes, which is a time-consuming procedure [56]. Figure 6 shows a schematic presentation of the green procedure for designing NPs from plant extract.

An excellent review of the biosynthesis of ZnO-NPs, its challenges, and potential solutions has recently been reported [56].

The green synthesis of TiO_2-NPs has been performed using several plant extracts (*Cicer arietinum* L. extract, Aloe Vera leaf extract, Annona squamosa fruit peel, and Acanthophyllum laxiusculum). These plants contain a rich source of metabolites, making them suitable for the rapid synthesis of NPs [57].

Bacterial extracts were employed to synthesize green TiO_2-NPs. Bacterial metabolites, similar to plant extracts, significantly impact the decrease and stability of TiO_2. *Lactobacillus* sp. and *Saccharomyces cerevisiae* mediated the biosynthesis of TiO_2-NPs with an NP size of 8–35 nm. A biosynthesis mechanism influenced by pH and redox potential is proposed [58].

A recent study utilized bitter olive seeds and a sol–gel approach to make a nanocomposite of ZnO−TiO_2 doped with iron ions [59]. The nano photocatalyst demonstrated high efficacy in treating industrial wastewater samples, with a dye degradation efficiency of over 75%. The device underwent successful semi-industrial testing, demonstrating its effectiveness under UV and visible light.

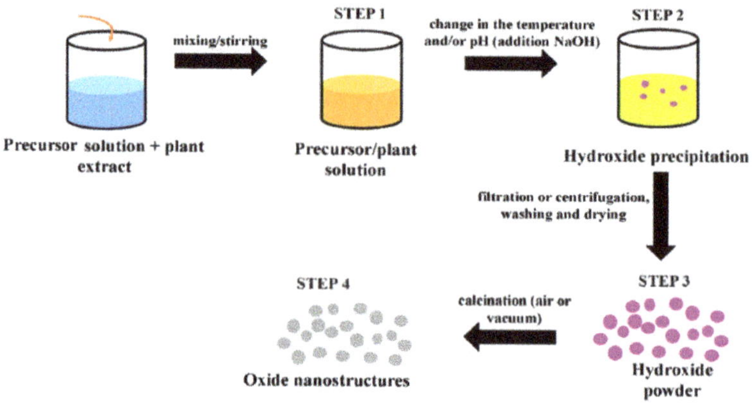

Figure 6. Schematic presentation of the sequential process of green synthesis for creating nanostructures from plant extract [56].

3. Most Common Methods for Enhancement of Photocatalytic Properties

3.1. Doping

Doping is the most frequently used approach for enhancing NMO electrical, optical, and structural properties and altering their electrical, catalytic, optical, structural, and magnetic properties to meet specific needs. It can be performed with metals and non-metals.

Many studies are based on doping the host lattice with rare earth element ions, primarily due to their exceptional conductivity, magnetism, electrical characteristics, electrochemical behavior, and luminescence. Among the many rare earth elements, ions like Sm, Nd, La, Pr, Ce, Gd, etc., have the unique characteristic of having a half-filled shell with 7f electrons, setting them apart from other elements. Rare earth elements exhibit high conductivity due to incompletely filled 4f highly localized orbitals and are insulated by the $5s^2$ and $5p^6$ orbitals in the outer shell. Jimkeli Singh and Chinnamuthu utilized the combustion approach to produce Ce-doped CuO nanoparticles [60]. They found that with an increased cerium concentration, crystallite size and band gap decreased. Ce dopants did not affect the peak positions but reduced intensity. Ce (4 mol.%)-doped CuO NPs showed the highest efficiency in the photocatalytic degradation of rhodamine and methylene blue under solar irradiation [60].

TiO_2's limited absorption in the visible range is a challenge to overcome. Its wide band gap absorbs only UV light, making photocatalysts for the whole solar spectrum exciting. Doping TiO_2 with transition metals, such as Fe, Sb, or Cr, can improve its response under visible light, enhancing photocatalytic efficiency [61]. Plasmonic composites combining TiO_2 with highly dispersed Au nanoparticles can generate OH• radicals under visible light illumination [62].

Non-metal doping offers advantages over metal doping, such as modifying photocatalytic NMOs without thermal instability or poor solubility. Incorporating nitrogen, carbon, sulfur, fluorine, or iodine, possibly as quantum dots [63], is a more efficient way to lower the band gap of TiO_2 and increase photoactivity [63]. The red shift of the S-doped TiO_2 absorption edge was explained with additional extrinsic electronic levels introduced by sulfur doping. This extends the absorption edge to the visible light region [63].

One innovative application is developing nanocomposites with TiO_2 and reduced graphene oxide doped with nitrogen, which shows excellent results in CO_2 photoreduction [64]. The enhancement of TiO_2-based composite with a low carbon-based component ratio for improved hydrogen generation via photocatalytic water splitting was reported [65]. Nitrogen-doped mesoporous TiO_2 (TiO_2-N) is a highly advanced photocatalyst with promising applications, showing a boosted visible light absorption due to shifting 2p orbital levels [66].

Recently, a trimetallic Cu-Mn-Fe silica-supported catalyst was prepared via direct synthesis by incorporating magnetic Fe_3O_4 nanoparticles and active catalytic/photocatalytic species (Cu and Mn) during the formation of the silica nanoparticles with interparticle mesoporosity. In the silica support, the Fenton-like activities of the catalyst result from the catalytic disproportionation of H_2O_2 by Cu^{2+} species and separated Mn species. The catalyst showed a high efficiency in the homolytic cleavage of H_2O_2 to hydroxyl radicals. Its effectiveness was further enhanced by the generation of hydroxyl radicals when exposed to visible light, making it operate as a catalyst similar to photo-Fenton. Magnetic Fe_3O_4 NPs facilitate the separation of the catalyst/photocatalyst following the reaction without affecting the catalytic/photocatalytic performance [67].

3.2. Heterojunction

Photocatalyst heterojunctions include amalgamating two or more distinct photocatalytic materials to create a junction that improves photocatalytic efficacy in destroying organic pollutants. The principal impetus for forming heterojunctions is to leverage the synergistic features of diverse materials, resulting in enhanced efficiency in light absorption, charge separation, and total photocatalytic efficacy. The fundamental principles of photocatalyst heterojunctions encompass charge carrier dynamics: electron–hole pairs are produced upon light absorption. The junction aids in segregating these charges, diminishing recombination rates and increasing the availability of charge carriers for chemical processes. The electronic band structures of the constituent components are essential. An advantageous band alignment (type II, p-n, etc.) can facilitate the effective transfer of electrons and holes between the elements. Materials frequently employed in forming photocatalyst heterojunctions comprise TiO_2, ZnO, CdS, g-C_3N_4, and metal oxides. Graphene and carbon nanotubes, as carbon-based materials, can improve charge conductivity. Various techniques can be utilized to fabricate heterojunctions, including hydrothermal or solvothermal procedures, sol–gel processes, and chemical vapor deposition.

Photocatalyst heterojunctions are promising for improving photocatalytic processes by combining semiconductors with different band gaps. They enhance light absorption, charge separation, and redox reaction kinetics. Type I heterojunctions promote electron–hole pairs, while type II heterojunctions create an internal electric field, preventing recombination and promoting redox reactions. Z-scheme heterojunctions combine type I and II, transferring electrons to holes through a redox mediator (Figure 7). Conventional type II heterojunctions reduce redox abilities, but Z-scheme heterojunctions have a distinct migration pattern, addressing these drawbacks while maintaining a high reactive capacity.

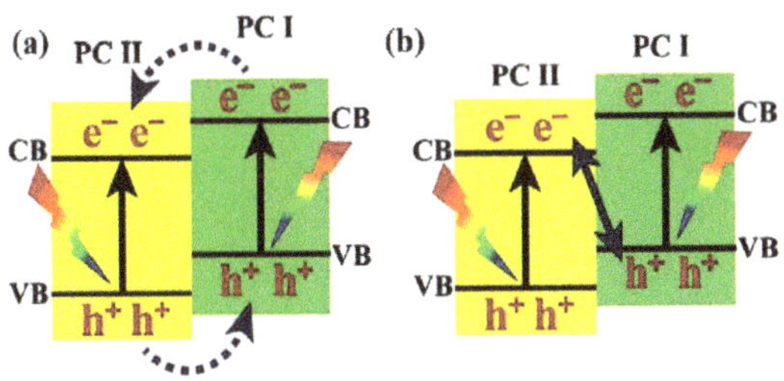

Figure 7. Schematic presentation of (**a**) type II-based and (**b**) direct Z-scheme heterojunctions [68].

Due to conventional type II and Z-heterojunction limitations, Yu et al. proposed a novel concept, the S-scheme heterojunction [69]. The S-concept consists of two n-type semiconductor photocatalysts, PC I, and PC II, representing oxidation and reduction photocatalysts, respectively (Figure 8b). A band-staggered combination of PC I and PC II forms the heterojunction. When the semiconductors come into contact, electrons flow spontaneously, bending photocatalysts B and A downward and upward. The internal electric field prevents electron migration, enhancing photocatalytic reaction efficiency [70].

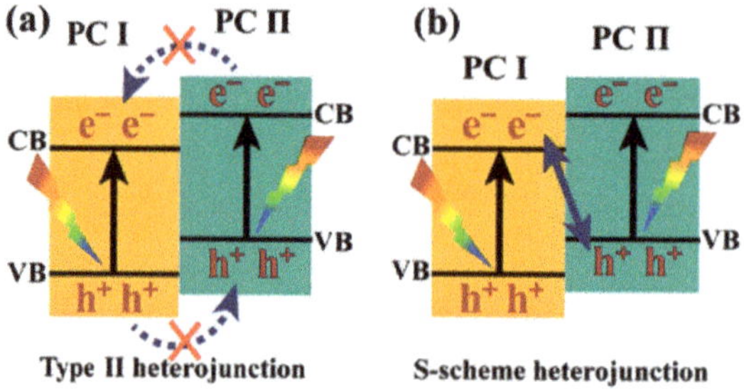

Figure 8. Schematic presentation of charge transfer in type II heterojunctions (**a**) and direct S-scheme (**b**) [69].

Hierarchical $CeO_2@Ni_{1-x}Co_xSe_2$ hollow spheres with an S-scheme heterojunction have been recently reported to exhibit rapid charge/mass transport, exceptional visible light absorption, superior activity, and excellent stability in photocatalytic CO_2 reduction with CO [71]. The photodegradation at the ZnO heterojunction was more efficient, occurring 300% and 33% faster than the individual Bi_2O_3 and ZnO catalysts, respectively, demonstrating a synergistic impact [72].

The selection of a synthesis technique for nano oxide particles is contingent upon the application's particular needs, encompassing particle size, morphology, distribution, purity, scalability, and cost factors. Researchers must evaluate these advantages and disadvantages to determine the suitable synthesis pathway for their intended use. Synthesis methods can vary widely, each offering advantages and disadvantages depending on the application, desired properties, and production scale. The pros and cons of the commonly applied methods can be summarized as follows. The sol–gel method offers controlled composition, homogeneity, and scalability for large-scale production. Still, it requires longer processing times and may require additional heat treatment. Coprecipitation is more straightforward, cost-effective, and quick but has limitations such as particle size control and potential contamination. Hydrothermal and solvothermal synthesis offers a high purity and crystalline quality but requires complex equipment and energy. Ball milling has scalability and versatility but has limitations like contamination, morphology control, and aggregation. Chemical vapor deposition produces high-purity and uniform materials but requires complex setups, high temperatures, and energy. It is suitable for thin-film applications but may require sophisticated setups and energy.

4. Metal Oxide Nanostructures and Photocatalytic Properties

4.1. TiO_2

Because of its cost-effectiveness and structural characteristics, TiO_2 (an *n*-type semiconductor) is the most intensively researched NMO for the photocatalytic breakdown of organic contaminants. With a broad energy band gap ranging from 3.0 to 3.2 eV, TiO_2 can only be stimulated by UV radiation. Hence, the utilization of visible light or sunshine is restricted.

The mechanisms of degradation of organic pollutants using TiO_2 as a photocatalyst proceed through the following steps: (a) adsorption onto the surface of TiO_2; (b) photocatalytic degradation by oxidation–reduction reactions, including photogenerated electrons, holes, and reactive species; and (c) desorption of the degradation outcomes. Organic dyes undergo degradation through two distinct pathways: (1) the indirect and (2) the direct pathway [73,74]:

1. The indirect pathway includes the indirect breakdown of dye molecules by generating potent oxidizing radicals following the absorption of UV radiation (Figure 9a). The degradation comprises several discrete steps. The first involves photon excitation of an electron from the valence band (VB) to the conduction band (CB) with energy equal to or

higher than the band gap energy (Eg), leading to an electron–hole (e^+/h^+) pair generation within the TiO_2:

$$TiO_2 + h\nu(UV) \rightarrow TiO_2(e^-(CB) + h^+(VB))$$

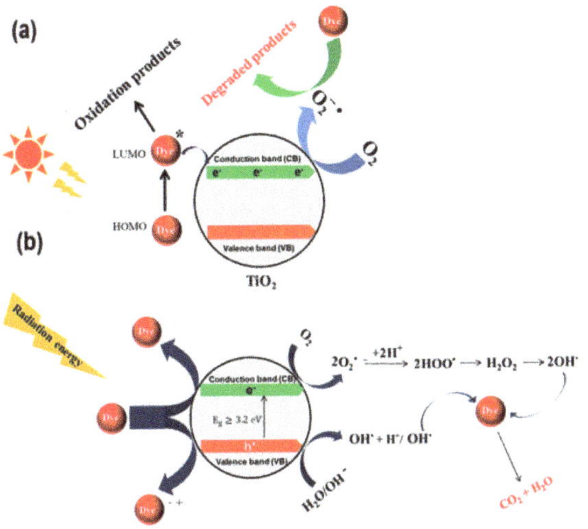

Figure 9. Schematic presentation of (**a**) indirect and (**b**) direct dye degradation mechanism on TiO_2 photocatalyst surface [74].

Moreover, photogenerated holes at VB produce OH• by reacting with adsorbed hydrogen formed by water ionization.

$$H_2O(ads) + h^+(VB) \rightarrow OH^{\cdot}(ads) + H^+(ads)$$

OH• radicals are potent oxidizing agents capable of chemically oxidizing organic molecules on the catalyst surface without selectivity, converting them into CO_2 and H_2O via mineralization. Then, electrons in the CB are captured by adsorbed oxygen, forming anionic superoxide radicals (O_2^-•).

$$O_2 + e^-(CB) \rightarrow O_2-(ads)$$

The involvement of the superoxide ion in the following oxidation processes relieves the recombination of electrons and holes, thereby preserving the electron neutrality of the TiO_2 material. Moreover, the superoxide ions react with hydrogen ions to produce hydroperoxyl radicals (HO_2•), which are then transformed into hydrogen peroxide (H_2O_2). The hydrogen peroxide subsequently decomposes to generate reactive OH•. Conventionally, oxidation and reduction occur on the surface of the photoexcited semiconductor photocatalyst, such as TiO_2.

$$2HOO^-(ads) \rightarrow H_2O_2(ads) + O_2$$

$$H_2O_2(ads) \rightarrow 2OH^-(ads)$$

$$Dye + OH^- \rightarrow CO_2 + H_2O \text{ (dye intermediates)}$$

$$Dye + h^+(VB) \rightarrow \text{oxidation products}$$

$$Dye + e^-(CB) \rightarrow \text{reduction products}$$

2. In the direct mechanism (Figure 9b), the excitation of dye molecules by visible light photons causes them to transition from their lowest energy state to an excited triplet state (refer to the equation below). The dye species that has been stimulated undergoes electron

transfer to the CV of TiO_2, converting it into a semi-oxidized radical cation (Dye + ϵ). Subsequently, the confined electrons react with the dissolved oxygen, forming superoxide radical anions (O_2^- ϵ), which produce hydroxyl radicals (OHϵ). The dominant role of these radicals is to oxidize organic molecules.

$$\text{Dye} + h\nu \rightarrow \text{Dye}^*$$

$$\text{Dye}^* + TiO_2 \rightarrow \text{Dye}^+ + TiO_2^-(\text{CB})$$

$$TiO_2^-(\text{CB}) + O_2(\text{ads}) \rightarrow TiO_2 + O_2^- \bullet \ (\text{ads})$$

$$\text{Dye} + O_2 - \bullet \ (\text{ads}) \rightarrow \text{Degraded products}$$

Despite its efficacy in photocatalysis, TiO_2 has some disadvantages, which are outlined as follows [75]: (1) limitations in the transfer of electron and hole pairs, with the likely recombination of these charge carriers within TiO_2 leading to reduced photocatalytic activity; (2) slow photocatalytic degradation rates due to the low adsorption of organic pollutants on the TiO_2 surface; (3) the aggregation of TiO_2 nanoparticles, resulting from their instability, which prevents effective light absorption at the active sites; (4) increased light scattering caused by TiO_2 nanoparticles, which reduces photocatalytic activity; (5) difficulties in the recovery and regeneration of the spent TiO_2; and (6) the high band gap energy of the initial TiO_2 photocatalyst, which affects its energy efficiency compared to standard heterostructure photocatalysts.

Diverse methodologies have been explored to augment its photocatalytic efficiency, such as introducing metals or non-metals or simultaneously introducing many elements. TiO_2 exhibits semiconducting properties because its CB and VB consist of Ti 3d and O 2p orbitals. Introducing dopants alters its electrical structure, wide band gap, and inherent characteristics, leading to an expanded photoresponse within the visible spectrum. Both metal ions and non-metal ions can enhance TiO_2 photolytic performance. The doping of TiO_2 with noble metal ions, rare earth elements, and transition metals has been usually reported. Nevertheless, there have been limited studies on doping TiO_2 with alkali metal ions, mainly due to their unexplored hydrophilicity and photocatalytic activity. Given the higher cost of transition metals than alkali metals, recent research increasingly focuses on alkali (Li, Na, K, Rb)-doped TiO_2.

Lithium doping decreases the diameters of TiO_2 NPs and alters their surface chemical morphologies and structures. As-prepared photocatalysts with varying LiBr concentrations show significantly improved methylene blue degradation efficiency and photocatalytic activity under UV light [76]. Moreover, Li-doped TiO_2 NPs exhibit enhanced photocatalytic activity in methyl orange decomposition, causing the rutile crystal phase formation, lowering the temperature needed to transition from the anatase to rutile phase, and triggering a mixed-crystal phenomenon [77]. Anatase TiO_2 nanotubes can be tailored by Na doping at different concentrations, resulting in reduced band gaps that match the visible solar spectrum. These nanotubes show the higher photodegradation efficiency of methylene blue dye. An increased Ti^{4+} to Ti^{3+} reduction plays an active role in improving the photocatalytic efficiency of the samples [78]. Doping TiO_2 with K enhanced its optical properties. The spectra demonstrated minimal absorption for pure TiO_2 film in the visible region. The band edge of the doped film was moved towards lower energy at 437 nm, increasing absorption. The decrease in band gap was caused by a shift in the optical absorption from UV to visible light and an increase in film thickness. TiO_2 absorption extended well into the visible area in the presence of a K dopant [79]. Rb doping enables a decrease in the NPs' size, improving the distribution of Rb-TiO_2 nanoflakes and inhibiting the recombination of photogenerated holes and electrons. These result in a degradation rate of up to 93% during one hour [80].

Doping with alkaline earth metals also enhances TiO_2 photocatalytic activity. Mg-doped NPs have superior activity in methylene blue degradation due to reduced band gap energy and charge recombination [81]. Oxygen vacancies increase with Mg doping, which is attributed to the difference in electronegativity and ionic radius between Ti and

Mg. These vacancies act as electron acceptors, reducing the recombination rate of electron–hole pairs [82]. Ca-doped TiO_2 nanofibers improved performance in photodegrading rhodamine B. The substitution of Ti^{4+} ions with Ca^{2+} ions, coupled with the introduction of oxygen vacancies, reduces the recombination of electron–hole pairs, thereby improving the photocatalytic efficiency of the nanofibers. Sr-doped TiO_2 NPs are efficient in Brilliant Green degradation under solar light. Sr doping increased the mesoporosity and surface area, suppressing electron–hole pair recombination.

Non-metals like nitrogen, carbon, sulfur, iodine, and fluorine have been used as dopants for TiO_2, with nitrogen being particularly suitable due to its atomic size and low ionization energy. Nitrogen doping reduces the band gap and mitigates electron and hole recombination [83,84]. Sulfur doping enhances degradation efficiency, but the choice of sulfur sources and preparation methods significantly influences S-doping efficiency [85–88]. Cationic doping replaces Ti with S^{4+} or S^{6+}, while anionic doping replaces oxygen with S^{2-}. In cationic form, hydroxyl groups and photoinduced holes contribute to photocatalytic activity, while in the anionic, electrons and holes contribute equally to the photocatalytic process [86]. Carbon doping enables a photocatalyst to be effective under visible light. Introducing carbon enhances photocatalytic activity by creating a new impurity energy level above the valence band, reducing the band gap and allowing the photocatalyst to absorb visible light.

The synergism between polyaniline and the TiO_2 nanocomposite was examined as an electrode system for use in dye-sensitized solar cells, both in the dark and under illumination with various metal phthalocyanine dyes. Cobalt phthalocyanine dye demonstrated a superior conversion efficiency compared to the other phthalocyanine dyes studied [89].

The immobilization of TiO_2 NPs onto suitable materials is also efficient in obtaining a more effective and stable photocatalyst; owing to their structural characteristics, such as a high surface area, stability, non-toxicity, and natural abundance, clay minerals are widely used as supports. The degradation mechanism involves the adsorption of pollutant molecules, followed by degradation through reactive oxygen species. The nanocomposites' high porosity and extensive surface area enable the rapid conversion and mineralization of the pollutants. The interlayer cations in the clay capture electrons while allowing holes to participate in the oxidation process, reducing the recombination rate and thereby improving photocatalytic activity compared to pure TiO_2. Furthermore, the clay enhances the reusability of the photocatalyst by facilitating its separation from the reaction mixture. Various clay minerals, montmorillonite, bentonite, kaolinite, smectite, and vermiculite, efficiently support TiO_2 NPs [90–94].

4.2. *ZnO*

ZnO (an n-type semiconductor with a band gap of 3.37 eV) is recognized as an advantageous photocatalyst due to its high stability, non-toxicity, environmental friendliness, and cost-effectiveness [95,96]. However, its wide band gap and rapid recombination rate of electron–hole pairs limit its photocatalytic effectiveness. Doping with metals or non-metals, incorporating noble metals, combining it with narrow band gap semiconductors, or supporting ZnO on various solid materials overcomes these limitations. Ba^{2+} ions introduced into the ZnO structure increased the surface area 14 times and improved hydrophilicity [97]. A positive correlation was observed between the surface hydrophilicity and the increased defectivity of the doped sample (Figure 10). The increased affinity with water is crucial for the better photocatalytic activity of the doped sample over the undoped one. Moreover, doping with Ba reduced photo corrosion and enhanced the stability of the sample when exposed to radiation.

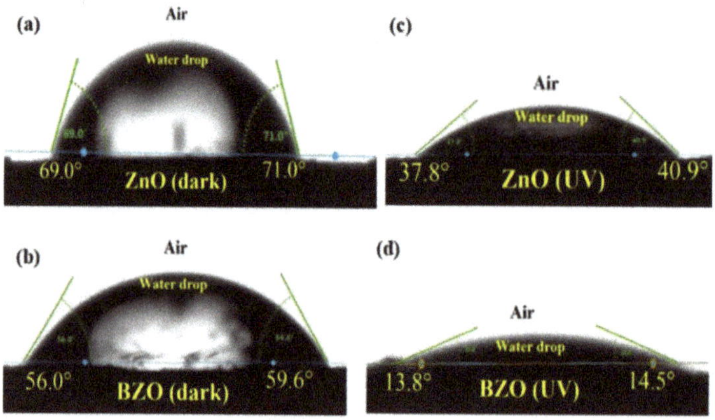

Figure 10. Shape of water drop on ZnO and Ba-doped ZnO (BZO) for static contact angle measurements in the dark (**a**,**b**) and after 10 min exposure to UV light irradiation (**c**,**d**) [97].

Doping ZnO with Fe creates sub-band states within its band gap, enhancing photocatalytic efficiency. The choice of doping element depends on electronegativity and the difference between dopant ionic radius and Zn^{2+} radius. Fe is a preferred element due to its chemical stability and ion radii. Fe-doped ZnO NPs showed efficacy as visible light-responsive photocatalysts [95,98]. Cr^{3+} has an ionic radius comparable to that of Zn^{2+}, which was utilized to synthesize highly efficient photocatalysts [99,100]. The observed efficacy is ascribed to the existence of Cr^{3+} and Cr^{2+}, which augment the surface oxygen vacancies and alter the band gap of ZnO [99,100]. A similar effect was reported for Mn^{2+} and Mn^{3+}, which substituted Zn^{2+} within the lattice of ZnO [101]. The presence of Ni^{2+} in the ZnO crystal lattice led to a distortion of the lattice due to the discrepancy in ionic radii between Ni and Zn cations. The distortion increased vacancy defects within the ZnO's structure, impacting its optical and photoluminescence properties [102].

A band gap analysis showed a notable band gap (Eg) reduction with increasing Ni concentrations in ZnO lattice ZnO [103]. The influence of Ni presence on the degradation of methylene blue under blue laser light irradiation was proved (Figure 11).

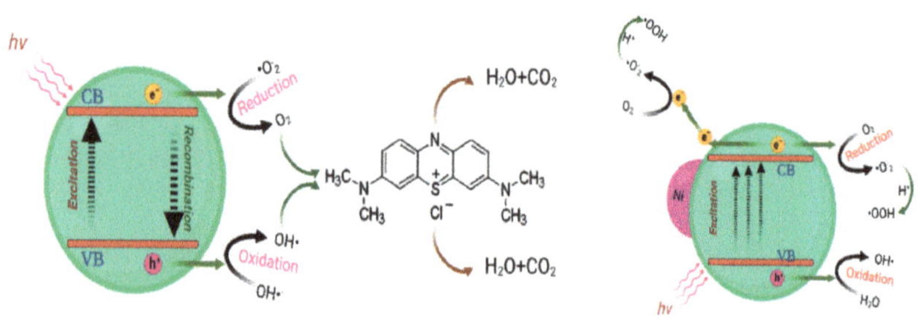

Figure 11. Diagram illustrating the impact of Ni doping on the ZnO nanostructure on the breakdown of methylene blue. Pure ZnO is on the left, and Ni-doped ZnO is on the right [103].

S-doped ZnO has an improved efficiency, which can be ascribed to an increased presence of oxygen vacancies. The photoactivity of ZnO doped with S and N was higher than that of C-doped ZnO in the degradation of methylene blue under UV–visible light. This behavior was attributed to the smaller crystallite size, lower band gap energy, and broader pore-size distribution resulting from S and N dopants [104].

The coupling of ZnO with other semiconductors to enhance its photocatalytic activity has also been explored. In this context, p-type oxides such as NiO and CuO are suitable for forming p–n heterojunctions with ZnO. This approach offers several benefits: (a) a reduced recombination rate of electron–hole pairs; (b) the creation of narrow band gaps, which enhances visible light absorption; and (c) improved reusability with minimal loss in performance [105]. The primary mechanism for the enhanced photocatalytic activity is attributed to the more efficient separation of photogenerated charge carriers facilitated by the internal electric field at the ZnO/NiO interface. Photocatalytic studies conducted by Chen et al. revealed the formation of an internal electric field at the core–shell structure of the n-ZnO/p-NiO interface, which significantly enhanced the photocatalytic activity for the degradation of methylene blue under UV irradiation [106]. CuO/ZnO (Figure 12) involves various radical reactions, including N-deethylation, decarboxylation, deamination, dealkylation, chromophore cleavage, and ring-opening in the degradation process of Rhodamine B. It was suggested that $O_2\bullet^-$ is a crucial reactive species in the degradation process [107].

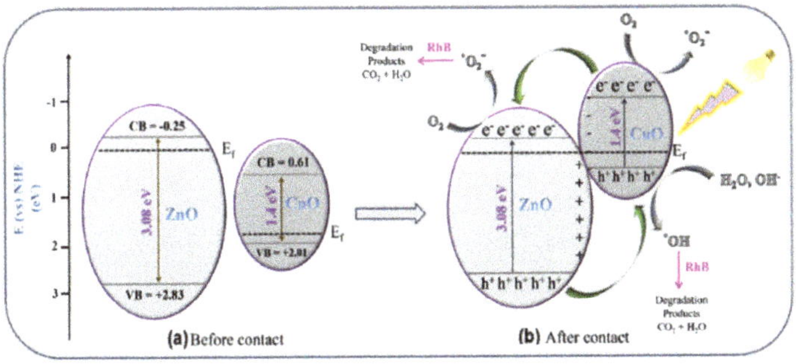

Figure 12. Schematic representation of the energy band diagram of the photodegradation process by CuO/ZnO nanocomposite [107].

The immobilization of ZnO NPs on porous materials like cellulose, clays, zeolites, and graphene oxides significantly enhances photocatalytic activity in reducing organic pollutants. The porous supports act as electron acceptors, improve adsorption, and generate oxygen vacancies in the ZnO lattice, extending its activity into the visible light spectrum. However, the ZnO content should be optimized to prevent aggregation [108–118].

4.3. CuO

CuO (a p-type semiconductor with a band gap of 1.2 to 2.1 eV) is effective in the visible range of the electromagnetic spectrum; however, its main drawback is the recombination between photogenerated electron–hole pairs, which reduces the efficiency of photocatalytic reactions. Zn-, Fe-, and Mn-doped CuO nanosheets are promising photocatalysts for the photodegradation of several organic dyes, since the doping induces a shift in the optical band gap, enhances visible light absorption, and reduces electron–hole recombination [119,120].

Combining CuO with other semiconductors yields an effective heterojunction material suitable for the photocatalytic degradation of different organic pollutants [121]. CuO/ZnO nanocomposites with a p–n heterojunction contribute to effectively separating the active charged ions, substantially reducing the recombination rate. A CuO/NiO photocatalyst with a p-p isotype heterojunction showed superior degradation activity under visible light due to superoxide and hydroxyl radicals, which are the primary reactive species involved in degradation [122,123].

CuO-supported graphene oxide has shown exceptional efficiency in reducing organic pollutants, thanks to synergistic effects with reduced graphene oxide properties [124]. The photocatalyst effectively degrades cationic methylene blue and anionic Congo red under visible light irradiation. Reduced graphene oxide creates a conductive network, enhancing charge transport. CuO-supported clays and clay minerals (bentonite, kaolinite, and montmorillonite) efficiently degrade several organics under sunlight [125–127]. Radicals

•O_2− and •OH are vital in the degradation process of cationic dye, methylene blue. The photocatalytic process involves dye adsorption by montmorillonite and CuO, facilitated by electrostatic attraction, Lewis's base interaction, and hydrogen bonding. Various cations in montmorillonite exchange with cationic dye, and metal complexation occurs between dye electron-rich functional groups and Al^{3+} from the montmorillonite skeleton (Figure 13).

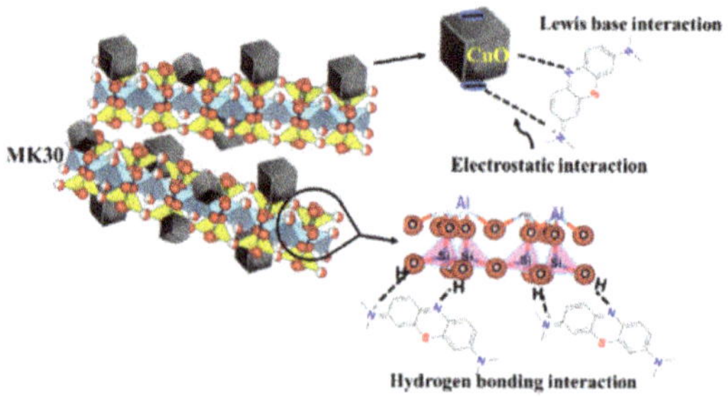

Figure 13. Schematic presentation of methylene blue degradation by CuO-supported montmorillonite [127].

4.4. Fe_2O_3

There are various kinds of iron oxide, including hematite (α-Fe_2O_3), maghemite (c-Fe_2O_3), and wustite (FeO). The most appealing mineral is hematite, α-Fe_2O_3. It has a small band gap energy of 2.1 eV and demonstrates excellent sensitivity to visible light. It is inexpensive, non-toxic, and chemically stable in water solutions. Its primary disadvantages are low electrical conductivity and absorptivity, leading to a high charge recombination and low photoactivity [128]. Additionally, the hydrophobic surface causes hydrophobic interactions between particles, promoting particle agglomeration and reducing the surface area [129]. Recently, it was reported that hematite impurities in clinoptilolite-rich tuff are responsible for the photolytic degradation of methylene blue under visible light [130].

Modifying the surface of Fe_2O_3 nanoparticles with Ag nanoparticles enhances their photocatalytic effectiveness in breaking down azo dyes under sunlight irradiation from 20% to 99%. Significantly, the Fe_2O_3 NPs doped with Ag are exceptionally stable and environmentally friendly, showing no signs of poisoning or photo-weathering [131]. The proposed photocatalytic process (Figure 14) involves the role of Ag NPs in capturing the photogenerated e− by acting as an electron absorber, therefore inhibiting charge conduction recombination.

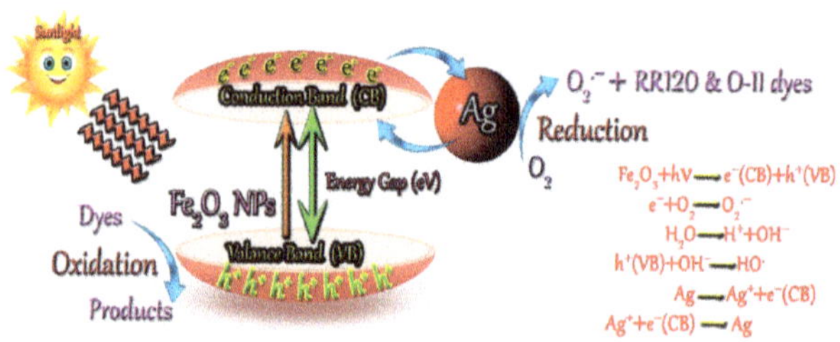

Figure 14. Mechanism of photocatalytic degradation of azo dyes using Ag–Fe_2O_3 NPs [131].

Ti- and Al-doped Fe_2O_3 exhibit significantly improved photocatalytic degradation of methyl orange compared to undoped Fe_2O_3 NPs [132]. The enhanced photocatalytic performance is attributed to grain refinement and the doping effect of Ti^{4+} and Al^{3+}. The Co(II) doping favors the growth of α-Fe_2O_3 and suppresses the growth of γ-Fe_2O_3. The doping narrows the band gap [133]. The doping of Cr(III) in α-Fe_2O_3 contributes to the ultrafast degradation of Congo red [134]. The doping of chromium influenced the significant factors responsible for the photocatalytic activity, the increase in range of absorbance, increased e−/h+ pair separation, improvement in the charge transfer process, and active site formation, which significantly enhanced the degradation process. The optical band gap energy of α-Fe_2O_3 NPs slightly decreased after V^{4+} doping ions. Weak ferromagnetic behavior due to canted surface spins and the reduced coercivity and remanence after doping of V^{4+} ions is attributed to the creation of oxygen defects. Doped NPs showed a maximum of 92% photocatalytic degradation efficiency within 180 min. This was made possible by traps or defect states induced by V substitution, in addition to the band gap reduction of α-Fe_2O_3 NPs [135]. Moreover, incorporating Y^{3+} ions triggered the formation of novel energy levels [136]. The presence of d-level electrons in hematite can elevate the valence top, decreasing the optical band gap and increasing the absorption intensity of visible light. Excited electrons can readily transition from the VB to the conduction band (CB) and can be immediately absorbed by the energy level produced by Y^{3+}. Such action will impede the recombination of photogenerated electrons and holes, expediting the electron and hole separation process. Furthermore, the electrons will be transported to the surface of the photocatalyst to undergo a reaction with O_2 and generate •O^{2-} (Figure 15).

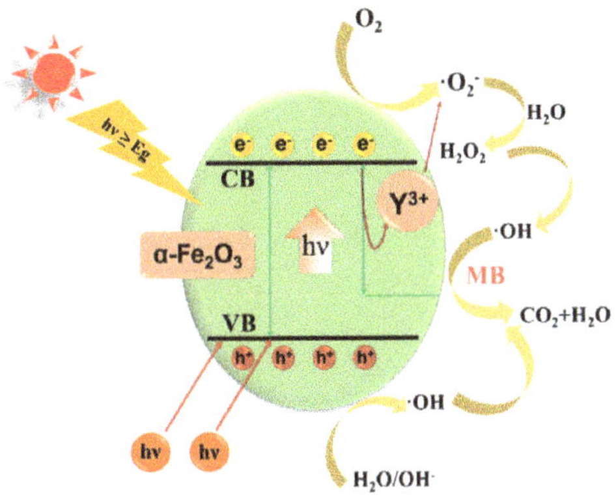

Figure 15. Photodegradation of methylene blue by Y-doped Fe_2O_3 [136].

In due course, more h+ ions will remain on VB to engage in reactions with H_2O molecules, resulting in the formation of •OH. This radical is widely recognized as the primary reactive species in the photocatalytic degradation process, improving the photocatalytic efficiency of hematite.

Recently, a theoretical study found that graphene-supported hematite with carbon vacancies has a nearly zero band gap, facilitating charge carrier transport to the surface. This improved carrier transport and catalysis benefit water splitting, a higher photocurrent density, and solar-driven water oxidation reactions [137].

4.5. Ag_2O

Silver(I) oxide (a p-type semiconductor) is receiving growing interest among semiconducting metal oxides. The narrow band gap (1.2 eV) of Ag_2O makes it well suited for application as a visible light photocatalyst. Yet, its sensitivity to light and fast recombination of electron–hole pairs generated by light significantly restrict its potential as a stable photocatalyst [138]. The morphology and dimensions of the particles modulate the photocatalytic characteristics of Ag_2O. The particle size can control the surface energy, resulting in a hydrophobic level or a superhydrophobic particle if the grain dimension meets roughness requirements. Ag_2O's wettability can be controlled by adjusting its crystal size during crystallization. Utilizing Ag_2O to photodecompose floating organic substances on the water surface is challenging. To overcome this, an attempt was made to prepare hydrophobic and oleophilic Ag_2O. The floatable superhydrophobic Ag_2O photocatalyst can effectively treat oil-based pollution and avoid switchable wettability under sunlight irradiation [139]. Its stable superhydrophobic status under illumination allows it to be used in various real-world scenarios, including as a high-performance photocatalyst for local oil degradation.

The crystal study of Ag_2O revealed that surfaces (100), (110), and (111) are especially susceptible to damage, and alterations in their surface shape had a substantial impact on the oxide's activity [140]. Its photosensitivity and unstable nature limit the photocatalytic activity of Ag_2O in light, which leads to photoreduction, producing O_2 and Ag metal, diminishing its functionality [141].

Ag_2O/rectorite nanocomposites showed remarkable stability in photocatalytic degradation under visible and NIR light. The radical-trapping experiment revealed that $\bullet O_2^-$ is an active radical species participating in the photocatalytic breakdown of contaminants when exposed to visible light and NIR irradiation. The short energy band gap of Ag_2O in the nanocomposites resulted in a significant absorption of visible and NIR light during the degradation of dyes. The rectorite integrated into the nanocomposites offered highly effective active sites for the photo-adsorption process, enhanced electron and hole separation characteristics, and successfully prevented charge recombination [140].

4.6. SnO_2

At ambient temperature, tin oxide is a particular n-type metal oxide semiconductor with a band gap of 3.6 eV. It combines a high electrical conductivity and optical transparency. Having its qualities adjustable by size and shape alteration, it offers a broad variety of applications. It is extensively used as a gas sensor material because of its sensitivity to different gases [142]. Recently, SnO_2 NPs have been used for photocatalysis because of their lack of toxicity, environmentally friendly nature, and chemical stability [143]. Nevertheless, the effective implementation of SnO_2 nanoparticles as a photocatalyst is hindered by their broad band gap and elevated electron and hole recombination rate. The limited duration of electron and hole propagation diminishes the effectiveness of SnO_2 photodegradation. Due to its broadband gap, it exhibits reduced sensitivity to visible light-induced photocatalytic activity.

A recent investigation indicated that the photoluminescent properties of small nanoparticles (approximately 6 nm) are affected by the heating regime used during their production. The properties of SnO_2 dried at ambient conditions and SnO_2 dried under vacuum exhibit noteworthy differences. The former demonstrates more significant UV–Vis absorption throughout the range of wavelengths, a larger Urbach energy, and a smaller band gap, suggesting a higher concentration of defects ('Sn' vacancies). Vacuum drying effectively enhances the crystallinity of SnO_2. Considering its small energy gap and many defects, the first material demonstrates satisfactory degrading properties of methyl orange dye when exposed to UV and visible light [144]. Experimental studies show that the structural, electrical, and optical characteristics of SnO_2 alter by introducing foreign atoms (X—Fe, Pt, Sb, Zn, Bi, Mg, F, Mn, Eu^{3+}, Al, Ti, Co). Doped with various metals and transition metal ions, SnO_2 exhibits exceptional electrical, optical, and electrochemical activities. Surface flaws occur as a consequence of dopants reducing the band gap. Profound variations in lattice characteristics and the absorption peak result in substantial differences in X-ray diffraction patterns between doped and undoped SnO_2. Doped SnO_2 exhibits improved absorption capacity in the visible light spectrum due to incorporating flaws and oxygen vacancies through doping [145]. X-doped SnO_2 exhibits a distinct band gap compared to both pure SnO_2 and the dopant (x) because of their conduction band (CB) and valence band (VB) interaction. The conduction band (CB) and valence band (VB) in dopant and SnO_2 exhibit a band disposition determined by the specific band gap. The band gap of the x-doped SnO_2 composite promotes the efficient partitioning and movement of the electrons and holes produced by light absorption. Under visible light, the photogenerated electrons in an x-doped SnO_2 nanocomposite are stimulated from the valence band of the dopant and moved to the conduction band of SnO_2 [145]. The photogenerated holes, in contrast, are stored in the VB. The extended lifespan of the excited electrons and holes is attributed to the transfer process (Figure 16B)). An expected consequence of electron transfer from the CB of SnO_2 to the CB of the doping material is inhibiting the reverse interaction between the photogenerated charge carriers (e^-/h^+). Recombination significantly reduces the photogenerated charge carriers, leading to photodegradation according to the method. The prevention of the recombination of photogenerated electron and hole pairs (e^-/h^+) can be achieved by doping SnO_2. This phenomenon enhances the photocatalytic efficiency of the photocatalyst when exposed to visible light. Upon rapid reaction, the electrons produce radicals that subsequently break down or oxidize the organic contaminants [145].

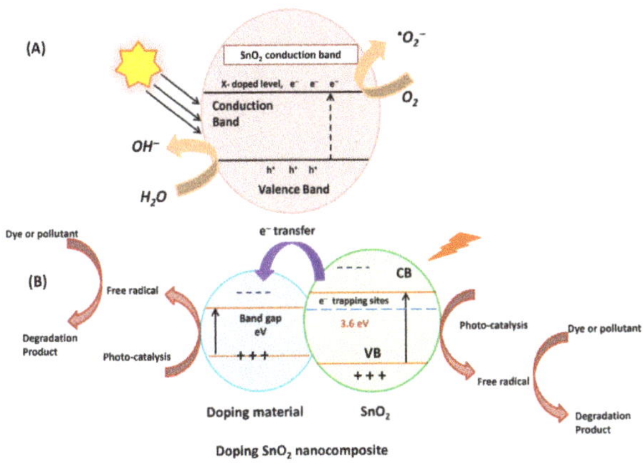

Figure 16. A schematic presentation of (**A**) the general mechanism of the SnO_2-based photocatalytic process and (**B**) the representation diagram of the electron transmission and the formation of degradation products by doped NPs [145].

For Fe-doped SnO_2, a recent report shows that doping causes (1) intermediate energy levels within the band gap created around the lower end of the conduction band; (2) this, in turn, enhances the separation of carriers and the dynamics of transformation; (3) the Fe-induced active sites play a crucial role in the adsorption and activation of the organic pollutant; (4) the oxygen vacancies have a role in both the adsorption and activation of O_2 and serve as dopant centers to decrease recombination and improve the separation of photoinduced electron–hole pairs [146].

Recent studies have shown that non-metal dopants, including nitrogen, carbon, sulfur, and fluorine, alter the band gap of SnO_2 by replacing the valence band, leading to an increase in oxygen vacancy defects on the nanoparticles' surface. Due to relatively small dimensions, C, F, O, and N diffuse through the lattice interstices and bind to the atoms via oxidation. Carbon is an ideal choice for non-metal dopants, particularly in semiconductors, due to its comparatively high mechanical strength, well-developed chemical resistance, and distinctive electrical characteristics. Moreover, doping can result in an increased generation of OH radicals, thereby enhancing the effectiveness of organic pollutant breakdown. This phenomenon can be elucidated because dopants will serve as electron hunters, thereby inhibiting the recombination of e^-/h^+ pairs. Consequently, the hole (h^+) of the photocatalyst will be liberated [147].

Doping the metal oxide with semiconductors of either a narrow or wide band gap can significantly improve its photocatalytic efficiency. Tin dioxide (SnO_2) semiconductors are extensively employed with TiO_2 semiconductors. A composite material consisting of SnO_2/TiO_2 nanotubes with different SnO_2 concentrations exhibited enhanced methylene blue absorption on the catalyst surface, leading to heightened metal oxide catalytic activity [148].

The surface modification of SnO_2 NPs can improve dispersion, luminescence, and mechanical properties. Chemical treatment with coupling agents affects the NPs' physical–chemical properties. Modified SnO_2 synthesized using grafting polymerization shows long-term durability and improved dispersion in the organic matrix. Ligand molecules are capping agents that prevent nanoparticle aggregation and control growth. Coating metal oxide nanoparticles with polymeric ligands improves their stability due to steric and electrostatic forces. However, the coating can also obstruct photocatalysis [149].

4.7. WO_3

Nano-structured WO_3 is a promising n-type semiconductor oxide because of its suitable band gap (2.6–2.8 eV), thus making it the second most searched photocatalyst activated under visible light. Due to its nontoxicity, affordability, high purity, long-lasting stability in different electrolytes, photosensitivity, and resistance to photo corrosion, it is of increasing interest. Compared to TiO_2 and ZnO, WO_3 offers benefits such as a more limited energy gap and a broader spectrum of light that can be absorbed. Its exceptional response properties to visible light, which constitutes over half of solar radiation energy, make it an ideal photocatalytic material. Metal–organic framework (WO_3) crystals are found in several forms, such as monoclinic, triclinic, orthorhombic, tetragonal, and hexagonal phases. At ambient temperature, monoclinic, triclinic, orthorhombic, and hexagonal phases are stable [150].

WO_3 has been extensively studied for its role in removing contaminants, reducing CO_2, and splitting water. However, the precise tuning of performance and morphology is necessary for specific applications [150]. Designing a WO_3-based photocatalyst tailored for each application could be essential for optimal effectiveness.

Modifying its morphology is crucial to enhancing WO_3 efficiency, and a comprehensive investigation is needed to understand the consequences of WO_3 morphological properties. Morphology plays an essential role in photocatalytic performance, as it exposes active crystal surfaces, increases surface area, and shortens transport distance. The regulation of morphology in photocatalytic materials is an effective policy that advances photocatalysis (Figure 17).

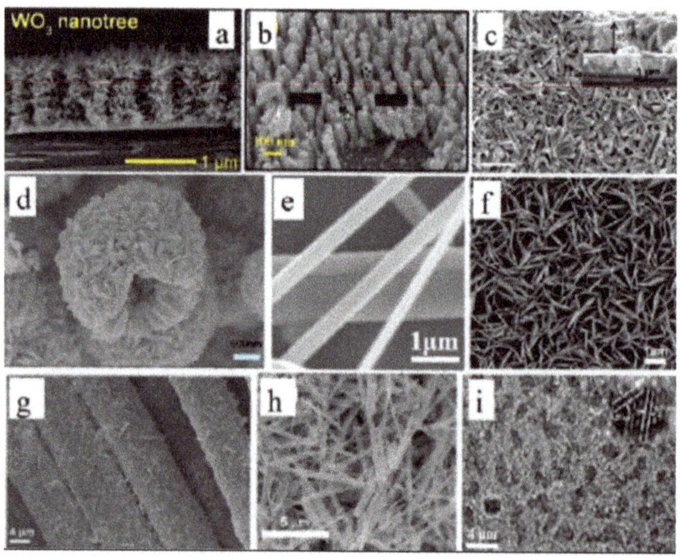

Figure 17. SEM images of different morphologies of WO_3 NPs: (**a**) nanotrees, (**b**) nanowires, (**c**) nanoplate array films, (**d**) hollow nest-like structure, (**e**) nanoroads, (**f**) nanosheets, (**g**) nanotubes, (**h**) needle-shaped nanorods, (**i**) quantum dots [150].

The surface properties significantly influence photocatalytic performance, with exposed crystal planes playing a crucial role in photocatalytic reactions. A near-perfect octahedron WO_3 was synthesized [151]. A thin tungstic acid layer covered this unique octahedron, significantly affecting its adsorption ability (Figure 18). The bounding {111} planes are likely not the lowest energy surfaces, and the tungstic acid covering helped stabilize the octahedron morphology. The as-prepared octahedra exhibits high visible-light-driven photocatalytic reducibility. They can remove Ag^+ ions in the photoprocessing of wastewater [151].

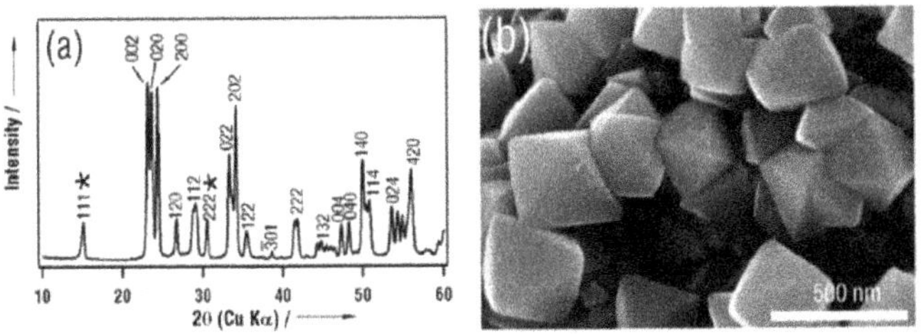

Figure 18. (**a**) XRD pattern and (**b**) SEM images of octahedral crystals (the asterisks represent diffractions corresponding to $H_2W_{1.5}O_{5.5}H_2O$) [151].

Moreover, doping is probably one of the most efficient approaches for enhancing WO_3 efficiency, improving light absorption, and reducing recombination rates. The electrocatalytic and photoelectrochemical activity of WO_3 changes non-monotonically with the Mn or V doping concentration due to local changes in the reduced nature of WO_3 and the formation of oxygen vacancies. The improved activity arises from fine-tuning the electronic structure and lessened free energy for atomic hydrogen adsorption. For photoelectrochemical water splitting, the photocurrent density increases from 0.61 mA cm^2 for the undoped WO_3 to about 1.38 mA cm^2 and 2.49 mA cm^2 for optimal Mn and V doping, respectively. Mn/V doping transforms WO_3 semiconductors into degenerate semiconducting materials with an improved metallic nature and suitable Gibbs-free energy [152].

Photocatalytic activity can also be enhanced by loading noble metals onto the WO_3. Although Au nanoparticles distributed on the WO_3 surface do not alter the selectivity of methylene blue photodegradation, they break methylene blue at a higher rate than the original WO_3 [68]. The increased photocatalytic activity has been ascribed to the improved separation of electron–hole pairs following the excitation of WO_3 at its band gap. Photogenerated holes in the CB of WO_3 are transported to Au, which has a lower Fermi level, under visible light. These holes directly contribute to methylene blue's oxidative degradation, boosting the photocatalyst's photocatalytic activity.

$g\text{-}C_3N_4$/CQDs/WO_3, a combination of graphite-like carbon nitride ($g\text{-}C_3N_4$), WO_3, and carbon quantum dots (CQDs) obtained by Z-scheme heterojunction, was tested for the depolymerization of lignin. The photocatalyst demonstrated a promising effect on depolymerizing four types of lignin, while effectively reducing the recombination rate of photogenerated carriers. The addition of CQDs expanded the light absorption range and improved the mobility of photogenerated carriers. An innovative approach for the highly selective cleavage of lignin C-C bonds was designed through heterojunction engineering [153].

A WO_3/Ag_3PO_4 composite showed improved photocatalytic performance and stability, ascribed to its direct Z-scheme heterojunction structure and the synergistic interaction between WO_3 and Ag_3PO_4. This combination enables the effective separation of electrons and holes and increases the conversion efficiency of light energy. Upon exposure to visible light, the photoexcited electrons in the CB of WO_3 and the retained holes in the valence band of Ag_3PO4 rapidly merge (Figure 19). Furthermore, the photogenerated holes in the VB of WO_3 greatly influence the oxidation reaction [154].

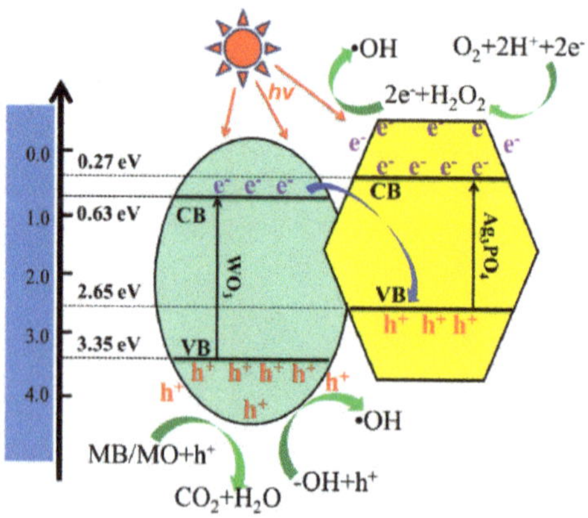

Figure 19. Proposed mechanism for the photodegradation of methylene blue (MB) and methyl orange (MO) in the direct Z-scheme charge carrier transfer process at the WO_3/Ag_3PO_4 composite interface [154].

Bare WO_3 displays a low photocatalytic performance because of the poor charge separation efficacy of its generated charge carriers. Graphene sheets are a promising material for charge transfer mechanisms. Reduced graphene oxide (RGO) enhances photogenerated electron transmission between graphene and MO semiconductors and enhances photocatalytic activity. A significant increase in O_2 production was reported for RGO/WO_3 compared to pristine WO_3 [155]. This improved the light absorption capability and utilization of electron–hole pairs in hybrid RGO/WO_3 nanoplates.

The photocatalytic effectiveness of the ternary photocatalyst (NH_2-GO/ZnO-WO_3) is superior in the degradation of several organic dyes when exposed to UV light. The improved photocatalytic efficiency mainly results from the low recombination rate of photoactive electron-hole pairs, the small energy band gap, and the extensive surface area of the ternary composite. Including lone pair electrons in the amino group forms a negatively charged surface composite. This, in turn, facilitates a favorable interaction with cationic dyes, which critically enhances the photocatalytic activity (Figure 20). The ternary nanocomposite NH_2-GO/ZnO-WO_3 has a suitable photocatalytic activity for degrading organic dyes often employed in industrial applications [156].

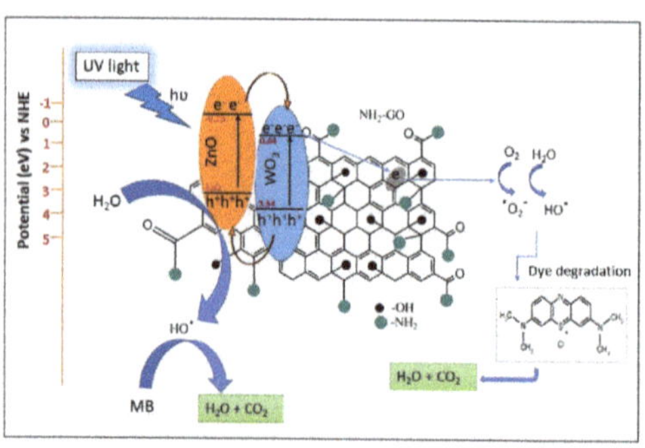

Figure 20. Schematic presentation of methylene blue (MB) photodegradation mechanism by NH_2-GO/ZnO-WO_3 composite [156].

A summary and comparison of the photocatalytic capabilities of various NMO-based photocatalysts are presented in Table 1.

Table 1. Photocatalytic properties of several NMO-based photocatalysts.

Photocatalyst	Model Pollutant	Operational Conditions (Light Type/C_0/Photocatalyst Amount)	Efficiency, %	Reaction Time, min	Reference
Rb-doped TiO_2	Methylene blue	UV light/10 ppm/0.4 g dm^{-3}	97	60	[80]
Ca-doped TiO_2	Rhodamine B	UV light/1 × 10^{-5} mol dm^{-3}/1.2 g dm^{-3}	95	100	[81]
TiO_2-pillared clay	Phenol	UV irradiation/50 ppm/2.0 g dm^{-3}	97.4	120	[93]
Fe-doped ZnO	Acid orange 7	Visible light/10 ppm/1.5 g dm^{-3}	52	180	[98]
Cu-Ni co-doped ZnO	Indigo carmine dye	Visible light/100 ppm/1 g dm^{-3}	93.3	60	[96]
CuO	Methylene blue Malachite green	Visible light/10 ppm/50 mg	62 70	270 30	[119]
Zn-CuO	Methylene blue Malachite green	Visible light/10 ppm/50 mg	66 85	270 30	[119]
Fe-CuO	Methylene blue Malachite green	Visible light/10 ppm/50 mg	66 90	270 30	[119]
Mn-doped CuO	Rhodamine B	Visible light/10 ppm/5 mg	93.8	90	[120]
CuO-ZnO	Tetracycline Ciprofloxacin	Natural sunlight/25 ppm/30 mg	94 93	50	[121]
CuO-NiO	Cefixime	Sunlight/15.22 ppm,/1 g dm^{-3}	90	180	[122]
CuO-montmorillonite 30	Methylene blue	Visible light/10 ppm/0.02 g dm^{-3}	98	40	[127]
Fe_2O_3	Orange-II Reactive red 120	Direct natural solar light irradiation/20 ppm/1 g dm^{-3}	23 20	120	[131]
Co-Fe_2O_3	Orange-II Reactive red 120	Direct natural solar light irradiation/20 ppm/1 g dm^{-3}	34 27	120	[131]
Cu–Fe_2O_3	Orange-II Reactive red 120 Orange-II Reactive red 120	Direct natural solar light irradiation/20 ppm/1 g dm^{-3}	81 71	120	[131]
Ag–Fe_2O_3	Orange-II Reactive red 120	Direct natural solar light irradiation/20 ppm/1 g dm^{-3}	97 99	120	[131]
Cr-Fe_2O_3	Congo red dye	Sunlight/30 ppm/30 mg dm^{-3}	95.2	15	[134]
Fe-SnO_2	Ciprofloxacin	UV light/10 ppm/50 mg	98.2	120	[146]
SnO_2-clinoptilolite	Methylene blue	Visible ligh/10 ppm/0.2 g dm^{-3}	45.0	180	[134]

4.8. Kinetics and Thermodynamics of Photodegradation

Kinetic studies on the photocatalytic degradation of organic compounds have been recorded in the literature, providing substantial evidence to clarify the process and its effectiveness. The Langmuir–Hinshelwood and pseudo-order kinetic models have frequently been assessed. The L–H model correlates effectively with the literature's experimental kinetic data about photocatalytic degradation. The photodegradation of methylene blue utilizing ZnO nanoparticles [157] indicated that the Langmuir–Hinshelwood model had a more favorable correlation with the experimental results than other models. The photodegradation of 2-chlorophenol using TiO_2 followed the Langmuir–Hinshelwood model, demonstrating a strong correlation coefficient (R^2 = 0.987) [158]. Moreover, the L–H model offered a more accurate representation than first-order kinetics for the photodegradation of amoxicillin using activated carbon-supported TiO_2 nanoparticles [159]. Tran et al. showed that only a limited number of research studies have successfully applied the pseudo-second model for photodegradation to date [160].

The thermodynamic driving force for photocatalysis is the alteration in Gibbs free energy (ΔG) of electron and hole systems in semiconductors due to light stimulation. Temperature does not influence the Gibbs free energy of semiconductors, so it cannot elicit the photocatalytic effect. A fundamental relationship between thermodynamics and kinetics is present in photocatalysis, with external factors such as organic species, light intensity, and temperature affecting the photocatalytic rate by influencing the thermodynamic driving force [161]. A recent investigation of the photodegradation of methylene blue by nanoparticles indicated that the reaction is spontaneous and endothermic [162], just as in the photodegradation of the dyes rhodamine B and alizarin R. The isotherm data were most accurately fitted to a Freundlich isotherm [163]. A recent study shows that photodegradation reactions facilitated by nickel manganite NPs are spontaneous and driven by entropy, advancing favorably at elevated temperatures [164].

5. Conclusions

Metal oxide-based photocatalyst systems have shown significant progress, but many challenges and opportunities remain to be explored. Most of the research on the photocatalysts is focused on powder materials. The synthesis of these powder materials in large quantities is complicated, and their respective recycling for real-time applications is also quite tricky. Moreover, the toxicity of the intermediates in the reaction is not easily predictable.

A comprehensive understanding of the charge transport process is essential to improving the activity of NMO-based photocatalysts. Heterogeneous photocatalysis is considered viable for degrading organic pollutants more effectively than conventional approaches. However, significant efforts are needed to overcome challenges such as understanding degradation mechanisms and nanostructure development.

NMO photocatalysts face challenges such as deactivation, cost, scalability, photostability, and charge carrier dynamics. Regeneration usually requires harsh conditions or chemical treatments, while synthesis can be expensive, especially for rare or precious metals. Long-term usability is affected by UV or visible light irradiation, and the fast recombination of charge carriers can lower efficiency. Accordingly, future perspectives must include material innovation, doping and defect engineering, advances in synthesis techniques, and integration with other technologies. Expanding our understanding of photocatalytic mechanisms at the nanoscale and addressing potential environmental impacts are also crucial for a successful application. A deep understanding of active sites, carrier transport paths, and carrier life on surfaces, interfaces, and nanostructures is also necessary.

Finally, a further challenge lies in translating laboratory-level research on NMO NP-anchored photocatalysts into commercially viable tools to meet the requirements for the practical use of solar energy in environmental cleaning and renewable energy.

Author Contributions: Conceptualization, N.R.; methodology, N.R., J.P. and N.N.T.; writing—original draft preparation, N.R., J.P. and N.N.T.; writing—review and editing, N.R. and J.P.; visualization, N.R., J.P. and N.N.T.; supervision, N.R. All authors have read and agreed to the published version of the manuscript.

Funding: This research was funded by the Ministry of Science, Technological Development, and Innovation of the Republic of Serbia (Project number: 451-03-66/2024-03/200011) and the Slovenian Research and Innovation Agency (Research program P1-0418) and NETPORE COST action CA20126.

Data Availability Statement: The data are contained within the article.

Conflicts of Interest: The authors declare no conflicts of interest.

References

1. Zan, L.; Amin, H.M.A.; Mostafa, E.; Abd-El-Latif, A.A.; Iqbal, S.; Baltruschat, H. Electrodeposited cobalt nanosheets on smooth silver as a bifunctional catalyst for OER and ORR: In situ structural and catalytic characterization. *ACS Appl. Mater. Interfaces* **2022**, *14*, 55458–55470. [CrossRef]
2. Suherman, A.L.; Zampardi, G.; Kuss, S.; Tanner, E.E.L.; Amin, H.M.A.; Young, N.P.; Compton, R.G. Understanding gold nanoparticle dissolution in cyanide-containing solution via impact-chemistry. *Phys. Chem. Chem. Phys.* **2018**, *20*, 28300–28307. [CrossRef] [PubMed]
3. Akbari, A.; Amini, M.; Tarassoli, A.; Eftekhari-Sis, B.; Ghasemian, N.; Jabbari, E. Transition metal oxide nanoparticles as efficient catalysts in oxidation reactions. *Nano-Struct. Nano-Objects* **2018**, *14*, 19–48. [CrossRef]
4. Awogbemi, O.; Ojo, A.A.; Adeleye, S.A. Advancements in the application of metal oxide nanocatalysts for sustainable biodiesel production. *Discov. Appl. Sci.* **2024**, *6*, 250. [CrossRef]
5. Fazio, E.; Spadaro, S.; Corsaro, C.; Neri, G.; Leonardi, S.G.; Neri, F.; Lavanya, N.; Sekar, C.; Donato, N.; Neri, G. Metal-oxide based nanomaterials: Synthesis, characterization and their applications in electrical and electrochemical sensors. *Sensors* **2021**, *21*, 2494. [CrossRef] [PubMed]
6. Tejashwini, D.M.; Harini, H.V.; Nagaswarupa, H.P.; Naik, R.; Deshmukh, V.V.; Basavaraju, N. An in-depth exploration of eco-friendly synthesis methods for metal oxide nanoparticles and their role in photocatalysis for industrial dye degradation. *Chem. Phys. Impact* **2023**, *7*, 100355. [CrossRef]
7. Malato, S.; Fernández-Ibánez, P.; Maldonado, M.I.; Blanco, J.; Gernjak, W. Decontamination and disinfection of water by solar photocatalysis: Recent overview and trends. *Catal. Today* **2009**, *147*, 1–59. [CrossRef]
8. Zhu, S.; Wang, D. Photocatalysis: Basic Principles, Diverse Forms of Implementations and Emerging Scientific Opportunities. *Adv. Energy Mater.* **2017**, *7*, 17100841. [CrossRef]
9. Yang, X.; Wang, D. Photocatalysis: From fundamental principles to materials and applications. *ACS Appl. Energy Mater.* **2018**, *1*, 6657–6693. [CrossRef]

10. Ren, G.; Han, H.; Wang, Y.; Liu, S.; Zhao, J.; Meng, X.; Li, Z. Recent Advances of Photocatalytic Application in Water Treatment: A Review. *Nanomaterials* **2021**, *11*, 1804. [CrossRef]
11. Tamirat, A.G.; Rick, J.; Dubale, A.A.; Su, W.N.; Hwang, B.J. Using hematite for photoelectrochemical water splitting: A review of current progress and challenges. *Nanoscale Horiz.* **2016**, *1*, 243–267. [CrossRef] [PubMed]
12. Nasrollahi, N.; Ghalamchi, L.; Vatanpour, V.; Khataee, A. Photocatalytic-membrane technology: A critical review for membrane fouling mitigation. *J. Ind. Eng. Chem.* **2021**, *93*, 101–116. [CrossRef]
13. Araújo, E.S.; Pereira, M.F.G.; da Silva, G.M.G.; Tavares, G.F.; Oliveira, C.Y.B.; Faia, P.M. A Review on the use of metal oxide-based nanocomposites for the remediation of organics-contaminated water via photocatalysis: Fundamentals, bibliometric study and recent advances. *Toxics* **2023**, *11*, 658. [CrossRef]
14. Srivastava, S.K. Recent advances in removal of pharmaceutical pollutants in wastewater using metal oxides and carbonaceous materials as photocatalysts: A review. *RSC Appl. Interfaces* **2024**, *1*, 340–429. [CrossRef]
15. Krishnan, A.; Swarnalal, A.; Das, D.; Krishnan, M.; Saji, V.S.; Shibli, S.M.A. A review on transition metal oxides based photocatalysts for degradation of synthetic organic pollutants. *J. Environ. Sci.* **2024**, *139*, 389–417. [CrossRef]
16. Sun, J.; Jiang, C.; Wu, Z.; Liu, Y.; Sun, S. A review on the progress of the photocatalytic removal of refractory pollutants from water by BiOBr-based nanocomposites. *Chemosphere* **2022**, *308*, 136107. [CrossRef]
17. Danks, A.E.; Hall, S.R.; Schnepp, Z. The evolution of "sol–gel" chemistry as a technique for materials synthesis. *Mater. Horiz.* **2016**, *3*, 91–112. [CrossRef]
18. Moothedan, M.; Sherly, K.B. Surface properties of nano mesoporous lanthanum oxide synthesized by sol-gel method. *Mater. Today Proc.* **2022**, *66*, 2338–2341. [CrossRef]
19. Radhakrishna, S.K.C.; Meda, U.S. Synthesis of silicon di oxide nano-particles by sol-gel method for applications in geopolymer composites. *ECS Trans.* **2022**, *7*, 5533–5541. [CrossRef]
20. John, A.K.; Palaty, S. Influence of solvent and pH on the synthesis of visible light active titanium dioxide nano particles. *J. Sol-Gel Sci. Technol.* **2018**, *87*, 391–399. [CrossRef]
21. Rafiq, F.; Govindsamy, P.; Periyasamy, S. Synthesis of a novel nanoparticle $BaCoO_{2.6}$ through sol-gel method and elucidation of its structure and electrical properties. *J. Nanomater.* **2022**, *2022*, 3877879. [CrossRef]
22. Hasson, S.S.; Alsammarraie, A.M. Synthesis of nickel oxide nanoparticles by sol-gel method. *Int. J. Health Sci.* **2022**, *6*, 6915–6924. [CrossRef]
23. Althomali, R.H.; Adeosun, W.A. Wet chemically synthesized metal oxides nanoparticles, characterization and application in electrochemical energy storage: An updated review. *Synth. Met.* **2023**, *298*, 117424. [CrossRef]
24. Sharma, N.; Jha, R.; Jindal, N. Hydrothermally synthesized stannic oxide nano-hexagons. *Mater. Today Proc.* **2018**, *5*, 13807–13815. [CrossRef]
25. Suman, S.; Mukurala, N.; Kushwaha, A.K. Annealing induced surface restructuring in hydrothermally synthesized gallium oxide nano-cuboids. *J. Cryst. Growth* **2021**, *554*, 125946. [CrossRef]
26. Goel, R.; Jha, R.; Bhushan, M.; Bhardwaj, R.; Ravikant, C. Hydrothermally synthesized nickel oxide (NiO) nano petals. *Mater. Today Proc.* **2022**, *48*, 687–689. [CrossRef]
27. Liu, J.L.; Fan, L.Z.; Qu, X. Low temperature hydrothermal synthesis of nano-sized manganese oxide for supercapacitors. *Electrochim. Acta* **2012**, *66*, 302–305. [CrossRef]
28. Youssef, W.B.; Nefzi, H.; Sediri, F. Controlled and environmentally friendly hydrothermal synthesis of nano-V_4O_9 plate-like for photocatalytic degradation of methyl orange under solar irradiation. *Solid State Sci.* **2023**, *137*, 107126. [CrossRef]
29. Ali, K.; Sajid, M.; Bakar, S.A.; Younus, A.; Ali, H.; Rashid, M.Z. Synthesis of copper oxide (CuO) via coprecipitation method: Tailoring structural and optical properties of CuO nanoparticles for optoelectronic device applications. *Hybrid Advan.* **2024**, *6*, 100250. [CrossRef]
30. Mahmood, N.B.; Saeed, F.R.; Gbashi, K.R.; Mahmood, U.S. Synthesis and characterization of zinc oxide nanoparticles via oxalate co-precipitation method. *Mater. Lett. X* **2022**, *13*, 100126. [CrossRef]
31. Wu, N.; Wang, Y.; Liu, H.; Xiong, X. The influence of pH value on the synthesis of ytterbium silicate nano-powders by cocurrent coprecipitation method. *J. Non-Cryst. Solids* **2023**, *614*, 122407. [CrossRef]
32. Tan, Y.; Wang, Y.; Liu, H.; Xiong, X.; Cheng, H. Synthesis and high-temperature phase transformation behavior of Dy_2O_3–Sc_2O_3 co-stabilized ZrO_2 powders prepared by cocurrent chemical coprecipitation. *Ceram. Int.* **2024**, *50*, 22395–22404. [CrossRef]
33. Bao, Z.; Li, K.; Wang, S.; Gao, K.; Zhang, D.; Li, M. Preparation and characterization of submicron-cerium oxide by hypergravity coprecipitation method. *Adv. Powder Technol.* **2021**, *32*, 1611–1618. [CrossRef]
34. Kromer, M.L.; Monzó, J.; Lawrence, M.J.; Kolodziej, A.; Gossage, Z.T.; Simpson, B.H.; Morandi, S.; Yanson, A.; Rodríguez-López, J.; Rodríguez, P. High-throughput preparation of metal oxide nanocrystals by cathodic corrosion and their use as active photocatalysts. *Langmuir* **2017**, *33*, 13295–13302. [CrossRef] [PubMed]
35. Lawrence, M.J.; Kolodziej, A.; Rodriguez, P. Controllable synthesis of nanostructured metal oxide and oxyhydroxide materials via electrochemical methods. *Curr. Opin. Electroche.* **2018**, *10*, 7–15. [CrossRef]
36. Yang, Y.; Li, Y.; Pritzker, M. Control of Cu_2O film morphology using potentiostatic pulsed electrodeposition. *Electrochim. Acta* **2016**, *213*, 225–235. [CrossRef]
37. Atta, N.F.; Amin, H.M.A.; Khalil, M.W.; Galal, A. Nanotube arrays as photoanodes for dye sensitized solar cells using metal phthalocyanine dyes. *Int. J. Electrochem. Sci.* **2021**, *6*, 3316–3332. [CrossRef]

38. Cheng, H.; Wang, C.; Qin, D.; Xia, Y. Galvanic replacement synthesis of metal nanostructures: Bridgingthe gap between chemical and electrochemical approaches. *Acc. Chem. Res.* **2023**, *56*, 900–909. [CrossRef]
39. Oh, M.H.; Yu, T.; Yu, S.H.; Lim, B.; Ko, K.T.; Willinger, M.G.; Seo, D.H.; Kim, B.H.; Cho, M.G.; Park, J.H.; et al. Galvanic replacement reactions in metal oxide nanocrystals. *Science* **2013**, *340*, 964–968. [CrossRef]
40. Fitri, M.A.; Ota, M.; Hirota, Y.; Uchida, Y.; Hara, K.; Ino, D.; Nishiyama, N. Fabrication of TiO_2-graphene photocatalyst by direct chemical vapor deposition and its anti-fouling property. *Mater. Chem. Phys.* **2017**, *198*, 42–48. [CrossRef]
41. Cuong, T.M.; Tuan, V.A.; Linh, B.H.; Phuong, D.T.; Hoa, T.T.K.; Tuyen, N.D.; Tuan, N.Q.; Kosslick, H. Novel method for doping of nano TiO_2 photocatalysts by chemical vapor deposition. *Stud. Surf. Sci. Catal.* **2010**, *175*, 497–500. [CrossRef]
42. Zhao, D.; Sathasivam, S.; Li, J.; Carmalt, C.J. Transparent and conductive molybdenum-doped ZnO thin films via chemical vapor deposition. *ACS Appl. Electron. MA* **2020**, *2*, 120–125. [CrossRef]
43. Nagirnyak, S.V.; Lutz, V.A.; Donstova, T.A.; Astrelin, I.M. Synthesis and characterization of tin(IV) oxide obtained by chemical vapor deposition method. *Nanoscale Res. Lett.* **2016**, *11*, 343. [CrossRef]
44. Yu, J.; Kim, D. The preparation of nano size nickel oxide powder by spray pyrolysis process. *Powder Technol.* **2013**, *235*, 1030–1037. [CrossRef]
45. Kim, D.H.; Seo, D.J.; Yu, J.K. Application of spray pyrolysis process for the preparation of nano sized cobalt oxide powder. *Korean J. Mater. Res.* **2014**, *24*, 25–32. [CrossRef]
46. Fernández-Barahona, I.; Muñoz-Hernando, M.; Herranz, F. Microwave-driven synthesis of iron-oxide nanoparticles for molecular imaging. *Molecules* **2019**, *24*, 1224. [CrossRef]
47. Hasanpoor, M.; Aliofkhazraei, M.; Delavari, H. Microwave-assisted synthesis of zinc oxide nanoparticles. *Procedia Mater. Sci.* **2015**, *11*, 320–325. [CrossRef]
48. Falk, G.S.; Borlaf, M.; López-Muñoz, M.J.; Fariñas, J.C.; Rodrigues Neto, J.B.; Moreno, R. Microwave-assisted synthesis of TiO_2 nanoparticles: Photocatalytic activity of powders and thin films. *J. Nanopart. Res.* **2018**, *20*, 23. [CrossRef]
49. Zhang, M.; Wang, M.; Liu, X.; Huang, H.; Zhao, S.; Ma, J.; Liao, Y.; Li, X. Microwave-hydrothermal synthesis of F and Fe co-doped CeO_2 photocatalysts for efficient removal of 2,4,6-TCP under visible light. *Mat. Sci. Semicon. Proc.* **2024**, *171*, 108032. [CrossRef]
50. Arellano-Cortaza, M.; Ramírez-Morales, E.; Castillo, S.J.; Lartundo-Rojas, L.; Zamudio-Torres, I.; López Alejandro, E.M.; Rojas-Blanco, L. Microwave-assisted hydrothermal synthesis of type II ZnSe/ZnO heterostructures as photocatalysts for wastewater treatment. *Ceram. Int.* **2023**, *49*, 24027–24037. [CrossRef]
51. Lin, Y.R.; Chang, Y.C.; Ko, F.H. One-pot microwave-assisted synthesis of In_2S_3/In_2O_3 nanosheets as highly active visible light photocatalysts for seawater splitting. *Int. J. Hydrogen Energy* **2024**, *52*, 953–963. [CrossRef]
52. El-Naggar, M.E.; Wassel, A.R.; Shoueir, K. Visible-light driven photocatalytic effectiveness for solid-state synthesis of ZnO/natural clay/TiO_2 nanoarchitectures towards complete decolorization of methylene blue from aqueous solution. *Environ. Nanotechnol. Monit. Manag.* **2021**, *15*, 100425. [CrossRef]
53. Hu, J.; Cao, Y.; Xie, B.J.; Jia, D. Simple solid-state synthesis and improved performance of $Ni(OH)_2$-TiO_2 nanocomposites for photocatalytic H_2 production. *Ceram. Int.* **2017**, *43*, 11109–11115. [CrossRef]
54. Jadoun, S.; Arif, R.; Jangid, N.K.; Meena, R.K. Green synthesis of nanoparticles using plant extracts: A review. *Environ. Chem. Lett.* **2021**, *19*, 355–374. [CrossRef]
55. Ghaffar, S.; Abbas, A.; Naeem-ul-Hassan, M.; Assad, N.; Sher, M.; Ullah, S.; Alhazmi, H.A.; Najmi, A.; Zoghebi, K.; Al Bratty, M.; et al. Improved photocatalytic and antioxidant activity of olive fruit extract-mediated ZnO nanoparticles. *Antioxidants* **2023**, *12*, 1201. [CrossRef]
56. Ali, J.; Bibi, S.; Jatoi, W.B.; Tuzen, M.; Jakhrani, M.A.; Feng, X.; Saleh, T.A. Green synthesized zinc oxide nanostructures and their applications in dye-sensitized solar cells and photocatalysis: A review. *Mater. Today. Commun.* **2023**, *36*, 106840. [CrossRef]
57. Verma, V.; Al-Dossari, M.; Singh, J.; Rawat, M.; Kordy, M.G.M.; Shaban, M. A Review on green synthesis of TiO_2 NPs: Photocatalysis and antimicrobial applications. *Polymers* **2022**, *14*, 1444. [CrossRef] [PubMed]
58. Jha, A.K.; Prasad, K.; Kulkarni, A.R. Synthesis of TiO_2 nanoparticles using microorganisms. *Colloid. Surf. B* **2009**, *71*, 226–229. [CrossRef]
59. Haghighizadeh, A.; Aghababai Beni, A.; Haghmohammadi, M.; Samie, M.S.; Farshad, A.S. Green synthesis of ZnO-TiO_2 nano-photocatalyst doped with Fe(III) ions using bitter olive extract to treat textile wastewater containing reactive dyes. *Water Air Soil Pollut.* **2023**, *234*, 366. [CrossRef]
60. Singh, S.J.; Chinnamuthu, P. Highly efficient natural-sunlight-driven photodegradation of organic dyes with combustion derived Ce-doped CuO nanoparticles. *Colloid. Surf. A* **2021**, *625*, 126864. [CrossRef]
61. Giuffrida, F.; Calcagno, L.; Leonardi, A.A.; Cantarella, M.; Zimbone, M.; Impellizzeri, G. Enhancing the photocatalytic properties of doped TiO_2 nanowires grown by seed-assisted thermal oxidation. *Thin Solid Film.* **2023**, *771*, 139783. [CrossRef]
62. Serra-Pérez, E.; Dražić, G.; Takashima, M.; Ohtani, B.; Kovačič, S.; Žerjav, G.; Novak Tušar, N. Influence of the surface structure of the TiO_2 support on the properties of the Au/TiO_2 photocatalyst for water treatment under visible light. *Catal. Today* **2024**, *437*, 114764. [CrossRef]
63. Cravanzola, S.; Cesano, F.; Gaziano, F.; Scarano, D. Sulfur-doped TiO_2: Structure and surface properties. *Catalysts* **2017**, *7*, 214. [CrossRef]
64. Lin, L.Y.; Nie, Y.; Kavadiya, S.; Soundappan, T.; Biswas, P. N-doped reduced graphene oxide promoted Nano TiO_2 as a bifunctional adsorbent/photocatalyst for CO_2 photoreduction: Effect of N species. *Chem. Eng. J.* **2017**, *316*, 449–460. [CrossRef]

65. Perović, K.; Kovačić, M.; Kraljić Roković, M.; Kušić, H.; Genorio, B.; Lavrenčič Štangar, U.; Novak Tušar, N.; Lončarić Božić, A. The development of ternary-based TiO_2-SnS_2/GO-RGO composite material for photocatalytic H2 production under Solar light irradiation. *Mater. Res. Bull.* **2023**, *167*, 112418. [CrossRef]
66. Sen, P.; Bhattacharya, P.; Mukherjee, G.; Ganguly, J.; Marik, B.; Thapliyal, D.; Verma, S.; Verros, G.D.; Chauhan, M.S.; Arya, R.K. Advancements in doping strategies for enhanced photocatalysts and adsorbents in environmental remediation. *Technologies* **2023**, *11*, 144. [CrossRef]
67. Jorge López, J.; Reya, A.; Vinuelas-Zahinos, E.; Álvarez, P.M. Preparation of a new green magnetic Fe_3O_4 @TiO_2-P25 photocatalyst for solar advanced oxidation processes in water. *J. Environ. Chem. Eng.* **2023**, *11*, 109999. [CrossRef]
68. Dutta, V.; Sharma, S.; Raizada, P.; Thakur, V.K.; Khan, A.A.P.; Saini, V.; Asiri, A.M.; Singh, P. An overview on WO_3 based photocatalyst for environmental remediation. *J. Environ. Chem. Eng.* **2021**, *9*, 105018. [CrossRef]
69. Fu, J.; Xu, Q.; Low, J.; Jiang, C.; Yu, J. Ultrathin 2D/2D WO_3/g-C_3N_4 step-scheme H_2-production photocatalyst. *Appl. Catal. B-Environ.* **2019**, *243*, 556–565. [CrossRef]
70. Wang, Y.; Du, K.; Xu, R.; Cui, D.; Shi, Y.; Hao, W.; Du, Y. Bismuth-based semiconductor heterostructures for photocatalytic pollution gases removal. *Curr. Opin. Green Sust.* **2023**, *41*, 100824. [CrossRef]
71. Han, W.; Chen, Y.J.; Jiao, Y.Z.; Liang, S.M.; Li, W.; Tian, G.H. Oxygen vacancy-rich S-scheme CeO_2@$Ni_{1-x}Co_xSe_2$ hollow spheres derived from NiCo-MOF for remarkable photocatalytic CO_2 conversion. *ACS Sustain. Chem. Eng.* **2023**, *11*, 7787–7797. [CrossRef]
72. Shahzad, R.; Muneer, M.; Khalid, R.; Amin, H.M.A. ZnO-Bi_2O_3 Heterostructured composite for the Photocatalytic degradation of orange 16 reactive dye: Synergistic effect of UV irradiation and hydrogen peroxide. *Catalysts* **2023**, *13*, 1328. [CrossRef]
73. Ajmal, A.; Majeed, I.; Malik, R.N.; Idriss, H.; Nadeem, A. Principles and mechanisms of photocatalytic dye degradation on TiO_2 based photocatalysts: A comparative overview. *RSC Adv.* **2014**, *4*, 37003–37026. [CrossRef]
74. Nur, A.S.M.; Sultana, M.; Mondal, A.; Islam, S.; Robel, F.R.; Islam, A.; Sumi, M.S.A. A review on the development of elemental and codoped TiO_2 photocatalysts for enhanced dye degradation under UV–vis irradiation. *J. Water Process Eng.* **2022**, *47*, 102728. [CrossRef]
75. Chen, D.; Cheng, Y.; Zhou, N.; Chen, P.; Wang, Y.; Li, K.; Huo, S.; Cheng, P.; Peng, P.; Zhang, R.; et al. Photocatalytic degradation of organic pollutants using TiO_2-based photocatalysts: A review. *J. Clean. Prod.* **2020**, *268*, 121725. [CrossRef]
76. Guo, D.; Yi, J.; Sun, Y.; Zhou, H. Recent advances in titanium-based electrode materials for stationary sodium-ion batteries. *Energy Environ. Sci.* **2016**, *9*, 2978–3006. [CrossRef]
77. Zou, F.; Hu, J.; Miao, W.; Shen, Y.; Ding, J.; Jing, X. Synthesis and characterization of enhanced photocatalytic activity with Li+-doping nanosized TiO_2 catalyst. *ACS Omega* **2020**, *5*, 28510–28516. [CrossRef]
78. Rahman, H.; Norbert, A.; Nair, P.S.; Joseph, J.A.; Shaji, S.; Deshpande, U.; Naduvath, J.; Shanu, A.S.; Philip, R.R. Influence of sodium doping on the material properties and photocatalytic activity of anatase titanium dioxide nanotubes prepared by anodization. *Opt. Mater.* **2022**, *134*, 113172. [CrossRef]
79. Amna, H.; Kayani, Z.N.; Anwar, M. Effect of Au ions on structural, optical, magnetic, dielectric, and antibacterial properties of TiO_2 dip-coated thin films. *J. Mater. Sci. Mater. Electron.* **2021**, *32*, 14398–14419. [CrossRef]
80. Jia, L.; Yang, L.M.; Wang, W.; Huang, S.T.; Xu, Z. Preparation and characterization of Rb-doped TiO_2 powders for photocatalytic applications. *Rare Met.* **2024**, *43*, 555–561. [CrossRef]
81. Minchi, L.; Cao, F.; Xinni, Z.; Youqiang, C.; Xuhua, L. Photocatalytic activity of Ca-TiO_2 nanofibers with different concentrations of calcium. *Chem. Phys. Lett.* **2019**, *736*, 1363807. [CrossRef]
82. Nair, S.B.; Aijo, J.K.; Joseph, J.A.; Shaji, S.; Philip, R.R. Role of magnesium doping for ultrafast room temperature crystallization and improved photocatalytic behavior of TiO_2 nanotubes. *Mater. Today Proc.* **2020**, *25*, 203–207. [CrossRef]
83. Ansari, S.A.; Khan, M.M.; Ansari, M.O.; Cho, M.H. Nitrogen-doped titanium dioxide (N-doped TiO_2) for visible light photocatalysis. *New J.Chem.* **2016**, *40*, 3000. [CrossRef]
84. Bakre, P.V.; Tilve, S.G.; Shirsat, R.N. Influence of N sources on the photocatalytic activity of N-doped TiO_2. *Arab. J. Chem.* **2020**, *13*, 7637–7651. [CrossRef]
85. Ohno, T.; Akiyoshi, M.; Umebayashi, T.; Asai, K.; Mitsui, T.; Matsumura, M. Preparation of S-doped TiO_2 photocatalysts and their photocatalytic activities under visible light. *Appl. Catal. A Gen.* **2004**, *265*, 115–121. [CrossRef]
86. Bakar, S.A.; Ribeiro, C. A comparative run for visible-light-driven photocatalytic activity of anionic and cationic S-doped TiO_2 photocatalysts: A case study of possible sulfur doping through chemical protocol. *J. Mol. Catal. A-Chem.* **2016**, *421*, 1–15. [CrossRef]
87. Piątkowska, A.; Janus, M.; Szymański, K.; Mozia, S. C-,N- and S-Doped TiO_2 Photocatalysts: A Review. *Catalysts* **2021**, *11*, 144. [CrossRef]
88. Ahmadpour, N.; Nowrouzi, M.; Avargani, V.M.; Sayadi, M.H.; Zendehboudi, S. Design and optimization of TiO_2-based photocatalysts for efficient removal of pharmaceutical pollutants in water: Recent developments and challenges. *J. Water Process Eng.* **2024**, *57*, 104597. [CrossRef]
89. Atta, F.N.; Galal, A.; Amin, H.M.A. Synthesis and photoelectrochemical behavior of a hybrid electrode composed of polyaniline encapsulated in highly ordered TiO_2 nanotubes array. *Int. J. Electrochem. Sci.* **2012**, *7*, 3610–3626. [CrossRef]
90. Zhang, Y.; Miao, B.; Chen, Q.; Bai, Z.; Cao, Y.; Davaa, B. Synthesis, structure, and photocatalytic activity of TiO_2-montmorillonite composites. *Catalysts* **2022**, *12*, 486. [CrossRef]

91. Liu, J.; Dong, M.; Zuo, S.; Yu, Y. Solvothermal preparation of TiO_2/montmorillonite and photocatalytic activity. *Appl. Clay Sci.* **2009**, *43*, 156–159. [CrossRef]
92. Chen, D.; Zhu, H.; Wang, X. A facile method to synthesize the photocatalytic TiO_2/montmorillonite nanocomposites with enhanced photoactivity. *Appl. Surf. Sci.* **2014**, *319*, 158–166. [CrossRef]
93. Wang, G.; Yan, Y.; Yang, Y.; Li, J.; Gao, H.; Wang, J.; Zheng, X.; Yang, J. In situ synthesis of TiO_2-pillared bentonite photocatalyst and its enhanced photocatalytic performance. *Water Air Soil Pollut.* **2022**, *233*, 494. [CrossRef]
94. Kassir, M.; Roques-Carmes, T.; Pelletier, M.; Bihannic, I.; Alem, H.; Hamieh, T.; Toufaily, J.; Villiéras, F. Adsorption and photocatalysis activity of TiO_2/bentonite composites. *Desalin. Water Treat.* **2017**, *98*, 196–215. [CrossRef]
95. Srinivasulu, T.; Saritha, K.; Ramakrishna Reddy, K.T. Synthesis and characterization of Fe-doped ZnO thin films deposited by chemical spray pyrolysis. *Mod. Electron. Mater.* **2017**, *3*, 76–85. [CrossRef]
96. Kumar, R.; Gedam, R.S. Synthesis and characterization of bi-functional Cu and Ni co-doped ZnO photocatalysts for organic pollutant degradation and antimicrobial activity. *Ceram. Int.* **2024**, *50*, 24716–24724. [CrossRef]
97. Hamrouni, A.; Moussa, M.; Fessi, N.; Palmisano, L.; Ceccato, R.; Rayes, A.; Parrino, F. Solar photocatalytic activity of Ba-doped ZnO nanoparticles: The role of surface hydrophilicity. *Nanomaterials* **2023**, *13*, 2742. [CrossRef]
98. Mancuso, A.; Sacco, O.; Mottola, S.; Pragliola, S.; Moretta, A.; Vaiano, V.; De Marco, I. Synthesis of Fe-doped ZnO by supercritical antisolvent precipitation for the degradation of azo dyes under visible light. *Inorg. Chim. Acta.* **2023**, *549*, 121407. [CrossRef]
99. Chang, C.J.; Yang, T.L.; Weng, Y.C. Synthesis and characterization of Cr-doped ZnO nanorod-array photocatalysts with improved activity. *J. Solid State Chem.* **2014**, *214*, 101–107. [CrossRef]
100. Mathai, J.; Jose, A.K.; Anjana, M.P.; Aleena, P.A.; Kunjumon, J.; Ittyachan, R.; Nair, S.S.; Vinithad, G.; Sajan, D. Substantial effect of Cr doping on the third-order nonlinear optical properties of ZnO nanostructures. *Opt. Mater.* **2023**, *142*, 114128. [CrossRef]
101. Aadnan, I.; Zegaoui, O.; El Mragui, A.; Daou, I.; Moussout, H.; Esteves da Silva, J.C.G. Structural, Optical and Photocatalytic properties of Mn doped ZnO nanoparticles used as photocatalysts for azo-dye degradation under visible light. *Catalysts* **2022**, *12*, 1382. [CrossRef]
102. Lins, A.; Jerônimo, A.G.; Barbosa, R.; Neves, L.; Trigueiro, P.; Almeida, L.C.; Osajima, J.A.; Pereira, F.A.; Peña-Garcia, R.R. Facile Synthesis of Ni-doped ZnO nanoparticles using cashew gum: Investigation of the structural, optical, and photocatalytic properties. *Molecules* **2023**, *28*, 7772. [CrossRef]
103. Zyoud, S.H.; Ganesh, V.; Che Abdullah, C.A.; Yahia, I.S.; Zyoud, A.H.; Abdelkader, A.F.I.; Daher, M.G.; Nasor, M.; Shahwan, M.; Zahran, H.Y.; et al. Facile synthesis of Ni-doped ZnO nanostructures via laser-assisted chemical bath synthesis with high and durable photocatalytic activity. *Crystals* **2023**, *13*, 1087. [CrossRef]
104. Joy, A.; Viswanathan, M.R.; Vijayan, B.K.; Silva, C.G.; Basheer, I.; Sugathan, S.; Mohamed, P.A.; Solaiappan, A.; Shereef, A. Solar photocatalysts: Non-metal (C, N, and S)-doped ZnO synthesized through an industrially sustainable in situ approach for environmental remediation applications. *RSC Adv.* **2024**, *14*, 21655–21667. [CrossRef] [PubMed]
105. Thampy, U.S.U.; Mahesh, A.; Sibi, K.S.; Jawahar, I.N.; Biju, V. Enhanced photocatalytic activity of ZnO–NiO nanocomposites synthesized through a facile sonochemical route. *SN Appl. Sci.* **2019**, *1*, 1478. [CrossRef]
106. Chen, Z.; Dedova, T.; Spalatu, N.; Maticiuc, N.; Rusu, M.; Katerski, A.; Acik, I.O.; Unold, T.; Krunks, M. ZnO/NiO heterostructures with enhanced photocatalytic activity obtained by ultrasonic spraying of a NiO shell onto ZnO nanorods. *Colloids Surf. A Physicochem. Eng. Asp.* **2022**, *648*, 129366. [CrossRef]
107. Jeevarathinam, M.; Asharani, I.V. Synthesis of CuO, ZnO nanoparticles, and CuO-ZnO nanocomposite for enhanced photocatalytic degradation of Rhodamine B: A comparative study. *Sci. Rep.* **2024**, *14*, 9718. [CrossRef]
108. Li, X.; Zhang, L.; Wang, Z.; Wu, S.; Ma, J. Cellulose controlled zinc oxide nanoparticles with adjustable morphology and their photocatalytic performances. *Carbohyd. Polym.* **2021**, *259*, 117752. [CrossRef]
109. Shi, C.; An, B.; Zhang, L.; Zai, Z.; Shi, Z.; Wang, Z.; Ma, J. Contribution of surface carboxyl of cellulose in the formation mechanism and interfacial catalysis activity of ZnO/cellulose nanocomposites. *Appl. Surf. Sci.* **2023**, *618*, 156633. [CrossRef]
110. Ehsan, M.F.; Barai, H.R.; Islam, M.M.; Susan, M.A.B.H.; Joo, S.W.; Miran, M.S. ZnO nanocomposites supported by acid-activated kaolinite as photocatalysts for the enhanced photodegradation of an organic dye. *Mater. Today Commun.* **2023**, *36*, 106563. [CrossRef]
111. Haounati, R.; Ighnih; Malekshah, R.E.; Alahiane, S.; Alakhras, F.; Alabbad, E.; Alghamdi, H.; Ouachtak, H.; Addi, A.A.; Jada, A. Exploring ZnO/Montmorillonite photocatalysts for the removal of hazardous RhB Dye: A combined study using molecular dynamics simulations and experiments. *Mater. Today Commun.* **2023**, *35*, 105915. [CrossRef]
112. Iazdani, F.; Nezamzadeh-Ejhieh, A. The photocatalytic rate of ZnO supported onto natural zeolite nanoparticles in the photodegradation of an aromatic amine. *Environ. Sci. Pollut. Res.* **2021**, *28*, 53314–53327. [CrossRef]
113. Nezamzadeh-Ejhieh, A.; Khodabakhshi-Chermahini, F. Incorporated ZnO onto nano clinoptilolite particles as the active centers in the photodegradation of phenylhydrazine. *J. Ind. Eng. Chem.* **2014**, *20*, 695–704. [CrossRef]
114. Bahrami, M.; Nezamzadeh-Ejhieh, A. Effect of the supported ZnO on clinoptilolite nano-particles in the photodecolorization of semi-real sample bromothymol blue aqueous solution. *Mater. Sci. Semicond. Process.* **2015**, *30*, 275–284. [CrossRef]
115. De Dios, F.S.; Morales, E.R.; Cortaza, M.D.A.; Hernández, G.P.; Mandujano, E.V.M.; Alejandro, E.M.L.; Blanco, L.R. Improvement of photocatalysis using ZnO/zeolite nanocomposites for contaminant removal in aqueous media. *Desalin. Water Treat.* **2023**, *312*, 79–88. [CrossRef]

116. Da Silvaa, P.L.; Nippes, R.P.; Macruz, P.D.; Hegetob, F.L.; Scaliante, M.H.N.O. Photocatalytic degradation of hydroxychloroquine using ZnO supported on clinoptilolite zeolite. *Water Sci. Technol.* **2021**, *84*, 764. [CrossRef]
117. Abd-Elrahim, A.G.; Chun, D.M. Room-temperature deposition of ZnO-graphene nanocomposite hybrid photocatalysts for improved visible-light-driven degradation of methylene blue. *Ceram. Int.* **2021**, *47*, 12812–12825. [CrossRef]
118. Mirikaram, N.; Pérez-Molina, Á.; Morales-Torres, S.; Salemi, A.; Maldonado-Hódar, F.J.; Pastrana-Martinez, L.M. Photocatalytic perfomance of ZnO-graphene oxide composites towards the degradation of vanillic acid under solar radiation and visible-LED. *Nanomaterials* **2021**, *11*, 1576. [CrossRef]
119. Tank, R.; Seling, T.R.; Katzbaer, R.R.; Thompson, K.L.; Aksoy, S.E.; Chitara, B.; Shringi, A.K.; Schaak, R.E.; Riaz, U.; Yan, F. Transition metal-doped CuO nanosheets for enhanced visible-light photocatalysis. *J. Photoch. Photobio A* **2024**, *448*, 115356. [CrossRef]
120. Tamam, N.; Aadil, M.; Hassanc, W.; Ejaz, S.R.; Najm, Z.M.; Alsafari, I.A.; Aman, S.; Trukhanov, A.V.; Al-Buriahi, M.S.; Boukhris, I. Surfactant assisted synthesis of nanostructured Mn-doped CuO: An efficient photocatalyst for environmental remediation. *Ceram. Int.* **2022**, *48*, 29589–29600. [CrossRef]
121. Bano, K.; Kaushal, S.; Lal, B.; Joshi, S.K.; Kumar, R.; Singh, P.P. Fabrication of CuO/ZnO heterojunction photocatalyst for efficient photocatalytic degradation of tetracycline and ciprofloxacin under direct sun light. *Environ. Nanotechnol. Monit. Manag.* **2023**, *20*, 100863. [CrossRef]
122. Ul Rahman, Z.; Shah, U.; Alam, A.; Shah, Z.; Shaheen, K.; Bahadar Khan, S.; Ali Khan, S. Photocatalytic degradation of cefixime using CuO-NiO nanocomposite photocatalyst. *Inorg. Chem. Commun.* **2023**, *148*, 110312. [CrossRef]
123. Muhambihai, P.; Rama, V.; Subramaniam, P. Photocatalytic degradation of aniline blue, brilliant green and direct red 80 using NiO/CuO, CuO/ZnO and ZnO/NiO nanocomposites. *Environ. Nanotechnol. Monit. Manag.* **2020**, *14*, 100360. [CrossRef]
124. Sagadevan, S.; Lett, J.A.; Weldegebrieal, G.K.; Garg, S.; Oh, W.-C.; Hamizi, N.A.; Johan, M.R. Enhanced photocatalytic activity of rGO-CuO nanocomposites for the degradation of organic pollutants. *Catalysts* **2021**, *11*, 1008. [CrossRef]
125. Sohrabnezhad, S.; Takas, M.E. Synthesis and characterization of porous clay heterostructure intercalated with CuO nanoparticles as a visible light-driven photocatalyst. *J. Iran. Chem. Soc.* **2019**, *16*, 45–55. [CrossRef]
126. Yang, C.; Xu, H.; Shi, J.; Liu, Z.; Zhao, L. Preparation and photocatalysis of CuO/Bentonite based on adsorption and photocatalytic activity. *Materials* **2021**, *14*, 5803. [CrossRef]
127. Chellapandi, T.; Madhumitha, G.; Avinash, J. Ultrasonication-assisted synthesis of CuO-decorated montmorillonite K30 nanocomposites for photocatalytic removal of emerging contaminants: A response surface methodology approach. *Environ. Res.* **2024**, *259*, 119574. [CrossRef]
128. Han, S.; Hu, L.; Liang, Z.; Wageh, S.; Al-Ghamdi, A.A.; Chen, Y.; Fang, X. One-step hydrothermal synthesis of 2D hexagonal nanoplates of α-Fe_2O_3/graphene composites with enhanced photocatalytic activity. *Adv. Funct. Mater.* **2014**, *36*, 5719–5727. [CrossRef]
129. Choi, Y.W.; Lee, H.; Song, Y.; Sohn, D. Colloidal stability of iron oxide nanoparticles with multivalent polymer surfactants. *J. Colloid. Interf. Sci.* **2015**, *443*, 8–12. [CrossRef] [PubMed]
130. Pavlović, J.; Šuligoj, A.; Opresnik, M.; Tušar, N.N.; Logar, N.Z.; Rajić, N. Studies of clinoptilolite-rich zeolitic tuffs from different regions and their activity in photodegradation of methylene blue. *Catalysts* **2022**, *12*, 224. [CrossRef]
131. Huang-Mu, L.; Devanesan, S.; Farhat, K.; Kim, W.; Sivarasan, G. Improving the efficiency of metal ions doped Fe_2O_3 nanoparticles: Photocatalyst for removal of organic dye from aqueous media. *Chemosphere* **2023**, *337*, 139229. [CrossRef]
132. Li, N.; Jiang, Y.; He, Y.; Gao, L.; Yi, Z.Z.; Zhai, F.; Chattopadhyay, K. Sustainable synthesis of multiple-metal-doped Fe_2O_3 nanoparticles with enhanced photocatalytic performance from Fe-bearing dust. *J. Mater. Res. Technol.* **2021**, *15*, 810–820. [CrossRef]
133. Suresh, R.; Giribabu, K.; Manigandan, R.; Mangalaraja, R.V.; Solorza, J.Y.; Stephen, A.; Narayanan, V. Synthesis of Co^{2+}-doped Fe_2O_3 photocatalyst for degradation of pararosaniline dye. *Solid State Sci.* **2017**, *68*, 39–46. [CrossRef]
134. Yadav, P.; Dhariwal, N.; Kumari, M.; Kumar, V.; Thakur, O.P. Enhanced degradation of Congo-red dye by Cr^{3+} doped α-Fe_2O_3 nano-particles under sunlight and industrial wastewater treatment. *Chemosphere* **2023**, *343*, 140208. [CrossRef]
135. Ramprasath, R.; Arul Pragasan, L.; Manikandan, V.; Sudha, S.; Cholan, S.; Alarfaj, A.A.; Hirad, A.H.; Gokul, B.; Sampath, S. Visible light photocatalytic and magnetic properties of V doped α-Fe_2O_3 (VFO) nanoparticles synthesized by polyol assisted hydrothermal method. *Chemosphere* **2022**, *307*, 135575. [CrossRef]
136. Huang, W.; Lu, X.; Jia, D.; Huang, J.; Li, Z.; Xie, H.; Wang, M.; Li, Y.; Zhang, D. Characterization of structural, optical and photocatalytic properties of yttrium modified hematite (α-Fe_2O_3) nanocatalyst. *Ceram. Int.* **2023**, *49*, 25602–25611. [CrossRef]
137. Bhattacharya, L.; Cohen, A.; Caspary Toroker, M. Theoretical understanding of graphene supported hematite photoanode for solar-driven water splitting applications. *Catal. Today* **2024**, *442*, 114912. [CrossRef]
138. Zia, J.; Riaz, U. Microwave-assisted degradation of paracetamol drug using polythiophene-sensitized Ag–Ag_2O heterogeneous photocatalyst derived from plant extract. *ACS Omega* **2020**, *5*, 16386–16394. [CrossRef]
139. Jiang, W.; Fu, H.; Zhu, Y.; Yue, H.; Yuan, S.; Liang, B. Floatable superhydrophobic Ag_2O photocatalyst without a modifier and its controllable wettability by particle size adjustment. *Nanoscale* **2018**, *10*, 13661–13672. [CrossRef] [PubMed]
140. Naing, H.H.; Wang, K.; Tun, P.P.; Zhang, G. Enhanced broad spectrum (vis-NIR) responsive photocatalytic performance of Ag_2O/rectorite nanoarchitectures. *Appl. Surf. Sci.* **2019**, *491*, 216–224. [CrossRef]
141. Kayed, K. The luminescence properties of individual silver nanoparticles in Ag/Ag_2O composites synthesized by oxygen plasma treatment of silver thin films. *J. Lumin.* **2021**, *237*, 118163. [CrossRef]

142. Kong, Y.; Li, Y.; Cui, X.; Su, L.; Ma, D.; Lai, T.; Yao, L.; Xiao, X.; Wang, Y. SnO_2 nanostructured materials used as gas sensors for the detection of hazardous and flammable gases: A review. *Nano Mater. Sci.* **2022**, *4*, 339–350. [CrossRef]
143. Šuligoj, A.; Pavlović, J.; Arčon, I.; Rajić, N.; Novak Tušar, N. SnO_2-containing clinoptilolite as a composite photocatalyst for dyes removal from wastewater under solar light. *Catalysts* **2020**, *10*, 253. [CrossRef]
144. Mahanta, R.; Chetri, P.; Bora, D. Investigation of visible light photocatalysis effect of SnO_2 nanoparticles. *Mater. Today Proc.* **2023**, *in press*. [CrossRef]
145. Roy, H.; Ur Rahman, T.; Riad Khan, M.A.J.; Al-Mamun, M.R.; Islam, S.Z.; Khaleque, M.A.; Hossain, M.I.; Hossain Khan, M.Z.; Islam, M.S.; Marwani, H.M.; et al. Toxic dye removal, remediation, and mechanism with doped SnO_2-based nanocomposite photocatalysts: A critical review. *J. Water Process. Eng.* **2023**, *54*, 104069. [CrossRef]
146. Huong, V.H.; Nguyen, V.C.; Pham, K.P.; Nguyen, T.B.; Ngac, A.B.; Loan, T.T. Construction dual active sites on SnO_2 via Fe doping for effective ciprofloxacin photodegradation. *J. Alloy. Compd.* **2024**, *1005*, 176020. [CrossRef]
147. Shabna, S.; Sahaya Jude Dhas, S.; Biju, C.S. Potential progress in SnO_2 nanostructures for enhancing photocatalytic degradation of organic pollutants. *Catal. Commun.* **2023**, *177*, 106642. [CrossRef]
148. Hou, L.r.; Yuan, C.Z.; Peng, Z. Synthesis and photocatalytic property of SnO_2/TiO_2 nanotubes composites. *J. Hazard. Mater.* **2007**, *139*, 310–315. [CrossRef]
149. Kango, S.; Kalia, S.; Celli, A.; Njuguna, J.; Habibi, Y.; Kumar, R. Surface modification of inorganic nanoparticles for development of organic–inorganic nanocomposites—A review. *Prog. Polym. Sci.* **2013**, *38*, 1232–1261. [CrossRef]
150. Gu, X.; Lin, S.; Qi, K.; Yan, Y.; Li, R.; Popkov, V.; Almjasheva, O. Application of tungsten oxide and its composites in photocatalysis. *Sep. Purif. Technol.* **2024**, *345*, 127299. [CrossRef]
151. Zhao, Z.G.; Liu, Z.F.; Miyauchi, M. Nature-inspired construction, characterization, and photocatalytic properties of single-crystalline tungsten oxide octahedra. *Chem. Commun.* **2010**, *46*, 3321–3323. [CrossRef]
152. Chandrasekaran, S.; Zhang, P.; Peng, F.; Bowen, C.; Huo, J.; Deng, L. Tailoring the geometric and electronic structure of tungsten oxide with manganese or vanadium doping toward highly efficient electrochemical and photoelectrochemical water splitting. *J. Matter. Chem. A* **2019**, *7*, 6161–6172. [CrossRef]
153. Jiang, Z.; Liu, X.; Zhang, X.; Cui, M.; Shi, M. Efficient cleavage of C-C bond in lignin adopting Z-scheme heterojunction photocatalyst by $g\text{-}C_3N_4/CQDs/WO_3$. *J. Environ. Chem. Eng.* **2024**, *12*, 113898. [CrossRef]
154. Lu, J.; Wang, Y.; Liu, F.; Zhang, L.; Chai, S. Fabrication of a direct Z-scheme type WO_3/Ag_3PO_4 composite photocatalyst with enhanced visible-light photocatalytic performances. *Appl. Surf. Sci.* **2017**, *393*, 180–190. [CrossRef]
155. Ke, J.; Zhou, H.; Liu, J.; Zhang, Z.; Duan, X.; Wang, S. Enhanced light-driven water splitting by fast electron transfer in 2D/2D reduced graphene oxide/tungsten trioxide heterojunction with preferential facets. *J. Colloid Interf. Sci.* **2019**, *555*, 413–422. [CrossRef]
156. Alam, R.; Hossain, M.A.; Elias, M. Amine functionalized graphene oxide decorated with ZnO-WO_3 nanocomposites for remediation of organic dye from wastewater. *J. Phys. Chem. Solids* **2024**, *190*, 112015. [CrossRef]
157. Irani, M.; Mohammadi, T.; Mohebbi, S. Photocatalytic degradation of methylene blue with ZnO nanoparticles; a joint experimental and theoretical study. *J. Mex. Chem. Soc.* **2016**, *60*, 218–225. [CrossRef]
158. Morales, G.V.; Sham, E.L.; Cornejo, R.; Farfan Torres, M.E. Photocatalytic degradation of 2-chlorophenol by TiO_2: Kinetic studies. *Lat. Am. Appl. Res.* **2013**, *43*, 325–328.
159. Basha, S.; Barr, C.; Keane, D.; Nolan, K.; Morrissey, A.; Oelgemöller, M.; Tobin, J.M. On the adsorption/photodegradation of amoxicillin in aqueous solutions by an integrated photocatalytic adsorbent (IPCA): Experimental studies and kinetics analysis. *Photochem. Photobiol. Sci.* **2011**, *10*, 1014–1022. [CrossRef]
160. Tran, H.D.; Nguyen, D.Q.; Do, P.T.; Tran, U.N.P. Kinetics of photocatalytic degradation of organic compounds: A mini-review and new approach. *RSC Adv.* **2023**, *13*, 16915–16925. [CrossRef]
161. Liu, B.; Zhao, X.; Terashima, C.; Fujishima, A.; Nakata, K. Thermodynamic and kinetic analysis of heterogeneous photocatalysis for semiconductor systems. *Phys. Chem. Chem. Phys.* **2014**, *16*, 8751–8760. [CrossRef]
162. Gharbani, P.; Mehrizad, A.; Mosavi, S.A. Optimization, kinetics and thermodynamics studies for photocatalytic degradation of methylene blue using cadmium selenide nanoparticles. *Npj Clean Water* **2022**, *5*, 34. [CrossRef]
163. Kanwal, A.; Shahzadi, T.; Riaz, T.; Zaib, M.; Khan, S.; Habila, M.A.; Sillanpaa, M. Photocatalytic degradation studies of organic dyes over novel Cu/Ni loaded reduced graphene oxide hybrid nanocomposite: Adsorption, kinetics and thermodynamic studies. *Molecules* **2023**, *28*, 6474. [CrossRef]
164. Fatima, S.; Iqbal, M.; Bhatti, H.N.; Alwadai, N.; Huwayz, M.A.; Nazir, A.; Igbal, M. Kinetics and thermodynamics studies of nickel manganite nanoparticle as photocatalyst and fuel additive. *Heliyon* **2022**, *10*, e33861. [CrossRef]

catalysts

MDPI

Review

Review on Recent Advances in the Removal of Organic Drugs by Advanced Oxidation Processes

Muhammad Umair, Tayyaba Kanwal, Vittorio Loddo, Leonardo Palmisano and Marianna Bellardita *

Engineering Department, University of Palermo, Viale dell Scienze Ed. 6, 90128 Palermo, Italy; muhammad.umair01@unipa.it (M.U.); tayyaba.kanwal277@gmail.com (T.K.); vittorio.loddo@unipa.it (V.L.); leonardo.palmisano@unipa.it (L.P.)

* Correspondence: marianna.bellardita@unipa.it

Abstract: In recent years, due to the high consumption of drugs both for human needs and for their growing use, especially as regards antibiotics, in the diet of livestock, water pollution has reached very high levels and attracted widespread attention. Drugs have a stable chemical structure and are recalcitrant to many treatments, especially biological ones. Among the methods that have shown high efficiency are advanced oxidation processes (AOPs) which are, among other things, inexpensive and eco-friendly. AOPs are based on the production of reactive oxygen species (ROS) able to degrade organic pollutants in wastewater. The main problem related to the degradation of drugs is their partial oxidation to compounds that are often more harmful than their precursors. In this review, which is not intended to be exhaustive, we provide an overview of recent advances in the removal of organic drugs via advanced oxidation processes (AOPs). The salient points of each process, highlighting advantages and disadvantages, have been summarized. In particular, the use of AOPs such as UV, ozone, Fenton-based AOPs and heterogeneous photocatalysis in the removal of some of the most common drugs (tetracycline, ibuprofen, oxytetracycline, lincomycin) has been reported.

Keywords: drug degradation; advanced oxidation processes (AOPs); reactive oxygen species (ROS); heterogeneous photocatalysis; Fenton-based AOPs; sulfate radical-based AOP

1. Introduction

Water is a fundamental need of human beings, and its main sources are rivers, lakes, aquifers and the desalination of seawater which, however, is exploited in a more limited way. With population growth and industrialization, almost all water sources are contaminated mainly by agricultural and industrial waste. As a result, one of the biggest problems facing humanity in the 21st century may be the sustainable use of water. For the remediation of water pollution, scientific society is attracted to the development of sustainable and green technologies [1].

Due to the rapidly growing demand for various products to treat humans and animals, pharmaceutical companies are expanding significantly. Molecules of pharmacological importance present in drugs are widespread in the environment [2,3]. In seawater, lakes, rivers, surface waters, urban wastewater and drinking water, their concentrations have been found to range between ng and μg per liter [4–6]. Furthermore, it must be kept in mind that, due to their presence, synergistic interactions can occur in these water systems which generally significantly increase ecotoxicity [7].

In this context, drug metabolites are not completely removed, which can increase the concentration of drugs in wastewater from treatment plants [8]. Drugs are difficult to remove from water systems through traditional wastewater treatment methods due to their particularly stable structure which hinders their complete degradation and their high hydrophilic properties [9,10]. Traditional methods such as biological treatment [11], adsorption [12], nanofiltration [13] and membrane bioreactors [14] have been used for their removal, but these methods are often not very efficient and/or give rise to a transfer of the pollutant rather than its total abatement. Therefore, alternative technologies have been developed.

Among the different methods for drug removal, advanced oxidation processes (AOPs) can be considered economical, flexible and highly efficient methods for destroying persistent organic molecules [15,16]. AOPs are generally known for in situ production of strongly oxidizing species in sufficient quantity and low selectivity such as hydroxyl radicals ($HO^\bullet$), O_3, H_2O_2 and superoxide anion radicals ($O_2^{\bullet-}$). These species almost always cause complete mineralization to CO_2, H_2O and inorganic compounds of the attacked molecule and, under certain operating conditions, could be preferred from an environmental point of view [17]. For example, they are highly effective in providing clean drinking water free of organic and inorganic substances and microorganisms [18]. The main advantage of AOPs compared to other available methods, however, is that they are completely ecological not only because they do not involve the removal of pollutants from one place to another (think for example of the precipitation of pollutants by chemical substances or their adsorption) but also because they do not produce large quantities of harmful waste [19,20].

Research progress on AOPs has increased significantly over the last 30 years, mainly due to the availability of a significant variety of technologies and numerous application areas. Among the main AOPs, we can mention ozonation, electrolysis, ultrasound, the use of Fenton reagents or various types of membranes, UV-based processes and heterogeneous photocatalysis using near-ultraviolet (UV) or visible light irradiation [21]. Less common but developing methods involve ionizing radiation, microwaves [22] and ferrate reagents. AOPs have been used for a variety of purposes, including odor control, groundwater purification, soil remediation and volatile organic compound treatment, but wastewater treatment is by far the most frequently studied and developed [23,24]. The application of AOPs, however, must be carefully evaluated considering their overall sustainability, chemical input, energy use and feasibility in real systems, comparing their effectiveness and cost with other traditional processes [25].

AOPs can be used alone or in combination with other biological and physico-chemical processes, depending on the properties of the wastewater to be treated and the purposes of the treatment. In principle, process coupling is advantageous because it generally increases the efficiency of the treatment. For example, AOPs could be used as a pre-treatment step to transform originally bio-recalcitrant molecules into other more easily biodegradable species, which would then be subjected to biological post-treatment. However, for drugs containing biodegradable substances, biological pre-treatment following chemical post-treatment might also be desirable since, even if biodegradable substances can be easily eliminated initially, the effectiveness of biodegradation is not comparable to that of a chemical oxidizing treatment [26].

All AOPs involve the in-situ production of the main oxidant species and the consequent combination of these species with contaminants. Reactor design and drug composition have an impact on the formation of reactive species, which are mainly radicals. In addition to radical scavenging, other factors including hydrodynamics and mass transfer of radicals are crucial for effectively destroying drugs [27]. Different types of AOPs are illustrated in Figure 1.

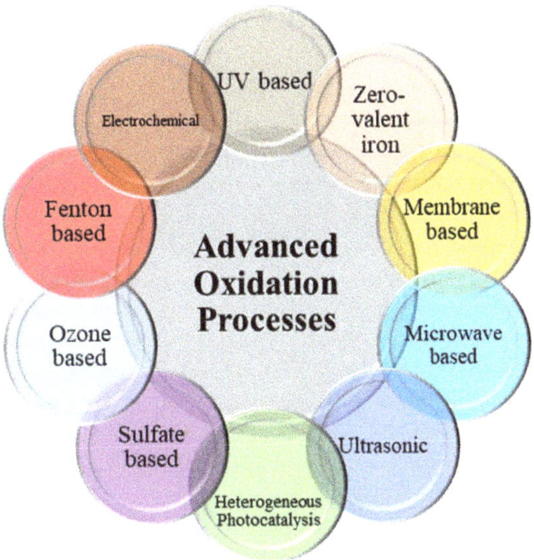

Figure 1. Different advanced oxidation processes.

1.1. UV-Based AOPs

UV-based AOPs are those processes that use ultraviolet (UV) light alone or in the presence of radical promoters for the degradation of organic compounds. Radical-based UV AOPs use only UV light to generate oxidant species, but other processes use Cl_2 (to generate radicals of the chlorine species and hydroxyl radicals), ozone, H_2O_2 and persulfate (to generate sulfate radicals).

Low-pressure Hg vapor lamps with a partial pressure of approximately 1 Pa are the most common UV radiation sources for UV-based AOPs. The efficiency of these lamps is 25–45% in the range of the emitted wavelength. The emission spectra of low-pressure Hg lamps show two distinct lines at approximately 254 and 185 nm. The line emitted at the wavelength of 254 nm is very useful for disinfection. In fact, UV light inactivates microbes which cause damage to DNA or RNA molecules, preventing their reproduction [28].

1.1.1. UV Irradiation

UV irradiation involves the direct interaction between UV light and a target pollutant and the induction of chemical reactions that can break down the pollutant into intermediate products whose subsequent decomposition eventually provides mineral end-products [29,30]. Traditionally, UV treatment has been used to disinfect drinking water with the benefit of limiting the creation of any regulated waste products of disinfection [31].

The homolysis (Equation (1)) and photochemical ionization (Equation (2)) of water are generated through UV light absorption:

$$H_2O + hv(185\ nm) \rightarrow HO^{\bullet} + H^{\bullet} \quad (1)$$

$$H_2O + hv(185\ nm) \rightarrow HO^{\bullet} + H^{+} + e^{-} \quad (2)$$

The main advantages of the AOP with UV irradiation are that it requires a relatively short time to treat, and the use of chemicals is not necessary, ensuring that no residual substances are produced during the process [1]. The degradation efficiency of drugs under UV light in water depends on several factors. Different types of buffer solutions can be used to modify the pH, but this generally has an important impact on the degradation of drugs as free radicals may be formed which are linked to the species constituting the buffer and interfere with the degradation mechanism [32].

1.1.2. UV/H_2O_2

UV/H_2O_2 is the most used AOP for the degradation of drugs. H_2O_2 present in solution gives rise to the production of two $HO^\bullet$ radical species through the photolytic cleavage of the O-O bond (initiation step) (Equation (3)).

$$H_2O_2 + hv \rightarrow 2HO^\bullet \quad (3)$$

After the formation of two hydroxyl radicals, a chain of reactions is formed (propagation and termination steps) (Equations (4)–(9)) [33].

$$HO^\bullet + H_2O_2 \rightarrow H_2O + HO^\bullet{}_2 \quad (4)$$

$$HO^\bullet + HO_2{}^- \rightarrow HO^\bullet{}_2 + OH^- \quad (5)$$

$$HO^\bullet{}_2 + H_2O_2 \rightarrow HO^\bullet + H_2O + O_2 \quad (6)$$

$$2HO^\bullet{}_2 \rightarrow H_2O_2 + O_2 \quad (7)$$

$$HO^\bullet + HO^\bullet{}_2 \rightarrow H_2O + O_2 \quad (8)$$

$$2HO^\bullet \rightarrow H_2O_2 \quad (9)$$

Some factors, i.e., pH, H_2O_2 concentration, organic compound structure, $HO^\bullet$ formation rate and water contents, influence the efficiency of UV/H_2O_2 AOP. At alkaline pHs, the absorption of CO_2 from the air increases, so the reaction should be carried out in a closed vessel to counteract this effect. The decrease in pH has a direct effect on the concentration of carbonate and bicarbonate ions resulting from the absorption of CO_2; therefore, the effectiveness of the process can increase as the amount of $HO^\bullet$ radicals in the solution increases [1].

1.1.3. UV/Chlorine

The UV/chlorine process is an interesting AOP because the chlorine (Cl_2) used in water is a common disinfectant and UV-activated chlorine radicals ($Cl^\bullet$) are formed. Chlorine dioxide (ClO_2) and the hypochlorite radical ($ClO^\bullet$) are the two main oxidizing species [34]. This method involves the addition of a sodium salt (Na^+ + ClO^-) to an aqueous solution, and ClO^- is also present as the protonated form HClO (pK_a = 7.52), depending on the pH values. This system involving an acid–base equilibrium is known as active chlorine (AC) [35]. The chlorine radical prefers to react with electron-rich molecules, so it is a more selective oxidant than the hydroxyl radical [36]. This method is generally useful for the treatment of wastewater with low pH values [37]. Indeed, pH significantly influences the molar absorption coefficient as the ratio between HOCl and ClO^- can change significantly.

Below are the main equations (Equations (10)–(12)) representing the process. Under UV light irradiation of an aqueous HClO solution, $HO^\bullet$ and $Cl^\bullet$ radicals are obtained [38,39] (Equation (10)). These highly reactive radicals subsequently interact with HClO to form the chlorine monoxide radical ($ClO^\bullet$) (Equations (11) and (12)).

$$HClO + hv \rightarrow HO^\bullet + Cl^\bullet \quad (10)$$

$$HO^\bullet + HClO \rightarrow ClO^\bullet + H_2O \quad (11)$$

$$Cl^\bullet + HClO \rightarrow ClO^\bullet + Cl^- + H^+ \quad (12)$$

Various types of active species such as $HO^{\bullet}$, AC, $Cl^{\bullet}$ and $ClO^{\bullet}$ coexist and often act in a complementary manner for efficient degradation of the pollutant. $HO^{\bullet}$, which is a selective oxidant, reacts at approximately the same rate with the organic species present [40]. $Cl^{\bullet}$, which is more selective, reacts with electron-rich organic components via H-abstraction, one-electron oxidation and addition to unsaturated C-C bonds. $Cl^{\bullet}$ is very reactive towards benzoic acid and phenol, compared to the $HO^{\bullet}$ radical [41]. In conclusion, the UV/chlorine AOP is more efficient than the UV/H_2O_2 AOP for the removal of drugs such as tolytriazole, iopamidole and benzotriazole [42].

1.1.4. UV/O_3

Ozone (O_3) in combination with UV light irradiation increases the concentration of $HO^{\bullet}$ radicals, improving drug removal efficiency. During the reaction, the by-product H_2O_2 is formed which, however, can in turn decompose into two $HO^{\bullet}$ radicals [43,44]. Equations (13) and (14) summarize these reactions:

$$O_3 + H_2O + hv \rightarrow O_2 + 2HO^{\bullet} \tag{13}$$

$$2HO^{\bullet} \rightarrow H_2O_2 \tag{14}$$

O_3 at extremely high concentrations can act as a radical scavenger and interact with them or give rise to secondary reactions by decomposing and inhibiting the oxidation process. It can react directly and electrophilically with the organic compounds to be broken down or indirectly through a radical reaction. However, the main reaction in the UV/O_3 system is the interaction of $HO^{\bullet}$ with the organic pollutant because the direct oxidation rate with molecular O_3 is slower [45].

1.1.5. UV/$SO_4{}^{\bullet -}$

$SO_4{}^{\bullet -}$ is a strong monoelectronic oxidant that shows higher degradation efficiency than $HO^{\bullet}$ under neutral and alkaline conditions due to its higher redox potential and longer lifetime [46]. These electrophilic radicals can rapidly oxidize some aromatic compounds through the abstraction of hydrogen atoms and the transfer of single electrons [47]. Compared to $HO^{\bullet}$, $SO_4{}^{\bullet -}$ can easily promote electron transfer but at a slower rate than the extraction and addition of hydrogen atoms [48]. pH is a crucial factor in oxidation in the presence of $SO_4{}^{\bullet -}$ for drug degradation. In fact, the production of $HO^{\bullet}$ and sulfate radicals increases with an increase in the pH, but it must be considered that the increase in these radicals involves their interaction with OH^- and causes an overall decrease in the reaction rate [49].

Sulfate radicals can be produced using peroxydisulfate (PS) or peroxymonosulfate (PMS). The radicals, in particular, can be generated from PS through its homolytic cleavage by UVC radiation (Equation (15)), in the presence of a photocatalyst which provides the photoproduced electron (Equation (16)) or heat (that can be also generated by microwaves) (Equation (17)).

$$S_2O_8{}^{2-} + hv \rightarrow 2SO_4{}^{\bullet -} \tag{15}$$

$$S_2O_8{}^{2-} + e^- \rightarrow SO_4{}^{2-} + SO_4{}^{\bullet -} \tag{16}$$

$$S_2O_8{}^{2-} + \Delta \rightarrow 2SO_4{}^{\bullet -} \qquad E_a = 33.5\ \text{kcal/mol} \tag{17}$$

1.1.6. Advantages and Disadvantages of UV-Based AOPs

In Table 1 the main advantages and disadvantages of UV-based AOPs are reported.

Table 1. Advantages and disadvantages of UV-based AOPs.

Process	Advantages	Disadvantages	References
UV	Absence of limitation of mass transfer Disinfection No bromate formation No need for off-gas treatment Potential to use sunlight	Cost- and energy-intensive Fouling of UV lamps UV light penetration decreases in presence of iron and nitrate Interference with chemical compounds	[50]
UV/H_2O_2	High stability of H_2O_2 Long-time storage ability Availability for drinking water treatment on full scale	UV light penetration is impacted by turbidity Special reactors are needed for UV light Residual H_2O_2 must be considered	[51]
UV/chlorine	$Cl^{\bullet}$ radical is a more selective oxidant than $HO^{\bullet}$ radical More efficient than UV/H_2O_2 Additional chlorine to quench residual H_2O_2 is not needed Cost-effective Favorable at low pH	Impact on efficiency of UV light due to suspended particles	[52] [53] [1]
UV/O_3	O_3 absorbs more UV light as compared to H_2O_2 The presence of UV light has a disinfectant effect The residual oxidant quickly deteriorates	Intensive energy required Cost-intensive Special reactors required Stripping of volatile compounds Blocking of UV light penetration due to turbidity Mass transfer limitation due to diffusion	[51]
$UV/SO_4^{\bullet-}$	Efficient at moderate pH conditions Efficient electron transfer reaction mechanism High standard reduction potential	Toxicity of by-products Presence of unreacted chemicals Metal contamination	[54] [55]

1.2. Ozone-Based AOPs

1.2.1. O_3/H_2O_2

The combination of O_3 with H_2O_2 is an efficient method for the degradation of organic drugs. This method is also known as peroxone in which the decomposition of O_3 takes place after the production of HO_2^- from H_2O_2 (Equations (18) and (19)) [56].

$$H_2O_2 \rightarrow HO_2^- + H^+ \tag{18}$$

$$HO_2^- + O_3 \rightarrow HO^{\bullet}{}_2 + O_3^{\bullet-} \tag{19}$$

It is important that neither an excessive quantity nor a too-low quantity of H_2O_2 is used because an excessive quantity can cause a decrease in $HO^{\bullet}$ species with the formation of $HO^{\bullet}{}_2$ (Equation (20)) and a too-low quantity can be insufficient for oxidizing organic drugs while achieving dissociation of the H_2O_2 molecule (Equation (21)):

$$HO^{\bullet} + H_2O_2 \rightarrow HO^{\bullet}{}_2 + H_2O \tag{20}$$

$$H_2O_2 \rightarrow H_2O + \frac{1}{2}O_2 \tag{21}$$

1.2.2. Catalytic Ozonation

In this process, catalysts react with O_3, increasing the degradation rate of drugs by generating $HO^{\bullet}$ radicals following the decomposition of O_3 [57]. Homogeneous catalytic ozonation involves the use of different transition metal ions (Mn^{2+}, Ni^{2+}, Co^{2+}, Cr^{2+},

Ag^{2+}, Zn^{2+}, Cd^{2+}, Fe^{2+}, Cu^{2+}) which function as catalysts for the degradation of organic drugs [58].

Ozone decomposition performance improves in the presence of highly stable heterogeneous catalysts which consequently can be recycled and reused without prior further treatment after the first use. Because of these advantages, heterogeneous catalytic ozonation is often used to treat aqueous effluents. In addition to the nature of the catalyst, in particular, its surface chemical–physical characteristics, and the pH of the solution, which influence the properties of the surface catalytic sites, the degradation processes of ozone in water is extremely important for the performance of catalytic ozonation [59]. The main heterogeneous catalysts used in coupling with O_3 are TiO_2 [60], MgO [61], MnO_2 [62], ZnO [63], SiO_2 [64] and $CuFe_2O_4$ [65].

1.2.3. Electro-Peroxone

In recent years, researchers have been attracted to electrochemical methods for water treatment. Advantages over traditional methods are, for example, the possibility of obtaining chemicals on site and often easy maintenance and reliable performance [66].

The electro-peroxone (E-peroxone) process is a new AOP that is a combination of traditional electrolysis and ozonation. H_2O_2 is produced in situ inside an electrolytic device which allows its quantity in the reaction medium to be controlled. This avoids the transport, management and storage of this chemical product which is dangerous and explosive [35]. The polluting drugs to be treated are present in a reactor equipped with a carbon-based cathode which converts the ozone (in reality it is a mixture of O_3 and O_2) coming from the generator and present in the effluent into H_2O_2 through an electrochemical reaction (Equation (22)). The H_2O_2 formed reacts with O_3 to produce $HO^\bullet$ radicals which play a major role in drug degradation (Equation (23)) [67–70].

$$O_2 + 2H^+ + 2e^- \rightarrow H_2O_2 \quad (22)$$

$$H_2O_2 + O_3 \rightarrow O_2 + HO^\bullet{}_2 + HO^\bullet \quad (23)$$

1.2.4. Advantages and Disadvantages of Ozone-Based AOPs

In Table 2 the main advantages and disadvantages of ozone-based AOPs are reported.

Table 2. Advantages and disadvantages of ozone-based AOPs.

Process	Advantages	Disadvantages	References
O_3/H_2O_2	Highly effective Highly efficient Handling of remediation Disinfection	Bromate formation Energy- and cost-intensive Excess H_2O_2 may need to be managed due to potential microbial growth The concentration of O_3-H_2O_2 must be properly controlled	[51]
Catalytic ozonation	Low operating cost No need for pH adjustment Complete mineralization Improved O_3 utilization efficiency Enhanced reaction kinetics Production of $HO^\bullet$ radicals at low pH	Challenge for selection of green, cost-effective and efficient catalysts Complex synthesis of ozone catalysts Reuse of catalysts Fouling of catalysts Challenge of residual toxicity	[71] [72]
Electro-peroxone	Economical, convenient and safe method Low sludge formation Highly efficient for low reactive ozone species No secondary pollution On-site production of H_2O_2	Energy-intensive Lower current efficiency Not particularly efficient for pesticide degradation	[73] [74] [75] [76]

1.3. Fenton-Based AOPs

1.3.1. Fenton-like Process

The Fenton process involves the formation of $HO^{\bullet}$ radicals through a series of reactions of H_2O_2 with Fe(II) salts. The reagent system is called the Fenton reagent [77,78] (Equations (24) and (25)).

$$Fe^{2+} + H_2O_2 \rightarrow Fe^{3+} + HO^{\bullet} + OH^{-} \quad \left(k = 63\text{–}76\ M^{-1}\ s^{-1}\right) \tag{24}$$

$$Fe^{3+} + H_2O_2 \rightarrow Fe^{2+} + HO_2^{\bullet} + H^{+} \quad \left(k = 0.001\text{–}0.01\ M^{-1}\ s^{-1}\right) \tag{25}$$

A hydrogen atom of the organic drug to be treated (R-H) is extracted by the $HO^{\bullet}$ radicals to produce an organic radical ($R^{\bullet}$) which gives rise to a chain reaction up to the final oxidation products (Equations (26)–(28)). H_2O_2 and Fe^{2+} should oxidize the drug to H_2O and CO_2 (and to inorganic species deriving from the possible presence of heteroatoms in the molecule) even without the presence of $HO^{\bullet}/R^{\bullet}$, radicals, obviously with different kinetics. However, it must be considered that the $HO^{\bullet}$ radicals produced can be further involved in subsequent reactions which have a negative effect on the oxidation reactions (Equations (29)–(31)):

$$RH + HO^{\bullet} \rightarrow H_2O + R^{\bullet} \tag{26}$$

$$R^{\bullet} + H_2O_2 \rightarrow ROH + HO^{\bullet} \tag{27}$$

$$R^{\bullet} + O_2 \rightarrow ROO^{\bullet} \tag{28}$$

$$Fe^{2+} + HO^{\bullet} \rightarrow Fe^{3+} + OH^{-} \quad \left(k = 3.2 \times 10^{8}\ M^{-1}\ s^{-1}\right) \tag{29}$$

$$H_2O_2 + HO^{\bullet} \rightarrow HO_2^{\bullet} + H_2O \quad \left(k = 3.3 \times 10^{7}\ M^{-1}\ s^{-1}\right) \tag{30}$$

$$^{\bullet}OH + HO^{\bullet} \rightarrow H_2O_2 \quad \left(k = 6 \times 10^{9}\ M^{-1}\ s^{-1}\right) \tag{31}$$

The optimal pH at which one must operate with the Fenton reagent must be between 3 and 5 as under neutral and near-neutral pH conditions, Fe^{3+} would react with $HO^{\bullet}$, producing insoluble ferric hydroxide which would separate from the solution. Consequently, the efficiency of the oxidation process would decrease, and continuous addition of Fe^{2+} ions should be necessary.

Finally, it should be noted that there is still a different hypothesis regarding the oxidizing species and their production. Without wanting to go into detail, referring to the specific papers, we only mention the possibility of the involvement of the high-valence ferryl-oxo species Fe(IV), instead of $HO^{\bullet}$ radicals, and in that case, Equation (24) can be replaced with the following Equation (32):

$$Fe^{2+} + H_2O_2 \rightarrow Fe^{IV}O^{2+} + H_2O \tag{32}$$

1.3.2. Photo-Fenton

In the photo-Fenton process (PFP), under UV irradiation, the reaction of H_2O_2 with Fe^{2+} occurs, producing $HO^{\bullet}$ radicals in quantities sufficient for oxidizing drugs. The operating mechanism of this process involves the photochemical regeneration of Fe^{2+} ions through the photoreduction of Fe^{3+} ions (Equation (33)). In particular, when all the Fe^{2+} ions are used in the Fenton reaction, the Fe^{3+} ions start to accumulate in the solution, and the reaction stops. The use of light, however, allows the cycle to continue as the Fe^{2+} ions which are necessary for the reaction with H_2O_2 are photochemically reformed [77,79,80].

$$FeOH^{2+} + hv \rightarrow Fe^{2+} + HO^{\bullet} \tag{33}$$

In the case of post-treatment of raw leachate, it should be noted that the concentration of total dissolved solids (TDSs) and the degree of turbidity significantly influence the performance of UV irradiation [81].

1.3.3. Electro-Fenton

The electro-Fenton process (EFP) is considered an efficient method for the degradation of drugs based on electrocatalytic in situ generation of hydroxyl radicals. It can be conducted in two different setups. In the first case, the ferrous ions are introduced into the reactor from the outside and H_2O_2 is produced at the cathode (Equation (34)). In the second possible configuration, also Fe^{2+} ions are produced in situ using cast iron sacrificial anodes (Equation (35)). EFP has some advantages over the traditional Fenton process; this method, in fact, allows better control of the process and does not require the transport or storage of H_2O_2 [82,83]. Furthermore, the electro-Fenton process is environmentally friendly and does not produce any harmful pollutants [84]. The main weakness of EFP is the operating cost, principally the chemical cost, when used on a practical scale.

$$O_2 + 2H^+ + 2e^- \rightarrow H_2O_2 \qquad \text{Cathode} \tag{34}$$

$$Fe^0 \rightarrow Fe^{2+} + 2e^- \qquad \text{Anode} \tag{35}$$

1.3.4. Photoelectro-Fenton

The photoelectro-Fenton process (PEFP) is a combination of photochemical and electrochemical processes with the Fenton process. This method involves UVA light and the electrochemical production of H_2O_2 which is used for the treatment of wastewater. The photoreduction of Fe(III) occurs, producing a high amount of $HO^\bullet$ radicals and Fe(II) (Equation (36)). Fe(III) can form complexes with some organic compounds present in solution that absorb in the near-UV and visible region. In this way, they can be decarboxylated under irradiation (Equation (37)) [85–87].

$$Fe(OH)^{2+} + hv \rightarrow Fe^{2+} + HO^\bullet \tag{36}$$

$$R(CO_2) - Fe(III) + hv \rightarrow R(^\bullet CO_2) + Fe(II) \rightarrow {}^\bullet R + CO_2 \tag{37}$$

1.3.5. Advantages and Disadvantages of Fenton-Based AOPs

In Table 3 the main advantages and disadvantages of Fenton-based AOPs are reported.

Table 3. Advantages and disadvantages of Fenton-based AOPs.

Process	Advantages	Disadvantages	References
Fenton-like process	Fe^{2+} is non-toxic and widely available No formation of chlorinated products No limitation of mass transfer H_2O_2 is easy to handle Complex formation enhances the coagulation of suspended solids	Sludge may formed Scavenging reactions may occur Regeneration of Fe^{2+} is very low	[88]
Photo-Fenton	High efficiency Wide pH range Low sludge formation	Cost- and energy-intensive Blocking of UV radiation due to turbidity of water Formation of oxalate layers on the surface of lamps Medium- or high-pressure lamps are required	[89] [90]

Table 3. *Cont.*

Process	Advantages	Disadvantages	References
Electro-Fenton	High oxidation efficiency High mineralization On-site production of reagents to generate H_2O_2 Ability to treat effluents with a wide range of concentrations and ease of handling	Cost-intensive Inefficient for treatment with large-scale volumes	[91] [90]
Photoelectro-Fenton	Regeneration of Fe^{2+} Production of $HO^{\bullet}$ radicals High mineralization No need to separate the catalyst Possibility of using solar light	No commercial availability of photoanodes Visible-light-active photoanodes are required Large-scale reactors are required Cost of artificial light irradiation	[92]

1.4. Ultrasonic Methods

In water, ultrasonic radiation that is emitted at values >20 kHz generates $HO^{\bullet}$ radicals that induce the degradation of pollutants by means of the so-called ultrasonic method (US). A process known as acoustic cavitation is used in the US and causes bubbles to form, grow and collapse in liquids due to the extremely high temperatures and pressures created within them [93]. The US is also known as sonolysis (Equation (38)) [94].

$$H_2O + \text{sonication} \rightarrow HO^{\bullet} + H^{\bullet} \quad (38)$$

Sonolysis is a fairly recent method for drug degradation, and therefore, it has not received much attention compared to other AOPs. There are very few publications on it. The degradation of many poorly soluble and very volatile organic drugs occurs rapidly, and consequently, this method could be useful for attacking pharmaceutical micropollutants. The efficiency of the US depends on several factors such as the type of drug, the intensity and frequency of the ultrasound, the temperature and the configuration of the reactor [26].

In Table 4 the main advantages and disadvantages of sonolysis are reported.

Table 4. Advantages and disadvantages of sonolysis.

Process	Advantages	Disadvantages	References
Sonolysis	High degradation efficiency Low energy required No need for chemicals No sludge waste Safe method Penetrability in aqueous medium Economical for small-volume treatments	Probe maintenance is required Turbidity of water Energy-intensive	[94] [95] [96]

1.5. Membrane-Based AOPs (M-AOPs)

Many organic micropollutants are not mineralized completely or to any great extent using a single traditional or advanced oxidation process. To improve degradation efficiency and successfully remove organic micropollutants, the combined use of different methods operating synergistically in hybrid systems has been proposed as an alternative approach [97,98]. Membrane processes such as distillation, nanofiltration and reverse osmosis can be applied for the removal of organic drugs from water or wastewater. However, in separation using membranes, contaminants are simply transferred to a concentrated phase but are not transformed or mineralized. Furthermore, during the filtration process, the membrane can deteriorate due to fouling and obtaining retentate in large quantities [99]. Taking this into account, the coupling of filtration processes and AOPs has recently been proposed as a strategy for the treatment of polluted effluents [51].

In an M-AOP, the AOP has the primary role of degrading the target drug into less hazardous contaminants. The membrane plays a secondary role because it simply allows the passage through it of the less dangerous species obtained after the action of the AOP that, moreover, can prevent fouling and keep the membrane in the best possible operating condition.

Combined M-AOP filtration systems can be classified into different categories, depending on the type of AOP used and how the formation and oxidizing action of $HO^\bullet$ radicals occur in the reaction medium. An interesting proposal involves using the AOP method to oxidize complex drug target molecules and then performing membrane filtration to separate the oxidized products. The type of membrane to be selected depends on the properties of the effluent and the size of the oxidized substrate to be separated, ultimately depending on the type of oxidizing reaction. The ability of membranes to couple with AOPs is still under discussion, considering in particular the polymeric ones that could be sensitive to heat, chemical attacks and irradiation, especially for long operating times [51].

In Table 5 the main advantages and disadvantages of Membrane-Based AOPs are reported.

Table 5. Advantages and disadvantages of membrane-based AOPs.

Process	Advantages	Disadvantages	References
Membrane-based AOP	Membrane captures the unoxidized contaminants and only allows safe, treated water to pass through The concentrated unoxidized contaminants on the membrane surface greatly accelerate their decomposition in the presence of AOPs By ozonating the membrane surfaces, membrane fluxes and permeability can be improved	Fouling of membranes Low stability of membranes (especially some polymeric membranes) for long irradiation times	[51] [100]

1.6. Electrochemical AOPs

Electrochemically based AOPs are called electrochemical advanced oxidation processes (EAOPs) and can be used for wastewater treatment. In EAOPs $HO^\bullet$ radicals are formed directly (anodic oxidation (AO)) or indirectly via the Fenton reagent (electro-Fenton (EF)). In AO, oxidation of water occurs and produces $HO^\bullet$ radicals at an anode having a high O_2 overvoltage (Equation (39)) [101].

$$H_2O \rightarrow HO^\bullet + H^+ + e^- \tag{39}$$

In turn, EF involves the reaction providing $HO^\bullet$ radicals between the Fenton reagent and H_2O_2 which is produced electrochemically at the cathode (Equation (40)) [102].

$$Fe^{2+} + H_2O_2 \rightarrow Fe^{3+} + HO^\bullet + OH^- \tag{40}$$

The type and quantity of reactive species created during an EAOP depend on various factors. The main ones are the composition of the polluted water, the material from which the electrodes are made and the applied potential [103]. The advantages and disadvantages of the method are shown in Table 6.

Table 6. Advantages and disadvantages of electrochemical AOP.

Process	Advantages	Disadvantages	References
Electrochemical AOPs	No need for light radiation Good energy efficiency No chemical required No waste produced Highly efficient as compared to other AOPs Eco-friendly Easy of handling Possibility to treat effluent with COD in the range 0.1–100 g L^{-1}	Cost- and energy-intensive Requires management of sludge-related electrocoagulant and indirect oxidation Limitation of mass transfer Poisoning effect	[51] [96] [104]

1.7. Zero-Valent Iron

Zero-valent iron (ZVI) is a very promising material, abundant, cheap, easy to produce and practically non-toxic. Therefore, it can be successfully used to degrade drugs [105,106]. The electrons that are produced directly by ZVI react with the molecules of the drug to be broken down and transform them into less dangerous contaminants. ZVI in the presence of dissolved oxygen (DO) can oxidize various organic pollutants because ZVI provides two electrons to O_2, producing H_2O_2 (Equation (41)) which can further react with two electrons also provided by ZVI and can transform into H_2O (Equation (42)). The strongly oxidizing $HO^{\bullet}$ radicals produced by the reaction of Fe^{2+} and H_2O_2 (Equation (43)), as in many other methods, are ultimately responsible for the attack on drugs and their degradation [107]:

$$Fe^0 + O_2 + 2H^+ \rightarrow Fe^{2+} + H_2O_2 \quad (41)$$

$$Fe^0 + H_2O_2 + 2H^+ \rightarrow Fe^{2+} + 2H_2O \quad (42)$$

$$Fe^{2+} + H_2O_2 \rightarrow Fe^{3+} + HO^{\bullet} + OH^- \quad (43)$$

In Table 7 the advantages and disadvantages of zero-valent iron are reported.

Table 7. Advantages and disadvantages of zero-valent iron.

Process	Advantages	Disadvantages	References
Zero-valent iron	No sludge formation Complete degradation No formation of undesirable by-products	Low stability Fast passivation Limited mobility	[108] [109]

1.8. Sulfate Radical-Based AOPs

As an alternative to $HO^{\bullet}$ radicals, sulfate radical ($SO_4^{\bullet-}$)-based AOPs have been extensively studied for wastewater treatment. $SO_4^{\bullet-}$ radicals are produced from two strong oxidants, namely persulfate (also called peroxydisulfate (PS)) and peroxymonosulfate (PMS), whose oxidation potentials are 2.1 and 1.82 eV, respectively. Their decomposition is illustrated in Equations (44) and (45) [54,110]:

$$S_2O_8^{2-} + 2e^- \rightarrow 2SO_4^{2-} \quad (44)$$

$$HSO_5^- + 2H^+ + 2e^- \rightarrow HSO_4^- + H_20 \quad (45)$$

Persulfate, in particular, is a promising oxidant as it is quite stable at room temperature and is not very selective for drug degradation. The activation of PS and PMS to produce $SO_4^{\bullet-}$ radicals can occur in various ways using heat, UV light, ultrasound, alkali, transitional ions or metal oxides [111,112].

In Table 8 the advantages and disadvantages of Sulfate-Based AOPs are reported.

Table 8. Advantages and disadvantages of sulfate-based AOPs.

Process	Advantages	Disadvantages	References
Sulfate-based AOPs	Wide pH range (2–8) Less need for reactants High selectivity High redox potential Easy availability of reactant On large scale, safe storage of oxidants	Cost-intensive Challenge of residual sulfate ions Potential to form toxic by-products	[110] [113]

1.9. Microwave-Based AOPs

Electromagnetic radiation within the frequency range of 0.3 to 300 GHz is referred to as microwave (MW) radiation. To prevent interference with cellular phone and telecommunication frequencies, all microwave reactors used for chemical synthesis and home kitchen microwave ovens operate at 2.45 GHz [114]. In the last few years, the use of MW irradiation has been widely studied in environmental applications such as wastewater and sewage treatment, soil remediation and biomedical applications [115–117] due to its rapid and uniform heating, "hot spot" effect and nonthermal effect. Although the use of microwaves alone was found to be not very efficient for drug removal, their combination with other AOPs is very promising because can it allow the mineralization of any organic pollutant [22,118–120]. MW-based AOPs (MW-AOPs) include mainly MW-enhanced photochemistry, MW-enhanced Fenton process, MW-activated persulfate, MW-assisted ultrasonic and MW-assisted ozone [121–124], and in these cases, MW radiation enhances the formation of ROS active species (such as $HO^{\bullet}$, $SO_4^{\bullet -}$, $O_2^{\bullet -}$), improving the process performance. The drawback is that MW-AOPs are currently only used on a laboratory scale, and furthermore, cost analyses have revealed their economic limitations [119].

1.10. Heterogeneous Photocatalysis

The heterogeneous photocatalysis AOP has been extensively studied over the past two decades and applied to drug degradation. This method appears to be more effective than other AOPs since some semiconductors are generally less expensive and non-selective, i.e., they can easily mineralize a large variety of toxic organic molecules [125,126]. The optoelectronic properties of a photocatalyst that behaves as a semiconductor depend on the energy of the conduction band (CB) and that of the valence band (VB). The energy difference between the two bands, called the bandgap, corresponds to the energy difference between the bottom of the conduction band and the top of the valence band. This value is between 1.0 and 4.0 eV in semiconductors, and therefore, thermal or light stimulation can increase the conductivity by transferring electrons from the valence band to the conduction band [1].

In heterogeneous photocatalysis, activation can be caused by irradiating the semiconductor with solar or artificial light. It must be considered that semiconductors and reagents are in two different phases. When a semiconductor is irradiated by light of an appropriate energy, the photoexcitation of electrons from the VB to the CB occurs, with the formation of positive holes in the VB. The photoproduced electrons and holes migrate to the surface of the catalyst and act as reducing and oxidizing agents, respectively. The holes can directly oxidize the species adsorbed on the surface of the photocatalyst or interact with the adsorbed water in the case of aqueous solutions producing $HO^{\bullet}$ radicals. The latter, in turn, attack the polluting molecules found near the surface or adsorbed on nearby sites. The molecules can be attacked in subsequent steps and are transformed into non-toxic products until complete mineralization into H_2O and CO_2 and any inorganic ions containing heteroatoms present in the degraded organic molecule [111]. The photocatalytic mechanism of drug degradation is reported as follows (Equations (46)–(52)) and in Figure 2:

$$\text{Semiconductor} + hv \rightarrow h^+ + e^- \quad (46)$$

$$H_2O + h^+ \rightarrow {}^\bullet HO^\bullet + H^+ \quad (47)$$

$$O_2 + e^- \rightarrow O_2{}^{\bullet -} \quad (48)$$

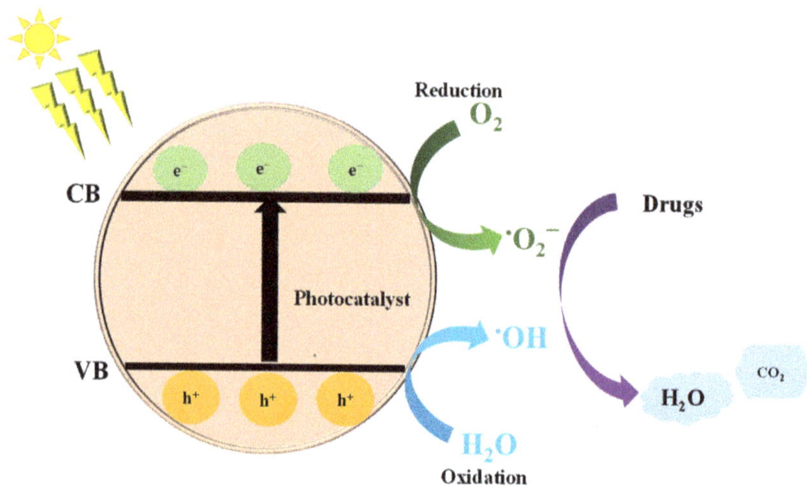

Figure 2. Photocatalytic mechanism of drug degradation.

$O_2{}^{\bullet -}$ and $HO^\bullet$ radicals react with a drug and transform it into other less harmful products.

$$O_2{}^{\bullet -} + H^+ \rightarrow HO^\bullet{}_2 \quad (49)$$

$$2HO^\bullet{}_2{}^\bullet \rightarrow O_2 + H_2O_2 \quad (50)$$

$$H_2O_2 + hv \rightarrow 2HO^\bullet \quad (51)$$

$$HO^\bullet + \text{drug} \rightarrow \text{drug}_{oxi} \quad (52)$$

TiO_2 is the most studied and used semiconductor for drug abatement due to its high chemical and thermal stability, wide availability, non-toxicity, cost-effectiveness and corrosion resistance [126]. However, TiO_2 has limitations due to its wide bandgap (3.0 eV and 3.2 eV in the case of the rutile and anatase polymorphs, respectively) and poor activity under the irradiation of sunlight (which contains only 4–5% of energetically effective photons for its excitation). Consequently, considerable efforts have been made to improve the performance of TiO_2-based photocatalysts for practical applications, mainly through modifications and/or combinations with other semiconductors [127,128].

In Table 9 the advantages and disadvantages of heterogeneous photocatalysis are reported.

Table 9. Advantages and disadvantages of heterogeneous photocatalysis.

Process	Advantages	Disadvantages	References
Heterogeneous photocatalysis	Active under the irradiation of near-UV light, compared to other AOPs which require shorter wavelengths Possibility for the use of sunlight as a clean and free energy source Mild operating conditions Non-toxicity of the catalysts Photochemical stability	Fouling of the catalyst The powdered photocatalyst needs to be recovered when employed as a slurry or suspension High recombination rates of photogenerated electrons and holes Mass transfer limitation Poor efficiency when using low lighting	[51] [95] [96]

1.11. Advanced Reduction Processes (ARPs)

Advanced reduction processes (ARPs) combine both activation methods and reducing agents to form highly reactive reducing radicals that can degrade oxidized contaminants. In order to determine the most efficient ARPs, different studies were carried out by applying several combinations of activation methods (ultraviolet light, ultrasound, electron beam and microwaves) and reducing agents (dithionite, sulfite, ferrous iron and sulfide) for the degradation of target contaminants at different pH-levels [129]. Many studies have been also carried out on reactions involving reducing free radicals [130], but only a few examples of the applications of reducing radicals to water treatment/contaminant degradation are present in the pertinent literature.

Ibrahim et al. [131], for the first time, expanded the concept of advanced reduction processes to green chemistry procedures. Indeed, they synthesized and characterized binary nanocomposites of TiO_2 nanotubes with $CoFe_2O_4$ ferrites and used them for the photocatalytic reduction of 4-nitrophenol to 4-aminophenol. Successively, ARPs have been used to degrade chlorinated organic contaminants such as 1,2-dichloroethane [132] and for the degradation of persistent pollutants such as per- and polyfluoroalkyl substances (PFASs) in water [133]. The degradation was achieved by coupling various advanced reduction processes that combine ultraviolet (UV) irradiation with various reagents (dithionite, sulfite, sulfide, ferrous iron). ARPs are capable of reducing the toxicity of the solution but do not lead to complete mineralization of the drugs.

Yu et al. [134] applied both advanced oxidation and reduction processes to the degradation of diclofenac and compared the reaction mechanism and the residual toxicity. Some of the intermediates formed by the two processes are different. Notably, with an AOP, a less efficient reduction in toxicity is accomplished, and with ARP, a higher irradiation dose is necessary. Similar results were found during the degradation of tetracyclines [135].

2. Factors That Affect the Degradation Efficiency

There are many factors that influence the degradation efficiency of drugs, such as pH, the initial concentration of the molecule to be degraded, the quantity of catalyst and the temperature.

2.1. pH

The pH is an important factor because varying the electrical charge of the functional groups on the surface of the catalyst due to the presence of an excess of $HO^{\bullet}$ (basic pH values) or an excess of H^+ (acidic pH values) in the solution produces various types of degradation products. In other words, therefore, changing the pH of the solution alters the surface charge of the semiconductor and shifts the potential of surface chemical reactions. In addition to the electrochemical/thermodynamic reasons, it must be considered that the reaction rate is also significantly influenced by the adsorption of drugs on the surface of the semiconductor. The degree of adsorption and the strength with which the reacting species adsorb depend greatly on the surface charges which in turn depend on the pH [136].

2.2. Initial Concentration of Drug

The initial concentration of the drug in the solution also plays an important role in its degradation during the photocatalytic reaction. In general, keeping the amount of catalyst constant, the degradation rate decreases as the drug concentration increases [137]. It must be kept in mind that an excessive concentration of the drug to be degraded could slow down the reaction because the molecules would not have sufficient sites available to adsorb and be attacked by oxidizing radicals.

2.3. Amount of Catalyst

The amount of photocatalyst also has an important influence on drug degradation. In heterogeneous photocatalysis, an increase in the degradation rate of the substrate is observed as the amount of catalyst increases. This is because a larger surface area has more active sites and therefore more $HO^{\bullet}$ radicals that can contribute to drug degradation. However, after a certain limit that depends on the type of solid and the solution in which it is suspended, the degradation efficiency begins to decrease as the catalyst increases. In fact, the radiation is partially shielded due to the excessive turbidity of the solution and the formation of particle aggregates with a consequent decrease in surface area [138].

2.4. Temperature

The effect of temperature has a limited effect on heterogeneous photocatalysis used to degrade drugs dissolved in aqueous effluents. An increase in it can increase the recombination rate of the photoproduced charges and favor the desorption of the reactive species adsorbed on the surface, causing a decrease in the photocatalytic efficiency [125]. However, a moderate increase could favor the desorption of some types of products, increasing the efficiency of the process (especially in gas–solid systems), and in this case, there is an optimal temperature even higher than the ambient one at which the experiments should be carried out.

2.5. Dosage of Reactive Oxygen Species (ROS)

During the application of AOPs, the concentration of ROS influences the extent of conversion/mineralization of the drugs, the reaction mechanism, and the reaction rate. The degradation of tetracycline in the presence of increasing amounts of peroxydisulfate (PS) showed a volcano-like trend [139]. First, the degradation rate increased with the PS concentration due to the enhanced production of $SO_4^{\bullet-}$ radicals, but when the PS concentration became high, the TC degradation rate began to decrease because some $SO_4^{\bullet-}$ radicals could be scavenged by PS.

The selection of an appropriate O_3 amount is very important for the degradation of drugs. The increase in ozone up to a certain amount generally has a beneficial effect because it improves the generation of $HO^{\bullet}$ radicals [56]. The behavior is not general and depends on the rate at which a drug is interacted with. Drugs can be easily degraded with a low dosage of ozone if they react with molecular ozone quickly, and a higher dosage is necessary if the reaction of a drug with ozone is difficult [140].

It is important also to use the optimum amount of H_2O_2 because an excess amount can lead to the formation of unwanted by-products and a low amount does not produce enough $HO^{\bullet}$ to obtain the complete mineralization.

2.6. Water Matrix

In an actual water matrix, different species such as natural organic matter (NOM), dissolved organic matter (DOM) and inorganic ions are abundantly present and affect the degradation of drugs in a positive or negative way [141,142]. When ionic species interact with drugs, they can be transformed into high-redox-potential radicals boosting the reaction rates, or they can function as free radical scavengers slowing down the degradation rate. Moreover, in the presence of a catalyst, they can compete with organic compounds for the same adsorption sites. Depending on the existence of promoting and inhibiting compounds, the water matrix may be crucial.

3. Strategies for Improving the Efficiency of AOPs: Use of Nanomaterials and/or Coupling with Conventional Techniques

Nanotechnology is very promising for treating polluted aqueous effluents and in general for environmental remediation through the development of AOPs. In fact, the notable reduction of toxic by-products is favored, and this is essential to meet water quality standards [143]. Furthermore, nanotechnology can offer economic advantages over conventional techniques, thanks to industrial production and new methods that use inexpensive raw materials and reduce energy use. In particular, the use of nanoparticles which increase the efficiency of the treatment is envisaged [144]. The beneficial role of the nanoparticles is mainly due to the increase in surface area, although the quantum size effect cannot be neglected, which can be summarized very briefly in the fact that if the particle size of a semiconductor decreases to the nanometric level, a widening of the bandgap occurs with a consequent shift of the light absorption to higher energy (blue shift) [145].

Dendritic polymers, metal/metal oxide nanoparticles, zeolites and carbon-based nanomaterials are essential for wastewater degradation, as they contain multi-branched chains and can more efficiently adsorb organic pollutants and heavy metals.

Nanomaterials combined with AOPs offer great potential to achieve a significant improvement in water treatment by not only removing contaminants but also transforming them into non-harmful compounds or compounds that can be easily degraded. However, most AOPs combined with nanomaterial-based methods are still under investigation, and further investigation and development are needed to increase their potential [146].

In the next paragraphs, we will report very briefly on some innovative oxidation processes for the treatment of contaminated water which may also involve the use of nanomaterials. They are characterized by relatively low costs and high efficiency [147].

3.1. *UV/H_2O_2 Processes*

The UV/H_2O_2 process in the presence of nanomaterials is a promising technology for the abatement of organic pollutants in water since the small size of the nanomaterials increases the degradation efficiency due to the increase in surface area [148].

In fact, the UV/H_2O_2 process, due to its low molar absorption coefficient, cannot effectively degrade pollutants as complete mineralization is often not achieved. Moreover, combining this process, for example, with TiO_2 nanoparticles, which function as a photocatalyst, can significantly improve efficiency [149]. Notably, the UV/H_2O_2/TiO_2 process, when combined with ZnO nanoparticles, increases the degradation rate even further with the production of a higher amount of active radical species [150].

It should be noted, however, that the presence of nanoparticles, which are difficult to separate from the system due to their small size, can complicate the execution of the process and can constitute a further element of pollution. Another drawback when ZnO is used in water is the anodic photo-oxidation to which this material is subjected which depends on the pH and causes the formation of soluble ionic species of zinc in the system to be purified.

3.2. *Persulfate-Based Processes*

Sulfate radicals are effective for removing organic pollutants from aqueous solutions. The use of magnetic iron oxide nanoparticles (MNPs) to obtain sulfate radicals from persulfate is a promising technology due to their wide availability and not only their magnetic characteristics but also their specific structural and catalytic properties. Furthermore, the excellent ferromagnetic behavior of Fe_3O_4 makes it easily separable from the solution [151]. Other nanomaterials such as ferrite-carbon aerogel, cobalt, iron, Co_3O_4/graphene oxide, $CoFe_2O_4$/titanate nanotubes, Co–MnO_4 and α-MnO_2 are proposed as promising heterogeneous catalysts for persulfate activation [144].

3.3. Coupling of AOPs with Conventional Water Treatment Techniques

While it is clear that several AOPs are effective at removing drugs, most AOPs are generally considered to be expensive techniques. To address this problem, the combination of advanced oxidation treatments with conventional water treatment technologies is suggested in some studies, although practical applications for large volumes of real effluent have not yet been seen [152].

Coupling of AOPs has been reported to improve the quality of the effluent prior to discharge to the environment, as demonstrated by a recent study demonstrating that the effluent is safer when ozone and sonolysis are coupled for the degradation of amoxicillin in water [152].

Furthermore, the coupling of heterogeneous photocatalysis with membrane technology constitutes another interesting example in terms of studies. The membrane, in fact, can perform the function of keeping the pollutant in the presence of the photocatalyst for longer during its degradation, preventing its permeation or the loss of the photocatalyst if the system is continuous. Another possibility is to allow the permeation of a useful intermediate with high added value before its further oxidation in the reagent system [153]. For this type of coupling, it is essential to appropriately choose the type of membrane and photocatalyst in relation to the process to be performed [154–156].

4. Use of AOPs for Drug Degradation

4.1. Tetracycline

The elimination of tetracycline (TC) by AOPs has been by far the most studied due to the widespread use of this molecule (it has been found in wastewater, surface water and groundwater at ng L^{-1} to μg L^{-1} levels) and its high stability and resistance in aqueous wastewater [157]. For these reasons, very often, the use of a single method is not effective in eliminating TC, and the coupling of different technologies is necessary [158,159]. The main problem related to the degradation of drugs is their partial oxidation to compounds which are often more harmful than their precursors. Unfortunately, in many papers, the degradation of drugs is evaluated but not their complete mineralization. The most used AOP methods are UV, UV/H_2O_2 treatments and heterogeneous photocatalysis.

The first photochemical oxidation of TC was reported in 1979 by Davies et al. [160], who studied the behavior of TC under UV light irradiation. The authors demonstrated that the process proceeds in several steps through the photo-deamination of TC followed by the interaction of the formed tetracycline radical with molecular oxygen forming a peroxyl radical, which abstracts a hydrogen atom, giving rise to a hydroperoxide. Finally, the hydroperoxide decomposes, losing water. Figure 3 presents a scheme showing the possible degradation route of TC and some of the intermediates that can be formed during its degradation. López-Peñalver et al. [161] investigated the aqueous degradation of TC in the presence of UV and UV/H_2O_2 by changing the initial concentration and initial solution pH and by adding H_2O_2. The degradation rate turned out to be dependent on the initial concentration and pH; TC degradation by UV radiation alone was low, and the addition of H_2O_2 before UV treatment increased the quantum yield of the reaction, also reducing the final TOC concentration and the toxicity of the by-products.

Photo-Fenton degradation of TC has been successfully carried out by using a magnetically recoverable $MnFe_2O_4$/MXene hierarchical heterostructure [159]. The uniform dispersion of $MnFe_2O_4$ nanoparticles within MXene nanosheets enhanced the visible light utilization and avoided the agglomeration of the $MnFe_2O_4$ particles. A TC degradation efficiency of 93.8% was reached at pH = 3 starting from a drug concentration of 10 mg L^{-1}.

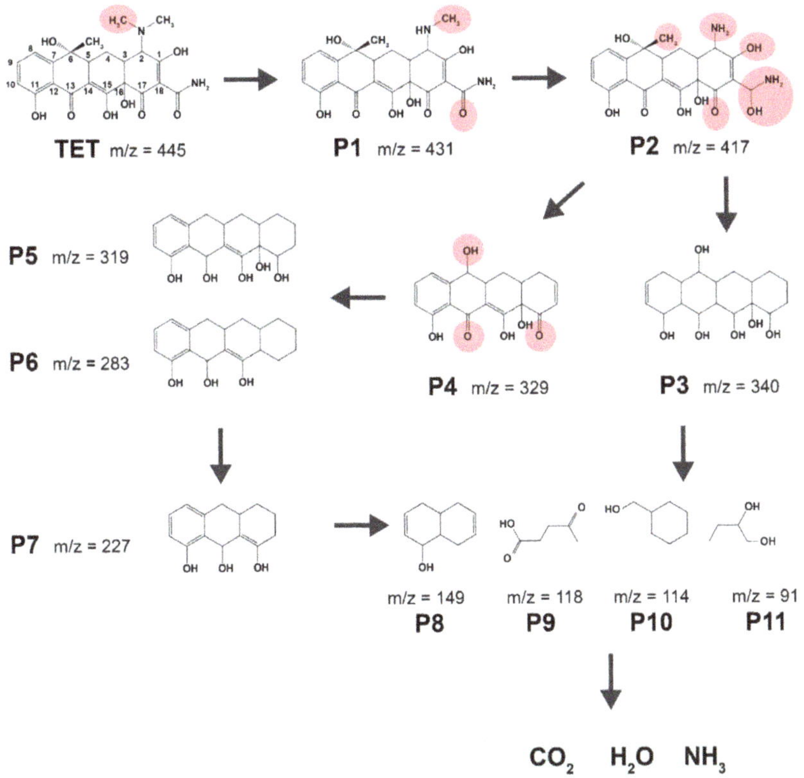

Figure 3. Scheme showing the possible degradation route of TC [162].

The photo-Fenton method has been used effectively for the complete mineralization of various types of antibiotics. As seen in more detail in Section 1.3.2, hydroxide radicals are produced due to the reaction between hydrogen peroxide and ferrous salt. In particular, this treatment allowed the removal of 24 mg L^{-1} of TC with a residual TOC concentration of 5 mg L^{-1} and 2 mg L^{-1} under black-light and solar irradiation, respectively, according to Bautiz and Nogueriav [163].

The photo-Fenton treatment of 40 mg L^{-1} of TC solution under UV irradiation in the presence of H_2O_2 (48% of stoichiometric dose) and Fe^{2+} (5 mg L^{-1}) allowed the total degradation of the drug and 77% TC mineralization [164]. An innovative 3D porous hydrogel composed of α-FeOOH/rGO (reduced graphene oxide) was able to generate reactive oxygen species in the absence of H_2O_2, eliminating 97.3% of TC in a Fenton-like process [165].

Liu et al. [166] compared the results of the degradation of TC obtained by UV irradiation, electro-Fenton and photoelectro-Fenton by using Pt gauze as an anode, a Fe_3O_4–graphite system as a cathode with an applied current density of 70 mA/cm^2 and Na_2SO_4 as electrolyte. A synergistic effect was noticed in the photoelectro-Fenton system (see Figure 4) between the two different technologies with a mineralization degree of ca. 84%. This finding demonstrates that the coupling of different technologies is an effective strategy for enhancing efficiency.

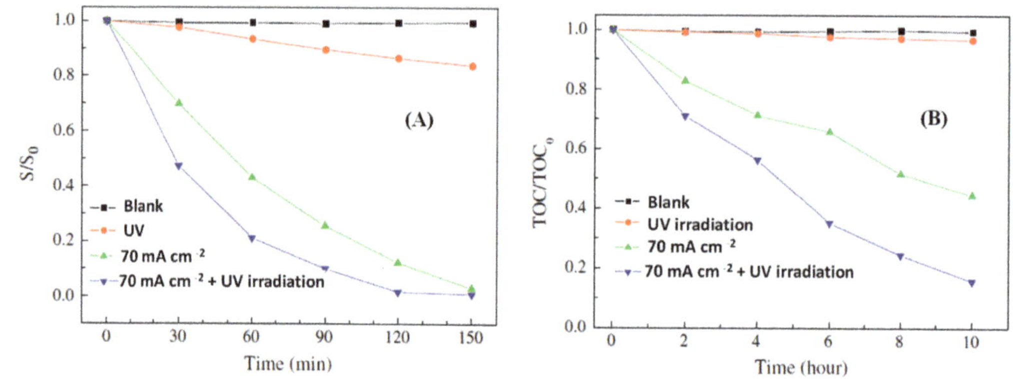

Figure 4. Effect of the different processes on tetracycline degradation (**A**) and mineralization (**B**) [166].

Ao et al. [167] applied the UV-activated peroxymonosulfate ($SO_4^{\bullet -}$) process to TC degradation both in synthetic home-prepared and real wastewater systems. PMS (HSO_5^-) was used to generate the $SO_4^{\bullet -}$ radicals under irradiation with a medium-pressure UV (MPUV) lamp, and the effect of PMS dose, pH and addition of some anions was evaluated together with the ecotoxicity and mutagenicity of the transformation products (TPs). As shown in Figure 5A, the genotoxicity of the solution first increases and then decreases as the irradiation is increased. Very low values are reached at the end of the treatment. The higher degradation rate in the real wastewater solutions with respect to the lab-prepared one (Figure 5B) has been attributed to the presence of anions like Cl^-, HCO_3^- and CO_3^{2-} in the former.

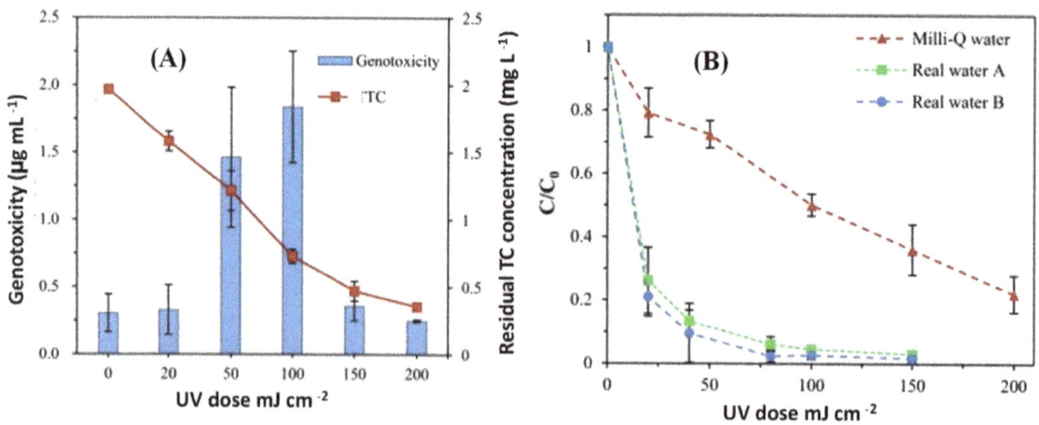

Figure 5. (**A**) Variation in genotoxicity of TC during MPUV/PMS process; (**B**) comparison of TC degradation in different systems [167].

Using 4 mM $S_2O_8^{2-}$ (PS) activated by ultrasound irradiation results in 96.5% removal of TC and 74% and 61.2% removal of chemical oxygen demand and total organic carbon, respectively [168]. Moreover, in this case, the TC degradation rate was higher in drinking water than in ultrapure water.

Ultrasound irradiation combined with Fe_3O_4 was very effective in the activation of PS for TC degradation, allowing 89% removal of the drug in just 90 min under the optimal operation conditions (TC initial concentration 100 mg L^{-1}, persulfate concentration 200 mM, initial pH 3.7, Fe_3O_4 concentration 1.0 g L^{-1} and ultrasound power at 80 W) [169]. Thermal activation of PS at 70 °C has also been described as a rapid and simple approach to activate the PS system, with almost complete elimination of TC within 30 min at 70 °C and ca. 70% at 40 °C within 240 min [170].

Natural bornite (Cu_5FeS_4), in which Cu(I) and Fe(III) ions are present abundantly, was efficient in persulfate activation for TC degradation [171]. The removal efficiency was 81.6% and the mineralization percentage was 48.7% in 180 min. Indeed, both of these ionic species can be used to efficiently trigger PS activation for TC degradation. Also, ferromanganese oxides (FMOs) displayed high activity in activating peroxymonosulfate (PMS) for TC degradation, allowing 94.3% of TC and ca. 55% of TOC removal after 30 min starting from an initial concentration of TC of 5 mg L^{-1} [172]. Electron spin resonance measurements revealed that Mn-oxides with active surface sites controlled by Fe are responsible for the generation of $SO_4^{\bullet-}$ radicals and the latter has a preponderant role compared to $HO^{\bullet}$ radicals in TC degradation. Similar results were found by using magnetic $Ni_{0.6}Fe_{2.4}O_4$ for activating PS: a TC removal of 86% in 35 min was achieved starting from a concentration of 20 mg L^{-1} [173].

Gao et al. [174] investigated the TC degradation mechanism of MW-activated PS. The effects of various experimental parameters such as TC and PS concentrations, initial pH and MW power were studied. Experiments carried out in the presence of scavengers revealed that sulfate radicals have a predominant role compared to hydroxyl radicals. By using the MW alone, the degradation of TC (initial concentration 20 mg L^{-1}) was low as its removal was only 10.3%; 99.4% of TC was instead degraded within 5 min in the MW-PS process. Moreover, compared with conventional heating processes, MW heating raised the degradation rates. When the MW power was varied from 500 to 700 W, the TC removal efficiency increased.

Ozonation is one of the most popular treatment methods because it can degrade complex substances into simpler by-products. However, due to the slow mass transfer rate of ozone from the gas phase to the liquid phase and high cost, it has some drawbacks. For this reason, ozonation is often used in combination with other processes, including O_3/H_2O_2, O_3/UV, O_3/ultrasound and catalytic ozonation. Using only ozonation, complete removal of TC (TC solution 0.5 mM) was achieved in just 4–6 min, but about 40% mineralization was reached after 2h [175].

After 20 min of ozonation in an internal loop-lift reactor, 2.08 mmol L^{-1} was almost totally converted [176]. Moreover, after 90 min of ozonation, 35% of the COD was removed, and practically no residual acute toxicity was detected. Complete mineralization and decreased toxicity of by-products were achieved during TC removal by applying ultrasound in the presence of ozone and a goethite catalyst (US/goethite/O_3) [177].

Combined processes including O_3/activated carbon, O_3/H_2O_2 and O_3/biological treatment were employed to achieve complete TC mineralization avoiding the formation of toxic intermediates [178]. When a US/Fe_3O_4/O_3 combined system was used, 100 mg L^{-1} of TC was nearly removed after 20 min, with a COD reduction of ca. 42% [179]. This COD reduction reached ca. 89% after 120 min, accompanied by a biological degradability ratio of 0.694. The system exhibited low energy consumption, excellent stability and reusability. Ultrasound-enhanced TC ozonation has been also studied using a rectangular air-lift reactor. The technique removes 91% of the COD and reduces the acute toxicity from initial values of 95% to 60% after 90 min of reaction [180].

The photocatalytic method has been widely employed for the treatment of wastewater containing TC. Figure 6 shows the number of papers published since 2000, the year the first paper appeared (source: Scopus, Elsevier (October 2023) relating to the entry "Photocatalytic degradation of tetracycline"). Due to the huge number of publications, only the most representative have been selected in this review.

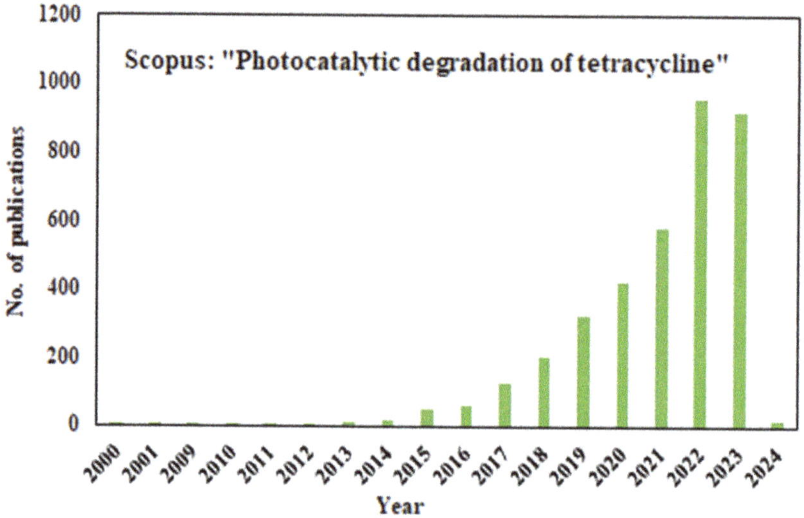

Figure 6. Number of papers published since 2000 on the photocatalytic degradation of tetracycline.

Different photocatalysts and setup configurations (irradiation sources, reactor geometry, experimental conditions, etc.) have been tested for TC removal by evaluating sometimes only the TC disappearing or also its mineralization. Table 10 reports a comparison of different photocatalytic systems for photocatalytic TC degradation. The reported results highlight the good efficiency of the reported systems in TC degradation under different experimental conditions. Only some papers reported data related to TC mineralization, and good TOC removal was found only in a few cases. A direct comparison of the results is not possible due to the different experimental conditions used by the authors.

Table 10. Results related to TC degradation with different photocatalytic systems.

Photocatalyst	Irradiation	TC Conc.	Photo Catalyst Conc.	TC Degradation	Ref.
TiO_2	Ultraviolet	32.44 mg L^{-1}	0.5 g L^{-1}	100% 50% mineralization	[181]
C–N–S tri-doped TiO_2	Visible light	5.0 mg L^{-1}	0.5 g L^{-1}	99%, 180 min 26% mineralization	[182]
α-Fe_2O_3/TiO_2	500 W halogen	29.9 mg L^{-1}	0.614 g L^{-1}	97.5%	[183]
Cu_2O–TiO_2 Cu_2O coupled with TiO_2 nanotubes	Visible light	100 mg L^{-1}	1.5 g L^{-1}	100%, 60 min	[184]
MWCNT/TiO_2 nano-composite	UVC irradiation	10 mg L^{-1}	0.2 g L^{-1}	100%, 100 min 37.3% mineralization	[185]
TiO_2@g-C_3N_4	Xenon lamp	20 mg L^{-1}	0.1 g L^{-1}	100%	[186]
Graphitic carbon nitride	Visible light	10 mg L^{-1}	1 g/L	77%, 120 min	[187]
5-PANI/$CuFe_2O_4$	UV-Vis	49.94 mg L^{-1}	0.1 g L^{-1}	86%, 120 min 95% mineralization	[188]
$Bi_2Ti_2O_7$ (BTO)	Visible light	25 mg L^{-1}	0.1 g L^{-1}	88.2%, 150 min	[189]
CuO/Fe_2O_3	UV	20 mg L^{-1}	0.05 g L^{-1}	88%, 50 min	[190]

Table 10. *Cont.*

Photocatalyst	Irradiation	TC Conc.	Photo Catalyst Conc.	TC Degradation	Ref.
ZnO/γ-Fe_2O_3	UV–visible light	30 mg L^{-1}	0.01 g L^{-1}	88.52%, 150 min	[191]
Black graphitic carbon nitride (CN-B)	UV-Vis	30 mg L^{-1}	0.05 g L^{-1}	92%, 120 min	[192]
$AgI/BiVO_4$	300 W xenon lamp	20 mg L^{-1}	0.03 g L^{-1}	94.91%, 60 min 90.5% mineralization	[193]
UiO-66-NDC/P–C_3N_4	UV–visible light	30 mg L^{-1}	1 g L^{-1}	95%, 120 min 49% mineralization	[194]
3% SnO_2/g-C_3N_4	Visible light	30 mg L^{-1}	0.5 g L^{-1}	95.90%, 120 min	[195]
$[(L1(Ag_4I_7)]CH_3CN$	Visible light	20 mg·L^{-1}	0.01 g L^{-1}	97.91%, 180 min	[196]
$ZnO/NiFe_2O_4/Co_3O_4$	Natural solar light	30 mg L^{-1}	0.02 g L^{-1}	98%, 20 min 90% mineralization	[197]
$Bi_2Sn_2O_7/Bi_2MoO_6$	300 W xenon lamp	20 mg L^{-1}	0.035 g L^{-1}	98.7%, 100 min 56.4% mineralization	[198]
MIL-88A	Visible light	200 mg L^{-1}	0.25 g L^{-1}	99.8%	[139]
UiO-66-NH_2@WO_3/CC	Visible light	20 mg L^{-1}	0.02 g L^{-1}	100%, 60 min	[199]

In addition to traditional TiO_2-based photocatalytic systems, new catalysts have been prepared with the aim of having more active samples under visible light irradiation. Below, along with some titania-based photocatalysts, some of the new systems used for tetracycline removal are described.

Palominos et al. [200] performed the degradation of TC in aqueous suspensions containing ZnO or TiO_2 under simulated solar light irradiation. The photocatalytic oxidation of the antibiotic tetracycline (TC) was performed in an aqueous suspension containing TiO_2 or ZnO under simulated solar light. ZnO showed a slightly higher activity than ZnO, and runs carried out in the presence of appropriate scavengers revealed that in the presence of TiO_2, the TC degradation occurs essentially by direct hole oxidation, and hydroxyl radicals played a secondary role, whilst when using ZnO, the oxidation is primarily driven by hydroxyl radicals.

Magnetic g-C_3N_4@$MnFe_2O_4$-graphene coupled systems with a relatively high specific surface area (SSA) and rapid separation of photoproduced e^-/h^+ pairs were able to remove 91.5% of TC under visible light illumination in the presence of persulfate in a photo-Fenton-like reaction [201]. The photocatalyst was recovered by applying an external magnetic field and reused numerous times without loss of activity.

Doping with Co^{2+} slows down the rapid recombination of the e^-/h^+ pair in TiO_2 nanosheets. Co-TiO_2/rGO nanocomposites synthesized by coating co-doped TiO_2 with rGO sheets using a one-pot hydrothermal method remove 60% of TC (initial concentration 30 mg L^{-1}) in 180 min under visible light with five-cycle repeatability [202]. A heterojunction core–shell structure consisting of a Co-TiO_2 nanofiber core and a g-C_3N_4 shell showed excellent photocatalytic performance with 90.8% TC removal and disinfection activity against E. coli [203]. This photocatalyst is advantageous due to its excellent chemical stability and non-toxicity of g-C_3N_4 with a moderate bandgap (2.7 eV). In a heterojunction with a flower shape, high-surface-area $BiOCl/TiO_2$ proves excellent for use in photocatalysis, resulting in 82% removal of TC during 10 min of illumination [204].

Black TiO_2 is considered an emerging photocatalyst different from white stoichiometric TiO_2 [16,205]. Black anatase-TiO_2 was effective for visible light photodegradation of TC, allowing ca. 66% removal after 240 min [16].

Calcite coating, $CaCO_3$ (CAL), is widely recommended to limit hole–electron recombination and improve the reusability of TiO_2 nanoparticles. A CAL/TiO_2 nanocomposite produced by the sol–gel method shows TC mineralization greater than 90% under UV irradiation [206]. Ilmenite, a mixture of NiO and TiO_2, is obtained by co-precipitation of $NiTiO_3$ and TiO_2 to form a nanocomposite with a heterojunction [207]. It has a visible band-gap energy and efficiently produces H_2 in addition to removing 58% of TC in just two hours. Mixed metal oxides (MMOs) showed increased visible light absorption and charge transfer. Zn/Fe-MMO, for example, is a composite with a layered double hydroxide structure and effectively removes 88% of TC in 2 h of visible light irradiation [208]. The TC removal ability of another MMO, namely TiO_2-Fe_2O_3, is 79.75% [209]. Ternary $Ag_xO/FeO_x/ZnO$ nanotubes were effective in the photocatalytic removal of TC under visible light irradiation [210]. The high activity has been ascribed both to interactions of light with the local magnetized domains in the Fe-containing composites and to efficient interfacial charge transfer between the different semiconductors. Moreover, the magnetic separation of the catalyst after the reaction reduces the cost of the separation of the photocatalyst from the reaction medium and makes the process advantageous compared to conventional methods.

The combination of Bi_2O_3 and g-C_3N_4 produced an effective core–shell material with a TC removal of 80.2% starting from an initial TC concentration of 10 mg L^{-1} with the use of a 250 W xenon lamp with a 420 nm cut-off filter [211]. Tun et al. [212] compared the photoactivity of Bi_2O_3, Bi_2O_3/montmorillonite and Ag-loaded Bi_2O_3/montmorillonite composites towards TC degradation under visible light irradiation. The ternary Ag-Bi_2O_3/montmorillonite sample exhibited superior outstanding activity due to the increased specific surface area, enhanced visible light absorption and photoproduced charge separation. Up to 90% TC removal efficiency was obtained in just 60 min of irradiation starting from a 20 mg L^{-1} TC initial concentration. The high activity has been ascribed to the surface plasmon resonance (SPR) which occurs on Ag nanoparticles. Similar results were found by Heidari et al. [213]. They compared the activity of $Bi_2Sn_2O_7$, gC_3N_4, $Bi_2Sn_2O_7$-C_3N_4, Ag/$Bi_2Sn_2O_7$, Ag/C_3N_4 and Ag/$Bi_2Sn_2O_7$-C_3N_4 photocatalysts towards TC degradation. The most active sample was Ag/$Bi_2Sn_2O_7$-C_3N_4, eliminating 89.1% of TC (20 mg L^{-1}) over 90 min under simulated solar light irradiation. The high activity has been attributed to both the formation of a type-II heterojunction between $Bi_2Sn_2O_7$ and gC_3N_4 and a surface plasmonic resonance on Ag nanoparticles. $Bi_{24}O_{31}Br_{10}$ nanosheets with controllable thickness exhibited high photocatalytic activity under solar light irradiation towards TC degradation under visible light irradiation [214]. Starting from an initial concentration of 20 mg L^{-1}, more than 95% of TC was degraded within 90 min. Ag/Ag_2S@BiOI nanowires were shown to be very effective in TC removal (100 mg L^{-1}), allowing its almost complete degradation within 60 min under simulated solar light irradiation [215]. Cytotoxicity tests revealed that this new catalyst is harmless or less harmful to humans after exposure in the visible region.

Reduced graphene oxide@ZnAlTi (rGO@ZnAlTi) photocatalysts have been applied for the oxidation of TC (10 mg L^{-1}) in the visible light range [216]. Graphene behaves like an electron donor and improves the adsorption of TC on the composite's surface. A removal efficiency of ca. 90% was realized in 120 min, along with a TOC abatement of ca. 80% within 270 min.

The photocatalytic treatment of TC may cause incomplete mineralization and is insufficient for higher pollutant concentrations. Its integration with adsorption technology or other AOPs can serve to obtain complete mineralization. The effectiveness of adsorption for the removal of antibiotics depends on the type of sorbent and properties such as SSA, porosity and pore diameter. Excellent TC removal is achieved with carbon-based materials, metal oxides, MOFs, clays, minerals, and composites [217].

Some Bi-containing semiconductors, such as BiOCl coupled to CdS nanoparticles, reveal both high adsorption capacity and photocatalytic removal efficiency under visible light towards TC (20 mg L^{-1}) degradation, resulting in 91.2% removal after 60 min [218]. The suggested mechanism is chelated TC adsorption followed by photocatalytic degradation.

Zhang et al. [139] coupled a Fe-based metal–organic framework (MIL-88A) with sulfate radical (SR)-based AOPs for TC degradation under visible light starting from an initial concentration of 100 mg L^{-1}. Thanks to a synergistic effect, TC was practically totally degraded within 80 min (Figure 7).

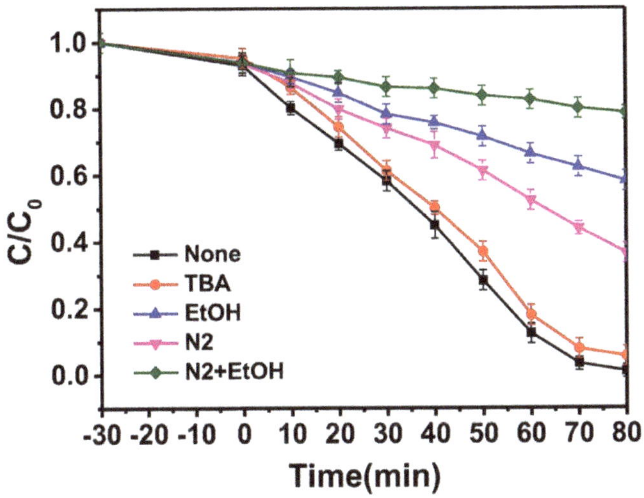

Figure 7. Degradation of TC by different processes. Reaction conditions: TC (200 mg L^{-1}), MIL-88A (0.25 g L^{-1}), PS (4.0 mM), initial pH 3.45 [139].

4.2. *Ibuprofen*

Ibuprofen (IBP), also known as 2-[4-(2-methylpropyl)phenyl] propanoic acid, is a non-steroidal anti-inflammatory drug typically used as an analgesic, anti-inflammatory and antipyretic. Different AOPs have been employed for its degradation in water solution.

The ultrasonic method (US) gave good results in IBP removal with an IBP degradation of 98% in 30 min starting from an initial concentration of 21 mg L^{-1} with an applied power of 80 W [219]. The degradation rate was higher in both the presence of air and oxygen, acid media being the most favorable. BOD_5 and COD measures indicated that, although some dissolved organic carbon remained, IBP was transformed into biodegradable by-products which could be destroyed in a subsequent biological step.

The activation of PDS, PMS and H_2O_2 at various ultrasonic frequencies was investigated [220]. A stock solution for IBP was prepared by dissolving 0.1 g of powder in methanol. Synthetic wastewater containing IBP was prepared by mixing deionized water with IBP stock solution. PS, PMS and H_2O_2 were added to the solution. Batch tests were conducted to determine the oxidative elimination of IBP. They were also used to investigate how IBP was degraded in the US-PS, US-PDS and US-H_2O_2 systems. It was reported that the IBP degradation followed pseudo-first-order kinetics regardless of the method used. The rate constant for IBP decomposition was found to be the highest at a frequency of 1000 kHz. US alone was efficient for IBP degradation, and the addition of PS, PMS and H_2O_2 improved the decomposition efficiency of IBP.

Direct photolysis by UV light (λ = 254 nm) and UV/H_2O_2 processes were assessed for the oxidation of IBP in synthetic solutions (initial concentration 9.90–34.24 mg L^{-1}) [221]. The effects of drug concentration, pH and H_2O_2 concentration were investigated (Figure 8). By varying the IBP initial concentration, an average efficiency of 82.63% was obtained after 270 min, and the oxidation was higher at pH 6. As pK_a = 4.9 is for IBU, it is mainly present in its dissociated form at pH 6, and this finding suggests its higher reactivity with respect to the protonated species. The optimum amount of H_2O_2 was 10 mg L^{-1}, which obtained 97.39% removal after 75 min. TOC and COD measurements revealed the partial mineralization of IBP both after UV and UV/H_2O_2 treatment.

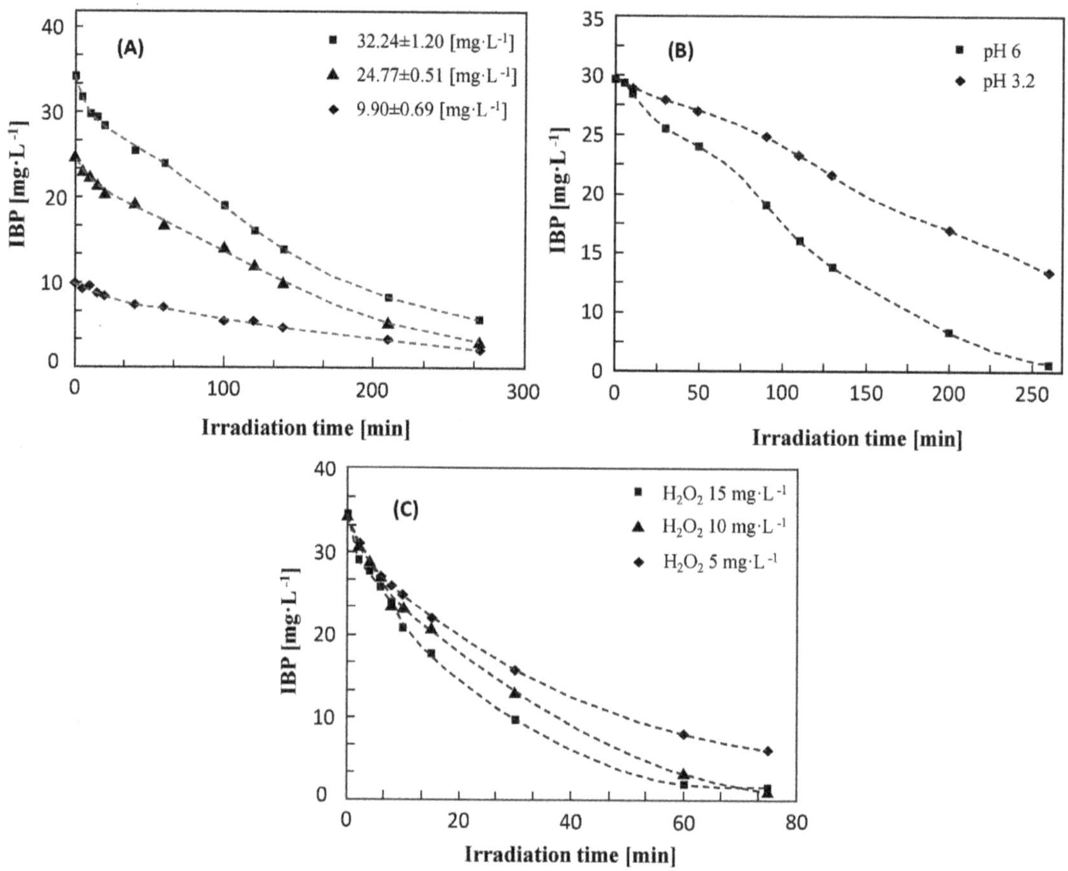

Figure 8. Influence of initial IBP concentration (**A**), pH (**B**) and H_2O_2 concentration (**C**) on IBP removal by UV/H_2O_2 process.

Adityosulindro et al. [222] used a heterogeneous Fenton process in the presence of a Fe-zeolite catalyst. The Fe-ZSM5 catalyst was effective in removing 20 mg L^{-1} IBP in pure water, with 88% degradation but only 27% TOC removal after 180 min at natural pH = 4.3. The drug decay followed a pseudo-first-order reaction with an activation energy of 53 kJ mol^{-1}. A few years later, the same research group [223] compared the degradation of IBP under UV, UV/H_2O_2, photo-Fenton and sono-photo-Fenton processes by varying the drug concentration and the irradiation source. In particular, two lamps were used: lamp L1, 254 nm, 6 W, and visible light lamp L2, 360–740 nm, 150 W. Applying the photo-Fenton process, by using both lamps, a complete degradation of IBP was obtained after 3 h while the mineralization was 82% with lamp L1 and 59% with lamp L2. The coupling of the L2-photo-Fenton process with ultrasound has a beneficial effect only at low Fe concentrations.

The degradation of IBP by the UV/chlorine and UV/H_2O_2 AOPs follows pseudo-first-order kinetics, with the UV/chlorine AOP having a rate constant 3.3 times higher than that of UV/H_2O_2 at pH 6. The degradation is sensitive to the dosage of chlorine and to the pH of the solution (decreasing at pH 9), but not to the temperature or the concentration of chloride ions. Increasing pH decreases the first-order rate constant and increases the contribution of reactive chlorine species [38].

The UV–Vis/H_2O_2 process was effective in the degradation of IBP (initial concentration 0.87 mM), allowing ca. 40% of degradation after 2 h [224]. With the addition of 1.2 mM of Fe(II) in the presence of 0.32 mM of H_2O_2, the complete degradation was reached with a mineralization degree of 40%.

Quero-Pastor et al. [225] studied the degradation of IBP (1 mg L^{-1}) by ozonation, also evaluating the residual toxicity of the solution after the treatment. Under the best operational condition, an almost complete conversion was reached after 20 min of treatment, but no mineralization was observed. The results of toxicity tests revealed that the intermediates are more toxic than the starting drug. Almeida et al. [226] evaluated the effects of single ozonation, oxidation in the presence of H_2O_2 and the combination of the two processes on IBP degradation (initial concentration 20 mg L^{-1}), mineralization and residual toxicity. When single ozonation was used, IBP was immediately removed, but no important TOC removal was reached. The addition of H_2O_2 did not present substantial enhancements; when O_3 and H_2O_2 were combined, a mineralization of 70% was accomplished after 180 min of reaction.

The investigation of IBP degradation by the UV/Fe^{3+}/Oxone process revealed that the efficiency depends on the operating parameters, and the best results were obtained at pH = 3 and with an optimal molar ratio of Fe^{3+}/Oxone/IBP equal to 2:2:1 [227]. Electro-peroxone treatment resulted in effective IBP removal, enhancing both its degradation and mineralization [228]. With an initial IBP concentration of 20 mg L^{-1}, O_3 gas phase concentration of 40 mg L^{-1}, current of 300 mA and pH = 7, IBP was completely degraded after 5 min and mineralized after 60 min.

The degradation of IBP by heterogeneous photocatalysis has been widely investigated in the presence of different photocatalysts by changing the experimental conditions and coupling photocatalysis with other technologies. In the presence of bare TiO_2, 0.03 g of photocatalyst was effective in totally degrading IBP (concentration 10^{-4} M) in 5 min at pH = 5 under UV irradiation [229].

Candido et al. [230] reported that UV light irradiation of TiO_2 suspended in 1 L of IBP solution (1.0 mg L^{-1}) at 25 °C and at pH = 7.8 produced, after 1 h, 92% and 78% removal of IBP and TOC in pure water and 64% and 35% in spring water, respectively. Ecotoxicity tests using some bioindicators of environmental conditions revealed that the solution had residual acute effects after the treatment.

Khalaf et al. [231] demonstrated that a synthetic solution of IBP (initial concentration 25 mg·L^{-1}) can be successfully treated under irradiation in the presence of photoactive glass coated with TiO_2. Moreover, the immobilization of TiO_2 on glass substrates avoided the recovery problems encountered when the photocatalyst is used as powder.

Agócs et al. [232] obtained 81% degradation of IBP (50 μM) in pure water after 60 min of treatment in the presence of nanometric TiO_2, stabilized with cyclodextrins, under UV irradiation. In tap water, the degradation was slower due to loss of efficiency of the oxidizing agents.

The activity of TiO_2 Degussa P25 under UV LED irradiation towards IBP degradation was evaluated in pure water and municipal and pharmaceutical wastewaters by measuring IBP degradation, mineralization and biotoxicity [233]. The process was effective in treating pure water and wastewater deriving from the pharmaceutical industry and less efficient in municipal wastewater, probably due to its complex composition. The degradation was higher at pH near 5.0 due to the enhanced electrostatic attractions between TiO_2 and IBP. For all the matrices, a reduction of 40% in acute toxicity was observed.

Jiménez-Salcedo et al. [234] compared the uses of TiO_2 nanoparticles under UV light and g-C_3N_4 nanosheets under visible light irradiation for IBP degradation. The authors observed a higher efficiency of TiO_2 with respect to g-C_3N_4, although the use of a higher-energy light must be considered. The initial IBP concentration was 5 μg mL^{-1}. With TiO_2, the complete degradation of IBP was achieved in 10 min, whilst more than 3 h was necessary with g-C_3N_4; in both cases, no complete mineralization was accomplished.

By using monoclinic $BiVO_4$ under simulated solar light irradiation [235], an IBP conversion of 90% was reached after 25 min starting from an initial concentration of the drug of 10 mg L^{-1}. No information about the mineralization has been reported by the authors.

By coupling ozonation with photocatalysis in the presence of $SrWO_4/ZnO$, an IBP removal efficiency of 93% and a 55% BOD elimination were obtained under UV irradiation starting from an initial drug concentration of 0.1 mg L^{-1} [236].

Fidelis et al. [237] studied the degradation of IBP by combining different methods: ozonation, photolytic ozonation, photocatalysis and photocatalytic ozonation (Figure 9). The single methods afforded a good IBP removal rate but a low mineralization degree. A synergistic effect was instead noticed in photocatalytic ozonation, with a complete degradation of IBP after 12 min and a 98% TOC reduction in 30 min of reaction.

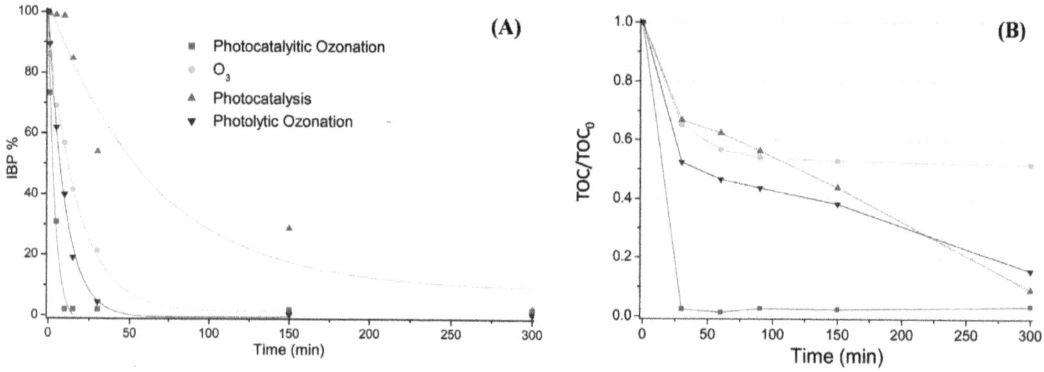

Figure 9. IBP (**A**) and TOC (**B**) removal with the different processes [237].

4.3. *Oxytetracycline*

The broad-spectrum antibiotic oxytetracycline (OTC) has notable biodegradability but has bioaccumulation and persistence properties and consequently is extremely harmful to human health.

Its abatement (initial concentration 5 mg L^{-1}) has been studied both in synthetic (ultrapure water) and real wastewater matrices using hybrid systems that combine microfiltration (MF) with photolysis (UVA/MF) or heterogeneous photocatalysis in the presence of a TiO_2-P25 photocatalyst [238]. A photocatalytic membrane reactor (PMR) has been tested using TiO_2-P25 nanoparticles both in suspension and immobilized on a nanoengineered membrane (NEM). A higher photocatalyst loading results in higher OTC removal efficiency (90% in 30 min), but a greater decrease in permeate flux because a denser TiO_2/P25 cake layer formed. The presence of NEM led to the improvement of the antifouling properties and also a decrease in the permeate flux.

Photo-Fenton catalytic activity for OTC degradation was tested with a $MnFe_2O_4$/g-C_3N_4 heterojunction composite which exhibited excellent catalytic activity as approximately 80.5% was removed in 10 min. OTC breakdown was primarily started by h^+ oxidation, with $HO^\bullet$ and $O_2^{\bullet-}$ playing only minor supporting roles [239]. The hypothesized reaction mechanism is reported in Figure 10. Through h^+ attacks, OTC molecules are oxidized, producing different intermediates by eliminating groups like -OH, $-NH_2$, $-CH_3$ and -$CONH_2$ and breaking the benzene ring. The intermediates are further degraded by h^+, $HO^\bullet$ and $O_2^{\bullet-}$ and form aliphatic compounds, tiny organic acids, CO_2 and water.

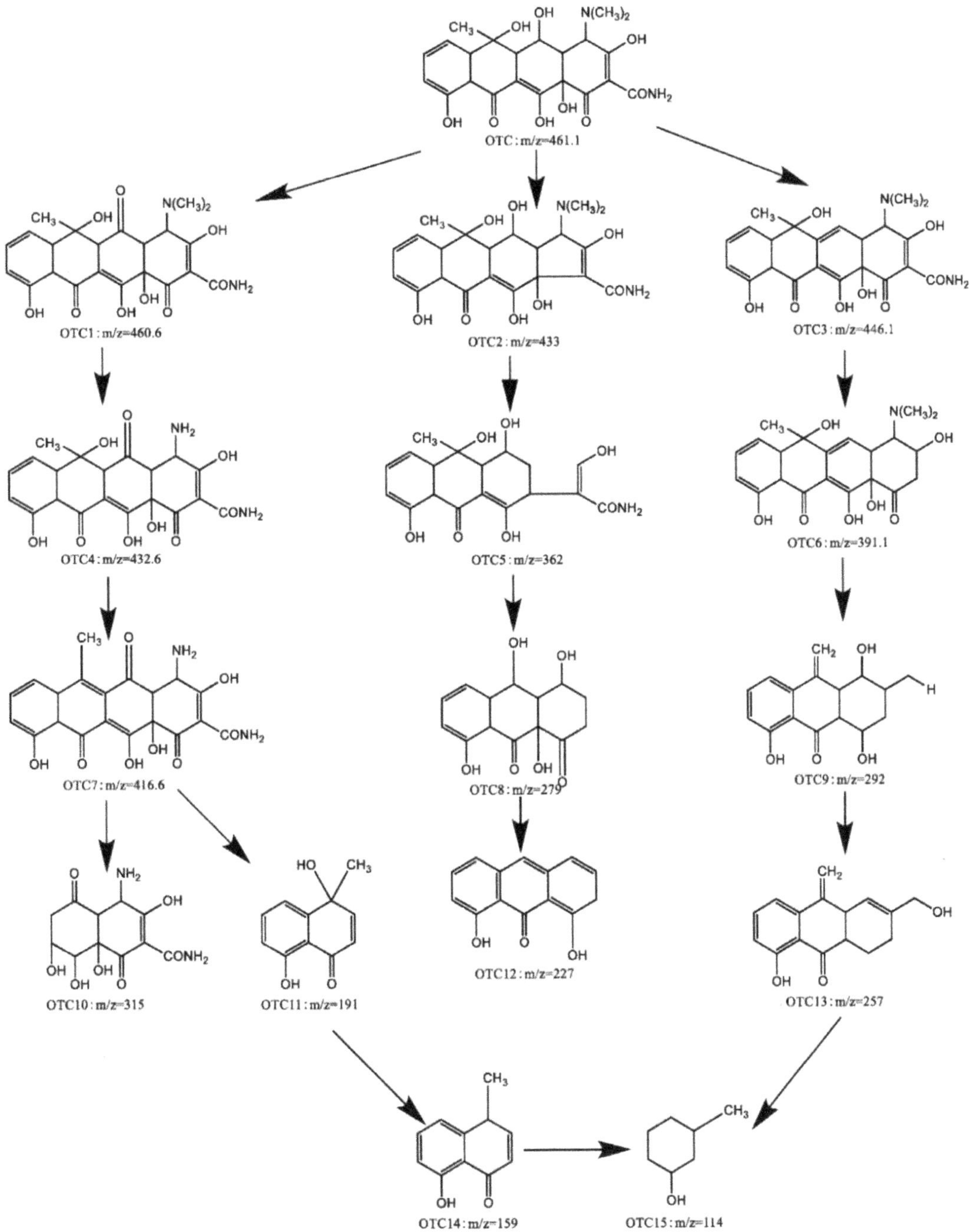

Figure 10. Possible degradation pathways of OTC in the photo-Fenton catalytic system [239].

The photocatalytic method using persulfate (PS) and Fe_3O_4/MIL-101(Fe) allowed the occurrence of 87.1% degradation of 70 mg L^{-1} of OTC in 60 min [240]. MWCNTs-$CuNiFe_2O_4$ nanomaterials were utilized to effectively remove OTC from an aqueous solution in the presence of persulfate following the mechanism shown in Figure 11 [241]. Excellent adsorption characteristics toward OTC were shown by the MWCNTs-$CuNiFe_2O_4$ combination, which also successfully activated potassium persulfate (KPS) for drug removal. By using a catalyst concentration of 10 mg L^{-1} with an initial concentration of OTC equal to 300 mg L^{-1}, 88.6% degradation was achieved. $SO_4^{\bullet-}$ and $HO^{\bullet}$ radical-capturing agents such as ethanol and isopropyl alcohol were used to investigate the reaction mechanism. The outcomes showed that the presence of these quenching agents reduced the removal effectiveness of MWCNTs-$CuNiFe_2O_4$, confirming the active role of these radicals in the degradation process.

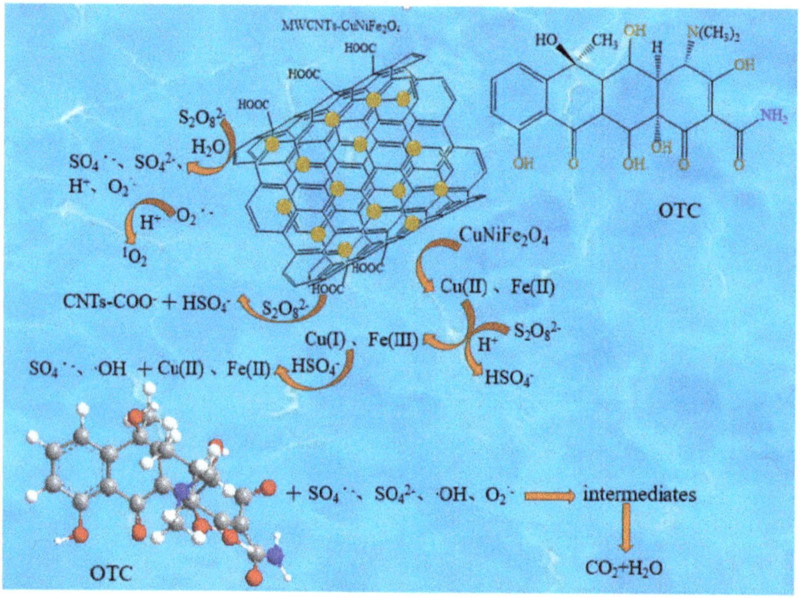

Figure 11. Mechanism of degradation of OTC by using MWCNTs-$CuNiFe_2O_4$/KPS system [241].

Ozonation was also proved to be beneficial for the degradation of OTC. Li et al. [242] investigated how ozonation affected OTC degradation at various pH levels. Chemical oxygen demand (COD), the concentration of oxytetracycline and the BOD_5/COD ratio were used to measure the effectiveness of the ozonation process. Using bioluminescence assays, the hazardous potential of OTC degradation was also investigated. The findings suggested that in pharmaceutical wastewater containing a high OTC concentration, ozonation as a partial step in a combined treatment concept can boost biodegradability.

4.4. Lincomycin

Lincomycin is an antibiotic found in water effluents. Its structure is shown in Figure 12.

TiO_2-based nanomaterials, including TNAs (TiO_2 nanotube arrays), TNWs/TNAs (TiO_2 nanowires on nanotube arrays), Au-TNAs and Au-TNWs/TNAs, were developed for enhanced photocatalytic degradation of antibiotics in aquaculture wastewater. TNWs/TNAs showed higher activity than TNAs due to larger surface area [243]. Au-TNWs/TNAs showed the highest activity under UV-VIS or VIS irradiation, exhibiting 100% efficiency in 20 min (lincomycin concentration 500 ng mL^{-1}) with reaction rates of 0.26 min^{-1} and 0.096 min^{-1}, respectively. The high activity of Au-TNWs/TNAs can be ascribed to the synergistic effects between the high surface area and the surface plasmonic effect of Au nanoparticles. Augugliaro et al. [244] reported that lincomycin is broken down by photocatalytic oxidation in aqueous suspensions of Degussa P25 polycrystalline TiO_2, using a hybrid system consisting of a solar photoreactor and a membrane module. The initial lincomycin concentration was 20 μM. The reported kinetics are pseudo-first-order with high membrane rejection for lincomycin and its oxidation products.

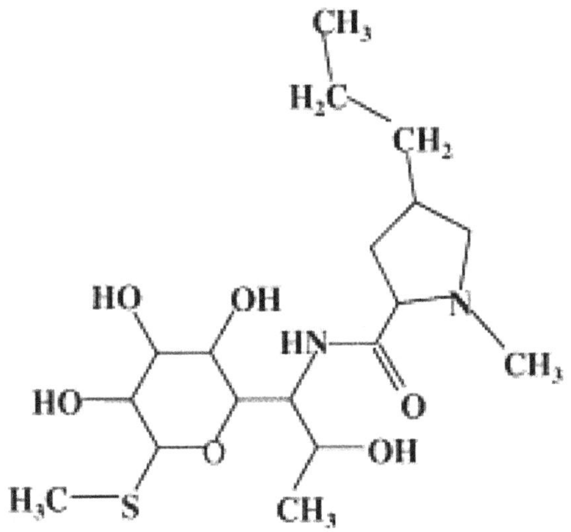

Figure 12. Structure of lincomycin.

As a highly effective metal-free photocatalyst, graphitic carbon nitride (g/C_3N_4) shows promising behavior for the degradation of drugs. Adjusting the energy band, improving charge extraction and adding a cocatalyst enhances the g/C_3N_4 photocatalytic activities and increases both the degradation rate and conversion degree of lincomycin under visible light irradiation [245]. Carbon quantum dots (CDs) were added as cocatalysts to improve the formation of $O_2^{\bullet-}$; graphene oxide (rGO) was employed to improve the charge mobility. The most active improved photocatalyst was CD-rGO-O-g/C_3N_4, which exhibited a tenfold increase in degradation rate when compared to the original g/C_3N_4. Starting from an initial concentration of the drug of 100 mg L^{-1}, 99% degradation was achieved after 180 min. The active species, such as $O_2^{\bullet-}$, $HO^{\bullet}$ and h^+, contribute differently to the degradation in each of the photocatalysts.

Metal–organic framework (MOF)-based photocatalytic treatment of lincomycin using Basolite F300 as a catalyst in the presence of two oxidants (H_2O_2, $S_2O_8^{2-}$) was reported by Kontogiannis et al. [246]. The results showed that the drug elimination was completed within 2 h, and the oxidant concentration (0–300 mg L^{-1}) affected the process rate. The best concentration for attaining maximum degradation was 100 mg/L for both oxidants. After 120 min, just 8% of MOFs were degraded according to photolysis studies. The photocatalyst Basolite® F300 was activated using H_2O_2 as an oxidant in order to study the photocatalytic process. MIL-100 surface iron sites showed Fenton-type activity, which synergistically increased degradation. When H_2O_2 was added, the rate of degradation increased to 95% in 90 min, indicating catalyst activation and a conceivable heterogeneous photo-Fenton reaction involving Fe-sites and H_2O_2. The mechanism of the reaction pathway is shown below in Figure 13.

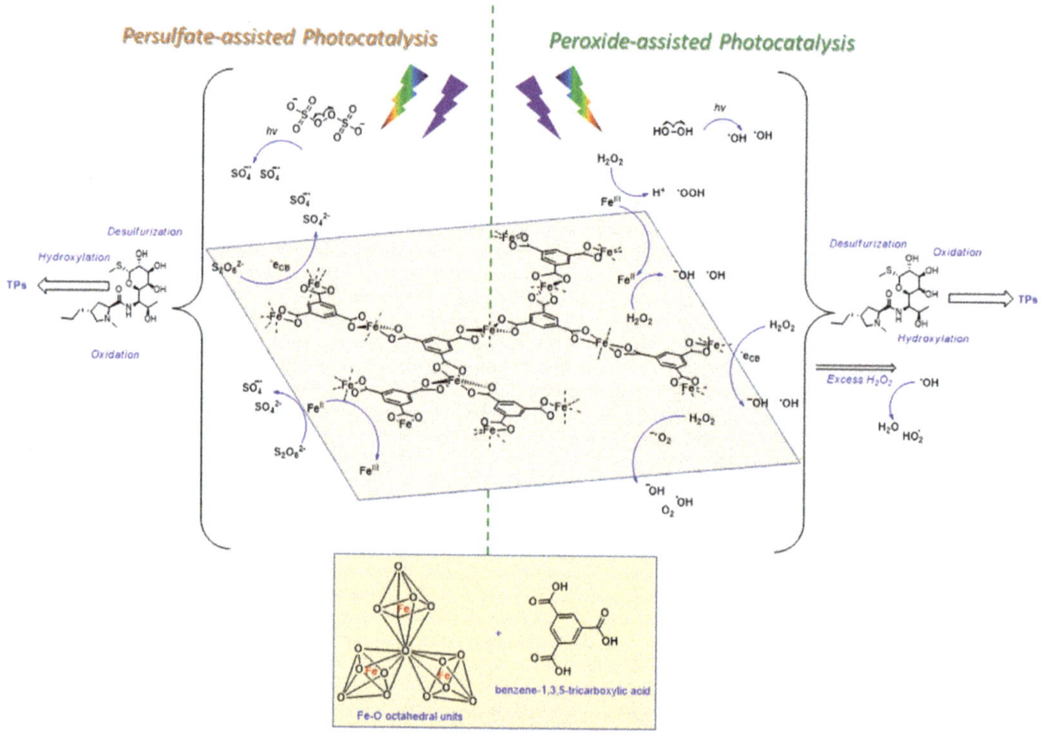

Figure 13. Mechanistic reaction pathways taking place on Basolite F300 as a catalyst activated by hydrogen peroxide or persulfate [246].

A Zr-based metal–organic framework (MOF), known as VNU-1, with increased pore size and improved optical properties, was prepared and used as a photocatalyst for the degradation of antibiotics [247]. The prepared sample was as active as well-known P25 in the removal of lincomycin, and a contemporary TOC decreasing to 95% was accomplished. After five cycles, the catalyst maintained its high photodegradation performance (e.g., 100% photodegradation in 10 min).

In perspective, good results can be obtained by combining a series of measures that enhance the photocatalytic activity, such as the doping of structured materials. Using nanowires on an Au-TiO_2 nanotube array causes 83% degradation of lincomycin in 20 min upon irradiation with visible light. This result can be attributed to the nanowires having a larger surface area than TiO_2 nanotubes [248].

4.5. Amoxicillin

Çağlar Yilmaz et al. [249] compared the activity of commercial TiO_2 P25 and lab-prepared bare and Co-doped TiO_2 towards the degradation of amoxicillin (AMX) under UV-C and visible irradiation. Co-TiO_2 was the most active photocatalyst, allowing the complete elimination of 100 mg L^{-1} of AMX after 240 min under UV-C irradiation and after 300 min under visible light. TiO_2 nanoparticles loaded on graphene oxide (GO/TiO_2) removed more than 99% of AMX (initial concentration 50 mg L^{-1}) under UV light irradiation using a lamp with an intensity of 36 W [250]. Photoproduced holes were the main active species in degrading AMX, and TOC analyses revealed a good mineralization degree, obtaining a COD and TOC removal of 91.25% and 89.7%, respectively. The photocatalysts showed good stability, recyclability and efficiency in reducing the initial toxicity of the solution.

Nanostructured photocatalysts consisting of titanium dioxide doped with iron and nitrogen (Fe^{3+}-$TiO_{2-x}N_x$) synthesized using the sol–gel (SG) method and microwave (MW) technology, were tested for their ability to break down amoxicillin (AMX) [251]. Higher activity was observed for the materials manufactured using the SG approach, and degradation efficiencies of 58.61% for SG and 46.12% for MW samples were observed after 240 min of visible light at pH 3.5.

Al-Musawi et al. [252] carried out the photocatalytic degradation of AMX by using a Fe_2O_3/bentonite/TiO_2 nanocomposite under both visible LED light and UV irradiation. The reaction rate followed a pseudo-first-order kinetics, and starting from an AMX concentration of 25 mg L^{-1}, under UV light, a complete degradation of the drug was obtained in 60 min, while under visible light, a removal percentage of 98.8% was observed in 90 min.

The electrophotocatalytic treatment of an AMX aqueous solution (100 mg L^{-1}) was studied by measuring the chemical oxygen demand (COD) [253]. The degradation of the drug was 79% after 120 min of irradiation.

Elmolla and Chaudhuri [254] compared Fenton, photo-Fenton, UV/ZnO and TiO_2 photocatalysis processes in the degradation of AMX. All methods were effective in AMX oxidation, and, except for UV/ZnO, an enhancement of the residual solution biodegradability was measured by the BOD_5/COD ratio evaluation. The best results were obtained by the photo-Fenton process.

4.6. *Erythromycin*

Erythromycin (ERY) is a penicillin medication that can remain in nature for up to a year, preserving its antibiotic activity. The inefficiency of conventional ERY degradation methods has prompted the development of cutting-edge technologies such as AOPs [255].

Chu et al. [256] studied ERY removal using PS activated by gamma radiation in different systems. The degradation rate follows the order deionized water > groundwater > secondary treated municipal wastewater, and in the deionized water, ERY was eliminated with a TOC removal of 25%.

Albornoz et al. [257] carried out ERY degradation by direct photolysis and by using photocatalysts such as TiO_2, $Ti_{1-x}Sn_xO_2$ and a commercial TiO_2 mesh under UV-A irradiation. Under direct photolysis, only a low degree of degradation and mineralization was observed with the formation of low-molecular-weight carboxylic acids (Figure 14). In the presence of photocatalysts, the best results (complete degradation after 4 h) were obtained with the sample $Ti_{1-x}Sn_xO_2$ due to the formation of a type-II heterojunction between TiO_2 and SnO_2.

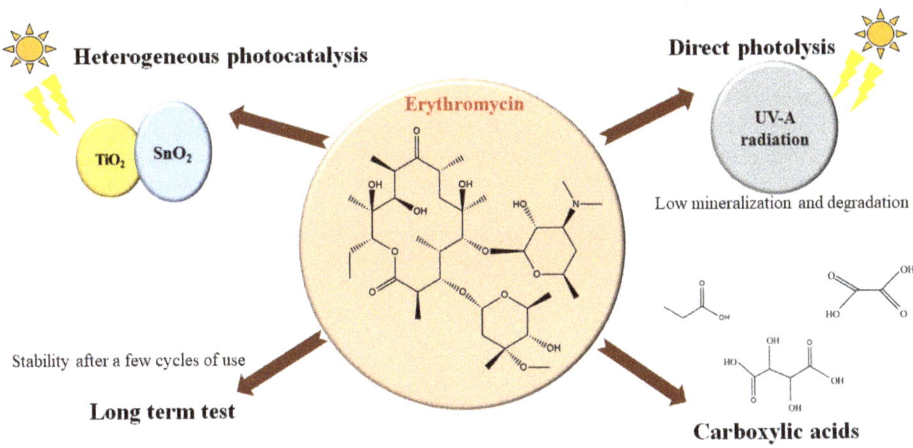

Figure 14. Photolytic and photocatalytic degradation of erythromycin.

Photocatalytic mineralization of ERY in aqueous TiO_2 suspensions using commercially available TiO_2 catalysts was also reported in the literature [258]. The most effective catalyst was Degussa P25, which reduced 90% of total organic carbon after 90 min of reaction with an ERY initial concentration of 10 mg L^{-1}.

Vignesh et al. [259] reported that when both zinc phthalocyanine and TiO_2 nanoparticles were used, a significant improvement in photocatalytic activity was demonstrated in comparison to pure TiO_2. A 74% degradation of ERY was achieved in 3 h by irradiation with visible light, while the undoped material degraded only 31.6% of the antibiotic under the same experimental conditions.

A composite g-C_3N_4/CdS photocatalyst showed good activity towards ERY degradation under simulated solar light irradiation [260]. Starting from an initial concentration of 50 mg L^{-1}, ca. 80% of the drug was çonverted in 1h. The high activity has been ascribed to the formation of a Z-scheme between g-C_3N_4 and CdS with an efficient charge separation.

The activity of $CaCO_3$ (calcite) towards ERY (initial concentration 30 mg L^{-1}) removal was explored under both UV and solar light irradiation by Mohsin et al. [261]. After 2 h of UV irradiation, 73% of conversion was reached, while 6 h was necessary under sunlight to remove 93% of ERY. Moreover, under sunlight irradiation, a reduction of 78.5%, 77.6% and 64.5% in COD, BOD and TOC was achieved, respectively. The photocatalyst showed good stability, maintaining its activity unchanged after three cycles.

γ-Fe_2O_3/SiO_2 composites allowed ca. 87% ERY degradation after 6 min of UV-C (λ = 254 nm) irradiation starting from a concentration of 6 mg L^{-1} of the drug [262]. The high activity is due to the high absorbent properties of silica combined with the good photocatalytic performance of γ-Fe_2O_3 under sunlight.

4.7. *Sulfonamide*

Sulfonamide antibiotics are widely used in human and veterinary medicine and as herbicides in agriculture [263]. They are very dangerous for the environment because they present a very stable structure that is difficult to break down with traditional processes.

P-doped TiO_2-αFe_2O_3 mixed oxide catalysts were used together with $K_2S_2O_8$ for photocatalytic degradation of a sulfonamide mixture containing sulfadiazine (SDZ), sulfamerazine (SMRZ) and sulfamethazine (SMTZ) (5 mg L^{-1} each) in a coupled process [264]. In 300 min under visible light irradiation, 69% mineralization was achieved.

Using 0.05 g Bi_2O_3-TiO_2/PAC catalysts, a 250 mL solution containing 20 mg L^{-1} of three different sulfonamides, namely sulfamethoxazole (SMX), sulfamethazine (SMT) and sulfadiazine (SDZ), was treated under solar light irradiation [265]. The degradation was ca. 100% for SMX after 30 min and ca. 100% for SDZ and 75% for SMT after 60 min of reaction. Tests carried out in the presence of trapping agents revealed that h^+ was the main active species in degrading SMX whilst $O_2^{\bullet-}$ played the main role in the oxidation of SMT and SDZ.

Batista and Nogueira [266] investigated the parameters influencing the photo-Fenton degradation of two antibiotics belonging to sulfonamide family, namely sulfadiazine (SDZ) and sulfathiazole (STZ). The addition of Fe(III)-oxalate improved the drugs' oxidation with respect to free iron, and at pH = 5 in the presence of H_2O_2, the complete degradation occurred in 8 min.

O_3, H_2O_2 and O_3/H_2O_2 processes were compared in the degradation of sulfonamides such as sulfamethoxazole (SMX), sulfasalazine (SSZ), metronidazole (MNZ) and sulfamethazine (SMT) [267].

The best results were found utilizing the O_3/H_2O_2 process due to a synergistic effect between the two methods, and a maximum degradation efficiency of 98.10%, 89.34%, 86.29% and 58.70%, was obtained for SSZ, SMX, SMT and MNZ, respectively, under optimum experimental conditions.

4.8. Ciprofloxacin

Ag_3PO_4 nanoparticles deposited onto TiO_2 nanotube arrays gave rise to a heterojunction that displayed high photocatalytic activity in degrading ciprofloxacin (CIP) under solar light irradiation [268]. Starting from an initial concentration of 10 mg L^{-1}, ciprofloxacin was destroyed by 78.4% in a 60 min period of irradiation. Liu et al. [269] compared the degradation of CIP (initial concentration 20 mg L^{-1}) by using TiO_2 P25 and lab-made TiO_2 nanosheets and composite $TiOF_2/TiO_2$ flower-shaped nanosheets under simulated solar light irradiation. Different $TiOF_2/TiO_2$ samples were prepared by the hydrothermal method at different temperatures: 140, 160, 180 and 200 °C. Composite nanosheets synthesized at 160 °C were the most active samples due to the formation of an efficient heterojunction allowing a decrease in the recombination rate of the photoproduced charges and an enhancement of the charge transfer (Figure 15). The $TiOF_2/TiO_2$ nanosheets allowed a 93.7% degradation of the antibiotic after 90 min of irradiation.

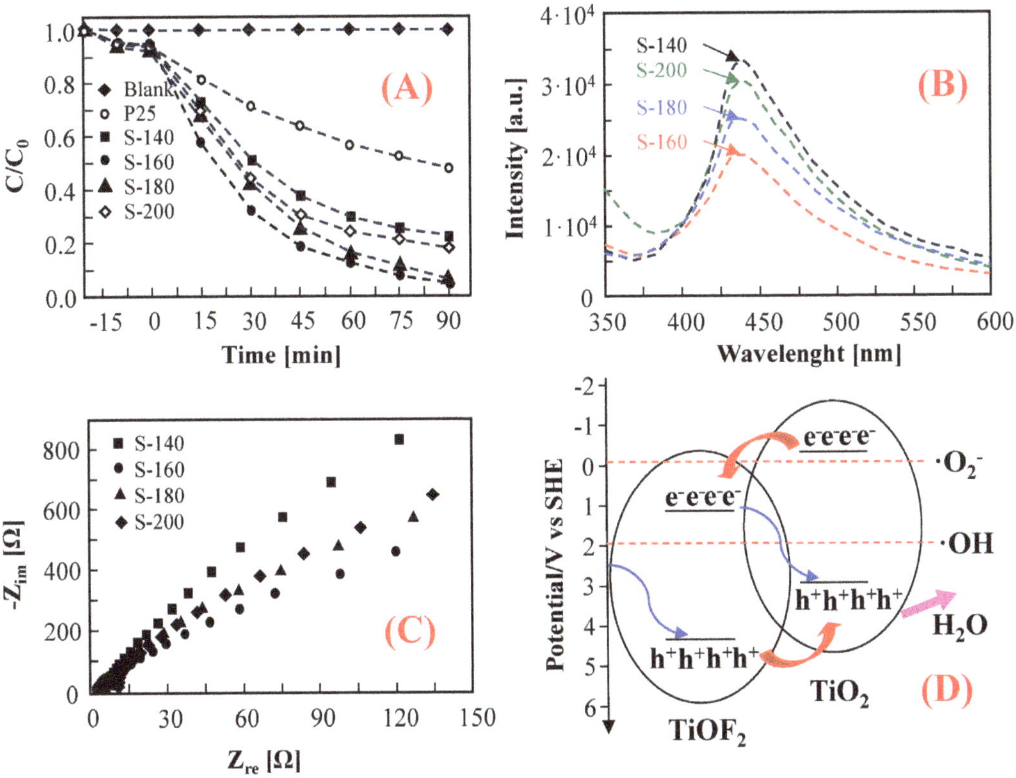

Figure 15. (**A**) Photocatalytic degradation of ciprofloxacin, (**B**) photoluminescence spectra, (**C**) electrochemical impedance spectroscopy measures and (**D**) heterojunction scheme.

El-Kemary et al. [270] investigated the photocatalytic activity of synthesized ZnO nanoparticles for the degradation of CIP under UV light irradiation in an aqueous solution at different PHs. The degradation process of CIP showed a pseudo-first-order reaction, and 50% degradation was observed after 60 min at pH = 10 and pH = 7.

Wen et al. [271] reported that upon exposure to visible light, CeO_2-Ag/AgBr composite photocatalysts with a Z-scheme arrangement demonstrated improved photocatalytic activity for the degradation of CIP. Accelerated interfacial charge transfer and better photogenerated electron–hole pair separation were credited with the improved performance.

Zn-doped Cu_2O synthesized by a solvothermal method exhibited excellent photoactivity towards the degradation of 20 mg L^{-1} of CIP, removing about 94.6% of it after 240 min [272]. Runs carried out in the presence of specific trapping agents revealed that the main active species responsible for the drug degradation were $HO^{\bullet}$ radicals and h^+.

Photo-Fenton activity of rGO-$ZnFe_2O_4$ towards CIP degradation was boosted by the thermal effect inducing H_2O_2 activation [273]. The integrated process showed excellent photoactivity thanks to a synergistic effect and displayed a superior performance when compared to a solely photo-Fenton or thermal-Fenton process in lowering the H_2O_2 activation barrier and speeding up the production and spread of radicals.

Ge et al. [274] reported that $LaFeO_3$/polystyrene (LFO/PS) photo-Fenton catalysts, prepared using ultrasound-assisted sol–gel and hydrothermal methods, were highly efficient (98.38%) for antibiotic degradation under the following experimental conditions: CIP concentration 10 mg L^{-1}, H_2O_2 5 mmol L^{-1} and pH = 9.00. Moreover, the TOC removal efficiency reached 76.44%. Finally, Gupta and Garg [275] carried out CIP degradation using a classical Fenton process in synthetic wastewater containing an initial concentration of the drug of 100 mg L^{-1}. Under the best experimental conditions, after 1 h, CIP degradation and TOC removal were 70% and 55%, respectively.

5. Challenges

As reported in this review, the use of a single AOP for the removal of drugs from wastewater may be, in many cases, insufficient to remove both the original drugs and their intermediates. Coupling of different AOPs can overcome this problem by allowing a more efficient decontamination of the effluents. Under appropriate experimental conditions, the use of integrated AOP systems is also convenient from an economical point of view because the pollutants can generally be mineralized, and less energy is needed. However, it is important to evaluate the quality of the treated water by considering its residual toxicity after the application of combined AOPs. Moreover, experiments in actual conditions must be incremented because the presence of specific chemical species can positively or negatively influence the performance of the process.

As the combined AOP treatments are still at the initial stage, in-depth research for cost reduction after a rigorous economic assessment deserves much attention and can provide positive surprises in the future. One possible benefit can be the use of solar light as an energy source in the activation of the different involved species.

6. Conclusions

The presence of pharmaceutical residues in wastewater is continuously growing due to increasing use, and their removal represents one of the emerging concerns regarding environmental protection and restoration. These compounds are present in a large variety, and they are extremely stable, very complex and highly persistent in the aquatic environment. Treatment using AOPs has been revealed to be effective for the removal of different drugs both in lab-prepared and real wastewater effluents due to the formation of highly reactive and unselective radicals which are the oxidizing species. In general, for almost all drugs, a good degradation efficiency was found in short treatment times with various AOPs. A weak point of some of these technologies is the low degree of mineralization of the drugs and the occurrence of only a partial oxidation giving rise to intermediates that are often more dangerous than the original compounds. Furthermore, many of the reported investigations have been conducted at laboratory or pilot scales, and large-scale application is still limited. This is probably due to the high operating (especially energetic) cost of most of the combined AOP processes.

Author Contributions: Writing—Original Draft Preparation, M.U.; Writing—Original Draft Preparation, T.K.; Funding Acquisition, Visualization, V.L.; Supervision, Writing—Review and Editing, L.P.; Conceptualization, Supervision, Writing—Review and Editing, M.B. All authors have read and agreed to the published version of the manuscript.

Funding: "SAMOTHRACE" (MUR, PNRR-M4C2, ECS_00000022), spoke 3-Università degli Studi di Palermo "S2-COMMs-Micro and Nanotechnologies for Smart & Sustainable Communities.

Data Availability Statement: No additional data are available.

Acknowledgments: The authors thank SiciliAn MicronanOTecH Research And Innovation CEnter for financial support.

Conflicts of Interest: The authors declare no conflict of interest.

References

1. Inamuddin; Asiri, A.M. *Correction to: Sustainable Green Chemical Processes and Their Allied Applications*; Springer Nature: Cham, Switzerland, 2020; ISBN 9783030422837.
2. Changotra, R.; Rajput, H.; Dhir, A. Natural Soil Mediated Photo Fenton-like Processes in Treatment of Pharmaceuticals: Batch and Continuous Approach. *Chemosphere* **2017**, *188*, 345–353. [CrossRef] [PubMed]
3. Wols, B.A.; Hofman-Caris, C.H.M.; Harmsen, D.J.H.; Beerendonk, E.F. Degradation of 40 Selected Pharmaceuticals by UV/H_2O_2. *Water Res.* **2013**, *47*, 5876–5888. [CrossRef] [PubMed]
4. Wang, J.; Chu, L.; Wojnárovits, L.; Takács, E. Occurrence and Fate of Antibiotics, Antibiotic Resistant Genes (ARGs) and Antibiotic Resistant Bacteria (ARB) in Municipal Wastewater Treatment Plant: An Overview. *Sci. Total Environ.* **2020**, *744*, 140997. [CrossRef] [PubMed]
5. Wang, J.; Wang, S. Removal of Pharmaceuticals and Personal Care Products (PPCPs) from Wastewater: A Review. *J. Environ. Manag.* **2016**, *182*, 620–640. [CrossRef] [PubMed]
6. Jodeh, S.; Abdelwahab, F.; Jaradat, N.; Warad, I.; Jodeh, W. Adsorption of Diclofenac from Aqueous Solution Using Cyclamen Persicum Tubers Based Activated Carbon (CTAC). *J. Assoc. Arab. Univ. Basic Appl. Sci.* **2016**, *20*, 32–38. [CrossRef]
7. Roberts, P.H.; Thomas, K.V. The Occurrence of Selected Pharmaceuticals in Wastewater Effluent and Surface Waters of the Lower Tyne Catchment. *Sci. Total Environ.* **2006**, *356*, 143–153. [CrossRef]
8. Mirzaei, R.; Yunesian, M.; Nasseri, S.; Gholami, M.; Jalilzadeh, E.; Shoeibi, S.; Mesdaghinia, A. Occurrence and Fate of Most Prescribed Antibiotics in Different Water Environments of Tehran. *Iran. Sci. Total Environ.* **2018**, *619–620*, 446–459. [CrossRef]
9. Szabó, R.K.; Megyeri, C.S.; Illés, E.; Gajda-Schrantz, K.; Mazellier, P.; Dombi, A. Phototransformation of Ibuprofen and Ketoprofen in Aqueous Solutions. *Chemosphere* **2011**, *84*, 1658–1663. [CrossRef]
10. Tran, H.T.T.; Kosslick, H.; Ibad, M.F.; Fischer, C.; Bentrup, U.; Vuong, T.H.; Nguyen, L.Q.; Schulz, A. Photocatalytic Performance of Highly Active Brookite in the Degradation of Hazardous Organic Compounds Compared to Anatase and Rutile. *Appl. Catal. B* **2017**, *200*, 647–658. [CrossRef]
11. Liu, W.; Sutton, N.B.; Rijnaarts, H.H.M.; Langenhoff, A.A.M. Anaerobic Biodegradation of Pharmaceutical Compounds Coupled to Dissimilatory Manganese (IV) or Iron (III) Reduction. *J. Hazard. Mater.* **2020**, *388*, 119361. [CrossRef]
12. Taoufik, N.; Elmchaouri, A.; Anouar, F.; Korili, S.A.; Gil, A. Improvement of the Adsorption Properties of an Activated Carbon Coated by Titanium Dioxide for the Removal of Emerging Contaminants. *J. Water Process Eng.* **2019**, *31*, 100876. [CrossRef]
13. De Souza, D.I.; Dottein, E.M.; Giacobbo, A.; Siqueira Rodrigues, M.A.; De Pinho, M.N.; Bernardes, A.M. Nanofiltration for the Removal of Norfloxacin from Pharmaceutical Effluent. *J. Environ. Chem. Eng.* **2018**, *6*, 6147–6153. [CrossRef]
14. Nguyen, L.N.; Hai, F.I.; McDonald, J.A.; Khan, S.J.; Price, W.E.; Nghiem, L.D. Continuous Transformation of Chiral Pharmaceuticals in Enzymatic Membrane Bioreactors for Advanced Wastewater Treatment. *Water Sci. Technol.* **2017**, *76*, 1816–1826. [CrossRef] [PubMed]
15. Taoufik, N.; Boumya, W.; Achak, M.; Sillanpää, M.; Barka, N. Comparative Overview of Advanced Oxidation Processes and Biological Approaches for the Removal Pharmaceuticals. *J. Environ. Manag.* **2021**, *288*, 112404. [CrossRef] [PubMed]
16. Wu, S.; Li, X.; Tian, Y.; Lin, Y.; Hu, Y.H. Excellent Photocatalytic Degradation of Tetracycline over Black Anatase-TiO_2 under Visible Light. *Chem. Eng. J.* **2021**, *406*, 126747. [CrossRef]
17. Khan, A.H.; Khan, N.A.; Ahmed, S.; Dhingra, A.; Singh, C.P.; Khan, S.U.; Mohammadi, A.A.; Changani, F.; Yousefi, M.; Alam, S.; et al. Application of Advanced Oxidation Processes Followed by Different Treatment Technologies for Hospital Wastewater Treatment. *J. Clean. Prod.* **2020**, *269*, 122411. [CrossRef]
18. Abdel-Raouf, N.; Al-Homaidan, A.A.; Ibraheem, I.B.M. Microalgae and Wastewater Treatment. *Saudi J. Biol. Sci.* **2012**, *19*, 257–275. [CrossRef] [PubMed]
19. Zhang, T.; Liu, Y.; Rao, Y.; Li, X.; Yuan, D.; Tang, S.; Zhao, Q. Enhanced Photocatalytic Activity of TiO_2 with Acetylene Black and Persulfate for Degradation of Tetracycline Hydrochloride under Visible Light. *Chem. Eng. J.* **2020**, *384*, 123350. [CrossRef]
20. Ayoub, K.; van Hullebusch, E.D.; Cassir, M.; Bermond, A. Application of Advanced Oxidation Processes for TNT Removal: A Review. *J. Hazard. Mater.* **2010**, *178*, 10–28. [CrossRef]
21. Loddo, V.; Bellardita, M.; Camera-Roda, G.; Parrino, F.; Palmisano, L. Heterogeneous Photocatalysis: A Promising Advanced Oxidation Process. In *Current Trends and Future Developments on (Bio-) Membranes: Photocatalytic Membranes and Photocatalytic Membrane Reactors*; Elsevier Inc.: Amsterdam, The Netherlands, 2018; pp. 1–43. ISBN 9780128135495.
22. Nascimento, U.M.; Azevedo, E.B. Microwaves and Their Coupling to Advanced Oxidation Processes: Enhanced Performance in Pollutants Degradation. *J. Environ. Sci. Health A Tox Hazard. Subst. Environ. Eng.* **2013**, *48*, 1056–1072. [CrossRef]

23. Deng, Y.; Zhao, R. Advanced Oxidation Processes (AOPs) in Wastewater Treatment. *Curr. Pollut. Rep.* **2015**, *1*, 167–176. [CrossRef]
24. Vogelpohl, A. Applications of AOPs in Wastewater Treatment. *Water Sci. Technol.* **2007**, *55*, 207–211. [CrossRef] [PubMed]
25. Wang, D.; Junker, A.L.; Sillanpää, M.; Jiang, Y.; Wei, Z. Photo-Based Advanced Oxidation Processes for Zero Pollution: Where Are We Now? *Engineering* **2023**, *23*, 19–23. [CrossRef]
26. Klavarioti, M.; Mantzavinos, D.; Kassinos, D. Removal of Residual Pharmaceuticals from Aqueous Systems by Advanced Oxidation Processes. *Environ. Int.* **2009**, *35*, 402–417. [CrossRef]
27. Pandis, P.K.; Kalogirou, C.; Kanellou, E.; Vaitsis, C.; Savvidou, M.G.; Sourkouni, G.; Zorpas, A.A.; Argirusis, C. Key Points of Advanced Oxidation Processes (AOPs) for Wastewater, Organic Pollutants and Pharmaceutical Waste Treatment: A Mini Review. *ChemEngineering* **2022**, *6*, 8. [CrossRef]
28. Blatchley, E.R.; Shen, C.; Scheible, O.K.; Robinson, J.P.; Ragheb, K.; Bergstrom, D.E.; Rokjer, D. Validation of Large-Scale, Monochromatic UV Disinfection Systems for Drinking Water Using Dyed Microspheres. *Water Res.* **2008**, *42*, 677–688. [CrossRef]
29. Doll, T.E.; Frimmel, F.H. Fate of Pharmaceuticals—Photodegradation by Simulated Solar UV-Light. *Chemosphere* **2003**, *52*, 1757–1769. [CrossRef]
30. Saritha, P.; Aparna, C.; Himabindu, V.; Anjaneyulu, Y. Comparison of Various Advanced Oxidation Processes for the Degradation of 4-Chloro-2 Nitrophenol. *J. Hazard. Mater.* **2007**, *149*, 609–614. [CrossRef]
31. Pereira, V.J.; Weinberg, H.S.; Linden, K.G.; Singer, P.C. UV Degradation Kinetics and Modeling of Pharmaceutical Compounds in Laboratory Grade and Surface Water via Direct and Indirect Photolysis at 254 nm. *Environ. Sci. Technol.* **2007**, *41*, 1682–1688. [CrossRef]
32. Chen, S.; Hou, Y.; Rong, Y.; Tu, L.; Yu, Z.; Sun, J.; Lan, D.; Li, Z.; Zhu, H.; Wang, S. Hydroxyl Radical and Carbonate Radical Facilitate Chlortetracycline Degradation in the Bio-Photoelectrochemical System with a Bioanode and a Bi_2O_3/CuO Photocathode Using Bicarbonate Buffer. *Chemosphere* **2022**, *296*, 134040. [CrossRef]
33. Bilal, M.; Rizwan, K.; Adeel, M.; Iqbal, H.M.N. Hydrogen-Based Catalyst-Assisted Advanced Oxidation Processes to Mitigate Emerging Pharmaceutical Contaminants. *Int. J. Hydrogen Energy* **2022**, *47*, 19555–19569. [CrossRef]
34. Guo, K.; Wu, Z.; Shang, C.; Yao, B.; Hou, S.; Yang, X.; Song, W.; Fang, J. Radical Chemistry and Structural Relationships of PPCP Degradation by UV/Chlorine Treatment in Simulated Drinking Water. *Environ. Sci. Technol.* **2017**, *51*, 10431–10439. [CrossRef]
35. Brillas, E. A Critical Review on Ibuprofen Removal from Synthetic Waters, Natural Waters, and Real Wastewaters by Advanced Oxidation Processes. *Chemosphere* **2022**, *286*, 131849. [CrossRef] [PubMed]
36. Fang, J.; Fu, Y.; Shang, C. The Roles of Reactive Species in Micropollutant Degradation in the UV/Free Chlorine System. *Environ. Sci. Technol.* **2014**, *48*, 1859–1868. [CrossRef] [PubMed]
37. Watts, M.J.; Linden, K.G. Chlorine Photolysis and Subsequent OH Radical Production during UV Treatment of Chlorinated Water. *Water Res.* **2007**, *41*, 2871–2878. [CrossRef] [PubMed]
38. Xiang, Y.; Fang, J.; Shang, C. Kinetics and Pathways of Ibuprofen Degradation by the UV/Chlorine Advanced Oxidation Process. *Water Res.* **2016**, *90*, 301–308. [CrossRef] [PubMed]
39. Kishimoto, N. State of the Art of UV/Chlorine Advanced Oxidation Processes: Their Mechanism, Byproducts Formation, Process Variation, and Applications. *J. Water Environ. Technol.* **2019**, *17*, 302–335. [CrossRef]
40. Lee, Y.; von Gunten, U. Oxidative Transformation of Micropollutants during Municipal Wastewater Treatment: Comparison of Kinetic Aspects of Selective (Chlorine, Chlorine Dioxide, FerrateVI, and Ozone) and Non-Selective Oxidants (Hydroxyl Radical). *Water Res.* **2010**, *44*, 555–566. [CrossRef]
41. Grebel, J.E.; Pignatello, J.J.; Mitch, W.A. Effect of Halide Ions and Carbonates on Organic Contaminant Degradation by Hydroxyl Radical-Based Advanced Oxidation Processes in Saline Waters. *Environ. Sci. Technol.* **2010**, *44*, 6822–6828. [CrossRef]
42. Wang, D.; Bolton, J.R.; Hofmann, R. Medium Pressure UV Combined with Chlorine Advanced Oxidation for Trichloroethylene Destruction in a Model Water. *Water Res.* **2012**, *46*, 4677–4686. [CrossRef]
43. Mota, A.L.N.; Albuquerque, L.F.; Beltrame, L.T.C.; Chiavone-Filho, O.; Machulek, A.; Nascimento, C.A.O. Brazilian Journal of Petroleum and Gas Advanced Oxidation Processes and Their Application in the Petroleum Industry: A Review. *Braz. J. Pet. Gas* **2008**, *2*, 122–142.
44. Poyatos, J.M.; Muñio, M.M.; Almecija, M.C.; Torres, J.C.; Hontoria, E.; Osorio, F. Advanced Oxidation Processes for Wastewater Treatment: State of the Art. *Water Air Soil Pollut.* **2010**, *205*, 187–204. [CrossRef]
45. Garoma, T.; Gurol, M.D. Degradation of Tert-Butyl Alcohol in Dilute Aqueous Solution by an O_3/UV Process. *Environ. Sci. Technol.* **2004**, *38*, 5246–5252. [CrossRef]
46. Ghauch, A.; Tuqan, A.M. Oxidation of Bisoprolol in Heated Persulfate/H_2O Systems: Kinetics and Products. *Chem. Eng. J.* **2012**, *183*, 162–171. [CrossRef]
47. Lee, J.; Von Gunten, U.; Kim, J.H. Persulfate-Based Advanced Oxidation: Critical Assessment of Opportunities and Roadblocks. *Environ. Sci. Technol.* **2020**, *54*, 3064–3081. [CrossRef] [PubMed]
48. Lutze, H.V.; Bircher, S.; Rapp, I.; Kerlin, N.; Bakkour, R.; Geisler, M.; Von Sonntag, C.; Schmidt, T.C. Degradation of Chlorotriazine Pesticides by Sulfate Radicals and the Influence of Organic Matter. *Environ. Sci. Technol.* **2015**, *49*, 1673–1680. [CrossRef] [PubMed]
49. Guan, Y.H.; Ma, J.; Li, X.C.; Fang, J.Y.; Chen, L.W. Influence of pH on the Formation of Sulfate and Hydroxyl Radicals in the UV/Peroxymonosulfate System. *Environ. Sci. Technol.* **2011**, *45*, 9308–9314. [CrossRef]
50. Fast, S.A.; Gude, V.G.; Truax, D.D.; Martin, J.; Magbanua, B.S. A Critical Evaluation of Advanced Oxidation Processes for Emerging Contaminants Removal. *Environ. Process.* **2017**, *4*, 283–302. [CrossRef]

51. Rosman, N.; Salleh, W.N.W.; Mohamed, M.A.; Jaafar, J.; Ismail, A.F.; Harun, Z. Hybrid Membrane Filtration-Advanced Oxidation Processes for Removal of Pharmaceutical Residue. *J. Colloid Interface Sci.* **2018**, *532*, 236–260. [CrossRef]
52. Khajouei, G.; Finklea, H.O.; Lin, L.S. UV/Chlorine Advanced Oxidation Processes for Degradation of Contaminants in Water and Wastewater: A Comprehensive Review. *J. Environ. Chem. Eng.* **2022**, *10*, 107508. [CrossRef]
53. Tian, F.X.; Ye, W.K.; Xu, B.; Hu, X.J.; Ma, S.X.; Lai, F.; Gao, Y.Q.; Xing, H.B.; Xia, W.H.; Wang, B. Comparison of UV-Induced AOPs (UV/Cl_2, UV/NH_2Cl, UV/ClO_2 and UV/H_2O_2) in the Degradation of Iopamidol: Kinetics, Energy Requirements and DBPs-Related Toxicity in Sequential Disinfection Processes. *Chem. Eng. J.* **2020**, *398*, 125570. [CrossRef] [PubMed]
54. Nidheesh, P.V.; Divyapriya, G.; Ezzahra Titchou, F.; Hamdani, M. Treatment of Textile Wastewater by Sulfate Radical Based Advanced Oxidation Processes. *Sep. Purif. Technol.* **2022**, *293*, 121115. [CrossRef]
55. Ricardo, I.A.; Paniagua, C.E.S.; Alberto, E.A.; Starling, M.C.V.M.; Agüera, A.; Trovó, A.G. A Critical Review of Trends in Advanced Oxidation Processes for the Removal of Benzophenone-3, Fipronil, and Propylparaben from Aqueous Matrices: Pathways and Toxicity Changes. *J. Water Process Eng.* **2022**, *49*, 102973. [CrossRef]
56. Rekhate, C.V.; Srivastava, J.K. Recent Advances in Ozone-Based Advanced Oxidation Processes for Treatment of Wastewater—A Review. *Chem. Eng. J. Adv.* **2020**, *3*, 100031. [CrossRef]
57. Qin, H.; Chen, H.; Zhang, X.; Yang, G.; Feng, Y. Efficient Degradation of Fulvic Acids in Water by Catalytic Ozonation with CeO_2/AC. *J. Chem. Technol. Biotechnol.* **2014**, *89*, 1402–1409. [CrossRef]
58. Guo, Y.; Yang, L.; Wang, X. The Application and Reaction Mechanism of Catalytic Ozonation in Water Treatment. *J. Environ. Anal. Toxicol.* **2012**, *2*, 1000150. [CrossRef]
59. Kasprzyk-Hordern, B.; Ziółek, M.; Nawrocki, J. Catalytic Ozonation and Methods of Enhancing Molecular Ozone Reactions in Water Treatment. *Appl. Catal. B* **2003**, *46*, 639–669. [CrossRef]
60. Cheng, S.W.; Li, Y.H.; Yuan, C.S.; Tsai, P.Y.; Shen, H.Z.; Hung, C.H. An Innovative Advanced Oxidation Technology for Effective Decomposition of Formaldehyde by Combining Iron Modified Nano-TiO_2 (Fe/TiO_2) Photocatalytic Degradation with Ozone Oxidation. *Aerosol Air Qual. Res.* **2018**, *18*, 3220–3233. [CrossRef]
61. Mashayekh-Salehi, A.; Moussavi, G.; Yaghmaeian, K. Preparation, Characterization and Catalytic Activity of a Novel Mesoporous Nanocrystalline MgO Nanoparticle for Ozonation of Acetaminophen as an Emerging Water Contaminant. *Chem. Eng. J.* **2017**, *310*, 157–169. [CrossRef]
62. Nawaz, F.; Xie, Y.; Cao, H.; Xiao, J.; Wang, Y.; Zhang, X.; Li, M.; Duan, F. Catalytic Ozonation of 4-Nitrophenol over an Mesoporous α-MnO_2 with Resistance to Leaching. *Catal. Today* **2015**, *258*, 595–601. [CrossRef]
63. Huang, W.J.; Fang, G.C.; Wang, C.C. A Nanometer-ZnO Catalyst to Enhance the Ozonation of 2,4,6-Trichlorophenol in Water. *Colloids Surf. A Physicochem. Eng. Asp.* **2005**, *260*, 45–51. [CrossRef]
64. Biard, P.F.; Werghi, B.; Soutrel, I.; Orhand, R.; Couvert, A.; Denicourt-Nowicki, A.; Roucoux, A. Efficient Catalytic Ozonation by Ruthenium Nanoparticles Supported on SiO_2 or TiO_2: Towards the Use of a Non-Woven Fiber Paper as Original Support. *Chem. Eng. J.* **2016**, *289*, 374–381. [CrossRef]
65. Qi, F.; Chu, W.; Xu, B. Comparison of Phenacetin Degradation in Aqueous Solutions by Catalytic Ozonation with $CuFe_2O_4$ and Its Precursor: Surface Properties, Intermediates and Reaction Mechanisms. *Chem. Eng. J.* **2016**, *284*, 28–36. [CrossRef]
66. Guo, Y.; Zhao, E.; Wang, J.; Zhang, X.; Huang, H.; Yu, G.; Wang, Y. Comparison of Emerging Contaminant Abatement by Conventional Ozonation, Catalytic Ozonation, O_3/H_2O_2 and Electro-Peroxone Processes. *J. Hazard. Mater.* **2020**, *389*, 121829. [CrossRef] [PubMed]
67. Turkay, O.; Ersoy, Z.G.; Barışçı, S. Review—The Application of an Electro-Peroxone Process in Water and Wastewater Treatment. *J. Electrochem. Soc.* **2017**, *164*, E94–E102. [CrossRef]
68. Yao, W.; Wang, X.; Yang, H.; Yu, G.; Deng, S.; Huang, J.; Wang, B.; Wang, Y. Removal of Pharmaceuticals from Secondary Effluents by an Electro-Peroxone Process. *Water Res.* **2016**, *88*, 826–835. [CrossRef]
69. Bakheet, B.; Yuan, S.; Li, Z.; Wang, H.; Zuo, J.; Komarneni, S.; Wang, Y. Electro-Peroxone Treatment of Orange II Dye Wastewater. *Water Res.* **2013**, *47*, 6234–6243. [CrossRef]
70. Wang, Y.; Yu, G.; Deng, S.; Huang, J.; Wang, B. The Electro-Peroxone Process for the Abatement of Emerging Contaminants: Mechanisms, Recent Advances, and Prospects. *Chemosphere* **2018**, *208*, 640–654. [CrossRef]
71. Priyadarshini, M.; Das, I.; Ghangrekar, M.M.; Blaney, L. Advanced Oxidation Processes: Performance, Advantages, and Scale-up of Emerging Technologies. *J. Environ. Manag.* **2022**, *316*, 115295. [CrossRef]
72. Shi, Y.; Xiong, Z.; Zhou, P.; Liu, Y.; Yu, S.; Xie, Z.; Wu, X.; Zheng, Y.; Shi, Y.; Xiong, Z. Review of Advanced Oxidation Processes for Treating Hospital Sewage to Achieve Decontamination and Disinfection. *Chin. Chem. Lett.* **2023**, *35*, 108714. [CrossRef]
73. Bavasso, I.; Montanaro, D.; Petrucci, E. Ozone-Based Electrochemical Advanced Oxidation Processes. *Curr. Opin. Electrochem.* **2022**, *34*, 101017. [CrossRef]
74. Shokri, A. Employing Electro-Peroxone Process for Degradation of Acid Red 88 in Aqueous Environment by Central Composite Design: A New Kinetic Study and Energy Consumption. *Chemosphere* **2022**, *296*, 133817. [CrossRef] [PubMed]
75. Dubey, S.; Joshi, A.; Trivedi, R.; Pal, D.; Prajapati, A.K. Electro-Peroxone Treatment of Rice Grain Based Distillery Biodigester Effluent: COD and Color Removal. *Water Resour. Ind.* **2021**, *25*, 100142. [CrossRef]
76. Rodríguez-Peña, M.; Mena, I.F.; Pérez, J.A.B.; Barrera-Díaz, C.E.; Rodrigo, M.A. Does Electro-Peroxonation Improve Performance of Electro-Ozonation? *J. Environ. Chem. Eng.* **2022**, *10*, 107578. [CrossRef]

77. Aramyan, S.M. Advances in Fenton and Fenton Based Oxidation Processes for Industrial Effluent Contaminants Control—A Review. *Int. J. Environ. Sci. Nat. Resour.* **2017**, *2*, 1–18. [CrossRef]
78. Santos, M.C.; Antonin, V.S.; Souza, F.M.; Aveiro, L.R.; Pinheiro, V.S.; Gentil, T.C.; Lima, T.S.; Moura, J.P.C.; Silva, C.R.; Lucchetti, L.E.B.; et al. Decontamination of Wastewater Containing Contaminants of Emerging Concern by Electrooxidation and Fenton-Based Processes—A Review on the Relevance of Materials and Methods. *Chemosphere* **2022**, *307*, 135763. [CrossRef]
79. Pacheco-Álvarez, M.; Picos Benítez, R.; Rodríguez-Narváez, O.M.; Brillas, E.; Peralta-Hernández, J.M. A Critical Review on Paracetamol Removal from Different Aqueous Matrices by Fenton and Fenton-Based Processes, and Their Combined Methods. *Chemosphere* **2022**, *303*, 134883. [CrossRef]
80. O'Dowd, K.; Pillai, S.C. Photo-Fenton Disinfection at near Neutral PH: Process, Parameter Optimization and Recent Advances. *J. Environ. Chem. Eng.* **2020**, *8*, 104063. [CrossRef]
81. Umar, M.; Aziz, H.A.; Yusoff, M.S. Trends in the Use of Fenton, Electro-Fenton and Photo-Fenton for the Treatment of Landfill Leachate. *Waste Manag.* **2010**, *30*, 2113–2121. [CrossRef]
82. Brillas, E.; Sirés, I.; Oturan, M.A. Electro-Fenton Process and Related Electrochemical Technologies Based on Fenton's Reaction Chemistry. *Chem. Rev.* **2009**, *109*, 6570–6631. [CrossRef]
83. Ismail, S.A.; Ang, W.L.; Mohammad, A.W. Electro-Fenton Technology for Wastewater Treatment: A Bibliometric Analysis of Current Research Trends, Future Perspectives and Energy Consumption Analysis. *J. Water Process Eng.* **2021**, *40*, 101952. [CrossRef]
84. Ghoneim, M.M.; El-Desoky, H.S.; Zidan, N.M. Electro-Fenton Oxidation of Sunset Yellow FCF Azo-Dye in Aqueous Solutions. *Desalination* **2011**, *274*, 22–30. [CrossRef]
85. Babuponnusami, A.; Muthukumar, K. Advanced Oxidation of Phenol: A Comparison between Fenton, Electro-Fenton, Sono-Electro-Fenton and Photo-Electro-Fenton Processes. *Chem. Eng. J.* **2012**, *183*, 1–9. [CrossRef]
86. GilPavas, E.; Dobrosz-Gómez, I.; Gómez-García, M.Á. Optimization of Solar-Driven Photo-Electro-Fenton Process for the Treatment of Textile Industrial Wastewater. *J. Water Process Eng.* **2018**, *24*, 49–55. [CrossRef]
87. Babuponnusami, A.; Muthukumar, K. A Review on Fenton and Improvements to the Fenton Process for Wastewater Treatment. *J. Environ. Chem. Eng.* **2014**, *2*, 557–572. [CrossRef]
88. Cuerda-Correa, E.M.; Alexandre-Franco, M.F.; Fern, C. Antibiotics from Water. *An Overview. Water* **2019**, *12*, 102.
89. Wang, B.; Song, Z.; Sun, L. A Review: Comparison of Multi-Air-Pollutant Removal by Advanced Oxidation Processes—Industrial Implementation for Catalytic Oxidation Processes. *Chem. Eng. J.* **2021**, *409*, 128136. [CrossRef]
90. Gautam, P.; Kumar, S.; Lokhandwala, S. Advanced Oxidation Processes for Treatment of Leachate from Hazardous Waste Landfill: A Critical Review. *J. Clean. Prod.* **2019**, *237*, 117639. [CrossRef]
91. Oturan, M.A. Outstanding Performances of the BDD Film Anode in Electro-Fenton Process: Applications and Comparative Performance. *Curr. Opin. Solid. State Mater. Sci.* **2021**, *25*, 100925. [CrossRef]
92. Nidheesh, P.V.; Trellu, C.; Vargas, H.O.; Mousset, E.; Ganiyu, S.O.; Oturan, M.A. Electro-Fenton Process in Combination with Other Advanced Oxidation Processes: Challenges and Opportunities. *Curr. Opin. Electrochem.* **2023**, *37*, 101171. [CrossRef]
93. Arefi-Oskoui, S.; Khataee, A.; Safarpour, M.; Orooji, Y.; Vatanpour, V. A Review on the Applications of Ultrasonic Technology in Membrane Bioreactors. *Ultrason. Sonochemistry* **2019**, *58*, 104633. [CrossRef] [PubMed]
94. Madhavan, J.; Theerthagiri, J.; Balaji, D.; Sunitha, S. Ultrasound: An Overview. *Molecules* **2019**, *24*, 1–18.
95. Zhou, N.; Liang, R.; Hu, A. *Nanotechnology for Water Treatment and Purification*; Springer Nature: Cham, Switzerland, 2014; Volume 22, ISBN 978-3-319-06577-9.
96. Zhang, Y.; Shaad, K.; Vollmer, D.; Ma, C. Treatment of Textile Wastewater Using Advanced Oxidation Processes—A Critical Review. *Water* **2021**, *13*, 3515. [CrossRef]
97. Grandclément, C.; Seyssiecq, I.; Piram, A.; Wong-Wah-Chung, P.; Vanot, G.; Tiliacos, N.; Roche, N.; Doumenq, P. From the Conventional Biological Wastewater Treatment to Hybrid Processes, the Evaluation of Organic Micropollutant Removal: A Review. *Water Res.* **2017**, *111*, 297–317. [CrossRef] [PubMed]
98. Mohapatra, D.P.; Kirpalani, D.M. Advancement in Treatment of Wastewater: Fate of Emerging Contaminants. *Can. J. Chem. Eng.* **2019**, *97*, 2621–2631. [CrossRef]
99. da Silva, T.L.; Costa, C.S.D.; da Silva, M.G.C.; Vieira, M.G.A. Overview of Non-Steroidal Anti-Inflammatory Drugs Degradation by Advanced Oxidation Processes. *J. Clean. Prod.* **2022**, *346*, 131226. [CrossRef]
100. Pan, Z.; Song, C.; Li, L.; Wang, H.; Pan, Y.; Wang, C.; Li, J.; Wang, T.; Feng, X. Membrane Technology Coupled with Electrochemical Advanced Oxidation Processes for Organic Wastewater Treatment: Recent Advances and Future Prospects. *Chem. Eng. J.* **2019**, *376*, 120909. [CrossRef]
101. Chaplin, B.P. Critical Review of Electrochemical Advanced Oxidation Processes for Water Treatment Applications. *Environ. Sci. Process. Impacts* **2014**, *16*, 1182–1203. [CrossRef]
102. Oturan, M.A.; Brillas, E. Electrochemical Advanced Oxidation Processes (EAOPs) for Environmental Applications. *Port. Electrochim. Acta* **2007**, *25*, 1–18. [CrossRef]
103. Ganiyu, S.O.; Martínez-Huitle, C.A.; Oturan, M.A. Electrochemical Advanced Oxidation Processes for Wastewater Treatment: Advances in Formation and Detection of Reactive Species and Mechanisms. *Curr. Opin. Electrochem.* **2021**, *27*, 100678. [CrossRef]
104. Sirés, I.; Brillas, E.; Oturan, M.A.; Rodrigo, M.A.; Panizza, M. Electrochemical Advanced Oxidation Processes: Today and Tomorrow. *A Review. Environ. Sci. Pollut. Res.* **2014**, *21*, 8336–8367. [CrossRef] [PubMed]

105. Cuervo Lumbaque, E.; Lopes Tiburtius, E.R.; Barreto-Rodrigues, M.; Sirtori, C. Current Trends in the Use of Zero-Valent Iron (Fe^0) for Degradation of Pharmaceuticals Present in Different Water Matrices. *Trends Environ. Anal. Chem.* **2019**, *24*, e00069. [CrossRef]
106. Li, Q.; Chen, Z.; Wang, H.; Yang, H.; Wen, T.; Wang, S.; Hu, B.; Wang, X. Removal of Organic Compounds by Nanoscale Zero-Valent Iron and Its Composites. *Sci. Total Environ.* **2021**, *792*, 148546. [CrossRef]
107. Fu, F.; Dionysiou, D.D.; Liu, H. The Use of Zero-Valent Iron for Groundwater Remediation and Wastewater Treatment: A Review. *J. Hazard. Mater.* **2014**, *267*, 194–205. [CrossRef] [PubMed]
108. Mahmoud, M.E.; Abdelwahab, M.S. One-Step Synthesis of Zero-Valent Sn Nanoparticles and Potential Microwave Remediation of Lead from Water. *Mater. Res. Bull.* **2021**, *134*, 111090. [CrossRef]
109. Guan, X.; Sun, Y.; Qin, H.; Li, J.; Lo, I.M.C.; He, D.; Dong, H. The Limitations of Applying Zero-Valent Iron Technology in Contaminants Sequestration and the Corresponding Countermeasures: The Development in Zero-Valent Iron Technology in the Last Two Decades (1994–2014). *Water Res.* **2015**, *75*, 224–248. [CrossRef]
110. Giannakis, S.; Lin, K.Y.A.; Ghanbari, F. A Review of the Recent Advances on the Treatment of Industrial Wastewaters by Sulfate Radical-Based Advanced Oxidation Processes (SR-AOPs). *Chem. Eng. J.* **2021**, *406*, 127083. [CrossRef]
111. Cardoso, I.M.F.; Cardoso, R.M.F.; Esteves da Silva, J.C.G. Advanced Oxidation Processes Coupled with Nanomaterials for Water Treatment. *Nanomaterials* **2021**, *11*, 2045. [CrossRef]
112. Xiao, R.; Luo, Z.; Wei, Z.; Luo, S.; Spinney, R.; Yang, W.; Dionysiou, D.D. Activation of Peroxymonosulfate/Persulfate by Nanomaterials for Sulfate Radical-Based Advanced Oxidation Technologies. *Curr. Opin. Chem. Eng.* **2018**, *19*, 51–58. [CrossRef]
113. Scaria, J.; Nidheesh, P.V. Comparison of Hydroxyl-Radical-Based Advanced Oxidation Processes with Sulfate Radical-Based Advanced Oxidation Processes. *Curr. Opin. Chem. Eng.* **2022**, *36*, 100830. [CrossRef]
114. Kappe, C.O. Controlled Microwave Heating in Modern Organic Synthesis. *Angew. Chem. Int. Ed.* **2004**, *43*, 6250–6284. [CrossRef] [PubMed]
115. Jones, D.A.; Lelyveld, T.P.; Mavrofidis, S.D.; Kingman, S.W.; Miles, N.J. Microwave Heating Applications in Environmental Engineering—A Review. *Resour. Conserv. Recycl.* **2002**, *34*, 75–90. [CrossRef]
116. Bose, S.; Kumar, M. Microwave-Assisted Persulfate/Peroxymonosulfate Process for Environmental Remediation. *Curr. Opin. Chem. Eng.* **2022**, *36*, 100826. [CrossRef]
117. Liu, X.; Huang, F.; Yu, Y.; Zhao, P.; Zhou, Y.; He, Y.; Xu, Y.; Zhang, Y. Ofloxacin Degradation over Cu–Ce Tyre Carbon Catalysts by the Microwave Assisted Persulfate Process. *Appl. Catal. B* **2019**, *253*, 149–159. [CrossRef]
118. Wang, F.; Wu, C.; Li, Q. Treatment of Refractory Organics in Strongly Alkaline Dinitrodiazophenol Wastewater with Microwave Irradiation-Activated Persulfate. *Chemosphere* **2020**, *254*, 126773. [CrossRef]
119. Verma, P.; Samanta, S.K. Microwave-Enhanced Advanced Oxidation Processes for the Degradation of Dyes in Water. *Environ. Chem. Lett.* **2018**, *16*, 969–1007. [CrossRef]
120. Gao, Y.; Zou, D. Efficient Degradation of Levofloxacin by a Microwave–3D $ZnCo_2O_4$/Activated Persulfate Process: Effects, Degradation Intermediates, and Acute Toxicity. *Chem. Eng. J.* **2020**, *393*, 124795. [CrossRef]
121. Cao, M.; Xu, P.; Tian, K.; Shi, F.; Zheng, Q.; Ma, D.; Zhang, G. Recent Advances in Microwave-Enhanced Advanced Oxidation Processes (MAOPs) for Environmental Remediation: A Review. *Chem. Eng. J.* **2023**, *471*, 144208. [CrossRef]
122. Wang, H.; Zhao, Z.; Zhang, X.; Dong, W.; Cao, Z.; He, L.; Wang, X. Rapid Decomplexation of Ni-EDTA by Microwave-Assisted Fenton Reaction. *Chem. Eng. J.* **2020**, *381*, 122703. [CrossRef]
123. Hu, L.; Zhang, G.; Wang, Q.; Wang, X.; Wang, P. Effect of Microwave Heating on Persulfate Activation for Rapid Degradation and Mineralization of P-Nitrophenol. *ACS Sustain. Chem. Eng.* **2019**, *7*, 11662–11671. [CrossRef]
124. Qi, C.; Liu, X.; Lin, C.; Zhang, X.; Ma, J.; Tan, H.; Ye, W. Degradation of Sulfamethoxazole by Microwave-Activated Persulfate: Kinetics, Mechanism and Acute Toxicity. *Chem. Eng. J.* **2014**, *249*, 6–14. [CrossRef]
125. Rauf, M.A.; Ashraf, S.S. Fundamental Principles and Application of Heterogeneous Photocatalytic Degradation of Dyes in Solution. *Chem. Eng. J.* **2009**, *151*, 10–18. [CrossRef]
126. Mahdi, M.H.; Mohammed, T.J.; Al-Najar, J.A. Advanced Oxidation Processes (AOPs) for Treatment of Antibiotics in Wastewater: A Review. *IOP Conf. Ser. Earth Environ. Sci.* **2021**, *779*, 012109. [CrossRef]
127. Dewil, R.; Mantzavinos, D.; Poulios, I.; Rodrigo, M.A. New Perspectives for Advanced Oxidation Processes. *J. Environ. Manag.* **2017**, *195*, 93–99. [CrossRef] [PubMed]
128. Banerjee, S.; Pillai, S.C.; Falaras, P.; O'shea, K.E.; Byrne, J.A.; Dionysiou, D.D. New Insights into the Mechanism of Visible Light Photocatalysis. *J. Phys. Chem. Lett.* **2014**, *5*, 2543–2554. [CrossRef] [PubMed]
129. Vellanki, B.P.; Batchelor, B.; Abdel-Wahab, A. Advanced Reduction Processes: A New Class of Treatment Processes. *Environ. Eng. Sci.* **2013**, *30*, 264–271. [CrossRef]
130. Neta, P.; Huie, R.E.; Harriman, A. One-Electron-Transfer Reactions of the Couple SO_2/SO_2^- in Aqueous Solutions. *Pulse Radiolytic and Cyclic Voltammetric Studies. J. Phys. Chem.* **1987**, *91*, 1606–1611.
131. Ibrahim, I.; Athanasekou, C.; Manolis, G.; Kaltzoglou, A.; Nasikas, N.K.; Katsaros, F.; Devlin, E.; Kontos, A.G.; Falaras, P. Photocatalysis as an Advanced Reduction Process (ARP): The Reduction of 4-Nitrophenol Using Titania Nanotubes-Ferrite Nanocomposites. *J. Hazard. Mater.* **2019**, *372*, 37–44. [CrossRef]
132. Liu, X.; Vellanki, B.P.; Batchelor, B.; Abdel-Wahab, A. Degradation of 1,2-Dichloroethane with Advanced Reduction Processes (ARPs): Effects of Process Variables and Mechanisms. *Chem. Eng. J.* **2014**, *237*, 300–307. [CrossRef]

133. Cui, J.; Gao, P.; Deng, Y. Destruction of Per- and Polyfluoroalkyl Substances (PFAS) with Advanced Reduction Processes (ARPs): A Critical Review. *Environ. Sci. Technol.* **2020**, *54*, 3752–3766. [CrossRef]
134. Yu, H.; Nie, E.; Xu, J.; Yan, S.; Cooper, W.J.; Song, W. Degradation of Diclofenac by Advanced Oxidation and Reduction Processes: Kinetic Studies, Degradation Pathways and Toxicity Assessments. *Water Res.* **2013**, *47*, 1909–1918. [CrossRef]
135. Jeong, J.; Song, W.; Cooper, W.J.; Jung, J.; Greaves, J. Degradation of Tetracycline Antibiotics: Mechanisms and Kinetic Studies for Advanced Oxidation/Reduction Processes. *Chemosphere* **2010**, *78*, 533–540. [CrossRef] [PubMed]
136. Akerdi, A.G.; Bahrami, S.H. Application of Heterogeneous Nano-Semiconductors for Photocatalytic Advanced Oxidation of Organic Compounds: A Review. *J. Environ. Chem. Eng.* **2019**, *7*, 103283. [CrossRef]
137. Macedo, L.C.; Zaia, D.A.M.; Moore, G.J.; de Santana, H. Degradation of Leather Dye on TiO_2: A Study of Applied Experimental Parameters on Photoelectrocatalysis. *J. Photochem. Photobiol. A Chem.* **2007**, *185*, 86–93. [CrossRef]
138. Wang, C.C.; Lee, C.K.; Lyu, M.D.; Juang, L.C. Photocatalytic Degradation of C.I. Basic Violet 10 Using TiO2 Catalysts Supported by Y Zeolite: An Investigation of the Effects of Operational Parameters. *Dye. Pigment.* **2008**, *76*, 817–824. [CrossRef]
139. Zhang, Y.; Zhou, J.; Chen, X.; Wang, L.; Cai, W. Coupling of Heterogeneous Advanced Oxidation Processes and Photocatalysis in Efficient Degradation of Tetracycline Hydrochloride by Fe-Based MOFs: Synergistic Effect and Degradation Pathway. *Chem. Eng. J.* **2019**, *369*, 745–757. [CrossRef]
140. Antoniou, M.G.; Hey, G.; Rodríguez Vega, S.; Spiliotopoulou, A.; Fick, J.; Tysklind, M.; la Cour Jansen, J.; Andersen, H.R. Required Ozone Doses for Removing Pharmaceuticals from Wastewater Effluents. *Sci. Total Environ.* **2013**, *456–457*, 42–49. [CrossRef]
141. Lado Ribeiro, A.R.; Moreira, N.F.F.; Li Puma, G.; Silva, A.M.T. Impact of Water Matrix on the Removal of Micropollutants by Advanced Oxidation Technologies. *Chem. Eng. J.* **2019**, *363*, 155–173. [CrossRef]
142. Li, D.; Feng, Z.; Zhou, B.; Chen, H.; Yuan, R. Impact of Water Matrices on Oxidation Effects and Mechanisms of Pharmaceuticals by Ultraviolet-Based Advanced Oxidation Technologies: A Review. *Sci. Total Environ.* **2022**, *844*, 157162. [CrossRef]
143. Domingues, E.; Fernandes, E.; Gomes, J.; Martins, R.C. Advanced Oxidation Processes Perspective Regarding Swine Wastewater Treatment. *Sci. Total Environ.* **2021**, *776*, 145958. [CrossRef]
144. Kurian, M. Advanced Oxidation Processes and Nanomaterials: A Review. *Clean. Eng. Technol.* **2021**, *2*, 100090. [CrossRef]
145. Henglein, A. Small-Particle Research: Physicochemical Properties of Extremely Small Colloidal Metal and Semiconductor Particles. *Chem. Rev.* **1989**, *89*, 1861–1873. [CrossRef]
146. Adeleye, A.S.; Conway, J.R.; Garner, K.; Huang, Y.; Su, Y.; Keller, A.A. Engineered Nanomaterials for Water Treatment and Remediation: Costs, Benefits, and Applicability. *Chem. Eng. J.* **2016**, *286*, 640–662. [CrossRef]
147. Cai, Z.; Sun, Y.; Liu, W.; Pan, F.; Sun, P.; Fu, J. An Overview of Nanomaterials Applied for Removing Dyes from Wastewater. *Environ. Sci. Pollut. Res.* **2017**, *24*, 15882–15904. [CrossRef]
148. Lei, S.; Wang, S.; Gao, B.; Zhan, Y.; Zhao, Q.; Jin, S.; Song, G.; Lyu, X.; Zhang, Y.; Tang, Y. Ultrathin Dodecyl-Sulfate-Intercalated Mg-Al Layered Double Hydroxide Nanosheets with High Adsorption Capability for Dye Pollution. *J. Colloid Interface Sci.* **2020**, *577*, 181–190. [CrossRef]
149. Janson, O.; Unosson, E.; Strømme, M.; Engqvist, H.; Welch, K. Organic Degradation Potential of a TiO_2/H_2O_2/UV–Vis System for Dental Applications. *J. Dent.* **2017**, *67*, 53–57. [CrossRef]
150. Massoudinejad, M.; Keramati, H.; Ghaderpoori, M. Investigation of Photo-Catalytic Removal of Arsenic from Aqueous Solutions Using UV/H_2O_2 in the Presence of ZnO Nanoparticles. *Chem. Eng. Commun.* **2020**, *207*, 1605–1615. [CrossRef]
151. Hu, L.; Wang, P.; Liu, G.; Zheng, Q.; Zhang, G. Catalytic Degradation of P-Nitrophenol by Magnetically Recoverable Fe_3O_4 as a Persulfate Activator under Microwave Irradiation. *Chemosphere* **2020**, *240*, 124977. [CrossRef]
152. Schröder, P.; Helmreich, B.; Škrbić, B.; Carballa, M.; Papa, M.; Pastore, C.; Emre, Z.; Oehmen, A.; Langenhoff, A.; Molinos, M.; et al. Status of Hormones and Painkillers in Wastewater Effluents across Several European States—Considerations for the EU Watch List Concerning Estradiols and Diclofenac. *Environ. Sci. Pollut. Res.* **2016**, *23*, 12835–12866. [CrossRef]
153. Sellaro, M.; Bellardita, M.; Brunetti, A.; Fontananova, E.; Palmisano, L.; Drioli, E.; Barbieri, G. CO_2 Conversion in a Photocatalytic Continuous Membrane Reactor. *RSC Adv.* **2016**, *6*, 67418–67427. [CrossRef]
154. Bellardita, M.; Camera-Roda, G.; Loddo, V.; Parrino, F.; Palmisano, L. Coupling of Membrane and Photocatalytic Technologies for Selective Formation of High Added Value Chemicals. *Catal. Today* **2020**, *340*, 128–144. [CrossRef]
155. Camera-Roda, G.; Santarelli, F.; Augugliaro, V.; Loddo, V.; Palmisano, G.; Palmisano, L.; Yurdakal, S. Photocatalytic Process Intensification by Coupling with Pervaporation. *Catal. Today* **2011**, *161*, 209–213. [CrossRef]
156. Molinari, R.; Lavorato, C.; Argurio, P. Recent Progress of Photocatalytic Membrane Reactors in Water Treatment and in Synthesis of Organic Compounds. *A Review. Catal. Today* **2017**, *281*, 144–164. [CrossRef]
157. Gopal, G.; Alex, S.A.; Chandrasekaran, N.; Mukherjee, A. A Review on Tetracycline Removal from Aqueous Systems by Advanced Treatment Techniques. *RSC Adv.* **2020**, *10*, 27081–27095. [CrossRef]
158. Emzhina, V.; Kuzin, E.; Babusenko, E.; Krutchinina, N. Photodegradation of Tetracycline in Presence of H_2O_2 and Metal Oxide Based Catalysts. *J. Water Process Eng.* **2021**, *39*, 101696. [CrossRef]
159. Shi, W.; Fu, Y.; Hao, C.; Guo, F.; Tang, Y. Heterogeneous Photo-Fenton Process over Magnetically Recoverable $MnFe_2O_4$/MXene Hierarchical Heterostructure for Boosted Degradation of Tetracycline. *Mater. Today Commun.* **2022**, *33*, 104449. [CrossRef]
160. Davies, A.K.; McKellar, J.F.; Phillips, G.O.; Reid, A.G. Photochemical Oxidation of Tetracycline in Aqueous Solution. *J. Chem. Soc. Perkin Trans. 2* **1979**, 369–375. [CrossRef]

161. López-Peñalver, J.J.; Sánchez-Polo, M.; Gómez-Pacheco, C.V.; Rivera-Utrilla, J. Photodegradation of Tetracyclines in Aqueous Solution by Using UV and UV/H_2O_2 Oxidation Processes. *J. Chem. Technol. Biotechnol.* **2010**, *85*, 1325–1333. [CrossRef]
162. Serrà, A.; Gómez, E.; Michler, J.; Philippe, L. Facile Cost-Effective Fabrication of Cu@Cu_2O@CuO–Microalgae Photocatalyst with Enhanced Visible Light Degradation of Tetracycline. *Chem. Eng. J.* **2021**, *413*, 127477. [CrossRef]
163. Bautitz, I.R.; Nogueira, R.F.P. Degradation of Tetracycline by Photo-Fenton Process-Solar Irradiation and Matrix Effects. *J. Photochem. Photobiol. A Chem.* **2007**, *187*, 33–39. [CrossRef]
164. Yamal-Turbay, E.; Jaén, E.; Graells, M.; Pérez-Moya, M. Enhanced Photo-Fenton Process for Tetracycline Degradation Using Efficient Hydrogen Peroxide Dosage. *J. Photochem. Photobiol. A Chem.* **2013**, *267*, 11–16. [CrossRef]
165. Zhuang, Y.; Liu, Q.; Kong, Y.; Shen, C.; Hao, H.; Dionysiou, D.D.; Shi, B. Enhanced Antibiotic Removal through a Dual-Reaction-Center Fenton-like Process in 3D Graphene Based Hydrogels. *Environ. Sci. Nano* **2019**, *6*, 388–398. [CrossRef]
166. Liu, S.; Zhao, X.R.; Sun, H.Y.; Li, R.P.; Fang, Y.F.; Huang, Y.P. The Degradation of Tetracycline in a Photo-Electro-Fenton System. *Chem. Eng. J.* **2013**, *231*, 441–448. [CrossRef]
167. Ao, X.; Sun, W.; Li, S.; Yang, C.; Li, C.; Lu, Z. Degradation of Tetracycline by Medium Pressure UV-Activated Peroxymonosulfate Process: Influencing Factors, Degradation Pathways, and Toxicity Evaluation. *Chem. Eng. J.* **2019**, *361*, 1053–1062. [CrossRef]
168. Nasseri, S.; Mahvi, A.H.; Seyedsalehi, M.; Yaghmaeian, K.; Nabizadeh, R.; Alimohammadi, M.; Safari, G.H. Degradation Kinetics of Tetracycline in Aqueous Solutions Using Peroxydisulfate Activated by Ultrasound Irradiation: Effect of Radical Scavenger and Water Matrix. *J. Mol. Liq.* **2017**, *241*, 704–714. [CrossRef]
169. Hou, L.; Zhang, H.; Xue, X. Ultrasound Enhanced Heterogeneous Activation of Peroxydisulfate by Magnetite Catalyst for the Degradation of Tetracycline in Water. *Sep. Purif. Technol.* **2012**, *84*, 147–152. [CrossRef]
170. Ji, Y.; Shi, Y.; Dong, W.; Wen, X.; Jiang, M.; Lu, J. Thermo-Activated Persulfate Oxidation System for Tetracycline Antibiotics Degradation in Aqueous Solution. *Chem. Eng. J.* **2016**, *298*, 225–233. [CrossRef]
171. Zhang, X.; Deng, H.; Zhang, G.; Yang, F.; Yuan, G.E. Natural Bornite as an Efficient and Cost-Effective Persulfate Activator for Degradation of Tetracycline: Performance and Mechanism. *Chem. Eng. J.* **2020**, *381*, 122717. [CrossRef]
172. Yang, Q.; Yang, X.; Yan, Y.; Sun, C.; Wu, H.; He, J.; Wang, D. Heterogeneous Activation of Peroxymonosulfate by Different Ferromanganese Oxides for Tetracycline Degradation: Structure Dependence and Catalytic Mechanism. *Chem. Eng. J.* **2018**, *348*, 263–270. [CrossRef]
173. Guan, R.; Yuan, X.; Wu, Z.; Wang, H.; Jiang, L.; Zhang, J.; Li, Y.; Zeng, G.; Mo, D. Accelerated Tetracycline Degradation by Persulfate Activated with Heterogeneous Magnetic $Ni_xFe_{3-x}O_4$ Catalysts. *Chem. Eng. J.* **2018**, *350*, 573–584. [CrossRef]
174. Gao, Y.; Cong, S.; He, Y.; Zou, D.; Liu, Y.; Yao, B.; Sun, W. Study on the Mechanism of Degradation of Tetracycline Hydrochloride by Microwave-Activated Sodium Persulfate. *Water Sci. Technol.* **2020**, *82*, 1961–1970. [CrossRef] [PubMed]
175. Khan, M.H.; Bae, H.; Jung, J.Y. Tetracycline Degradation by Ozonation in the Aqueous Phase: Proposed Degradation Intermediates and Pathway. *J. Hazard. Mater.* **2010**, *181*, 659–665. [CrossRef] [PubMed]
176. Wang, Y.; Zhang, H.; Zhang, J.; Lu, C.; Huang, Q.; Wu, J.; Liu, F. Degradation of Tetracycline in Aqueous Media by Ozonation in an Internal Loop-Lift Reactor. *J. Hazard. Mater.* **2011**, *192*, 35–43. [CrossRef]
177. Wang, Y.; Zhang, H.; Chen, L. Ultrasound Enhanced Catalytic Ozonation of Tetracycline in a Rectangular Air-Lift Reactor. *Catal. Today* **2011**, *175*, 283–292. [CrossRef]
178. Gómez-Pacheco, C.V.; Sánchez-Polo, M.; Rivera-Utrilla, J.; López-Peñalver, J. Tetracycline Removal from Waters by Integrated Technologies Based on Ozonation and Biodegradation. *Chem. Eng. J.* **2011**, *178*, 115–121. [CrossRef]
179. Hou, L.; Zhang, H.; Wang, L.; Chen, L. Ultrasound-Enhanced Magnetite Catalytic Ozonation of Tetracycline in Water. *Chem. Eng. J.* **2013**, *229*, 577–584. [CrossRef]
180. Wang, Y.; Zhang, H.; Chen, L.; Wang, S.; Zhang, D. Ozonation Combined with Ultrasound for the Degradation of Tetracycline in a Rectangular Air-Lift Reactor. *Sep. Purif. Technol.* **2012**, *84*, 138–146. [CrossRef]
181. Safari, G.H.; Hoseini, M.; Seyedsalehi, M.; Kamani, H.; Jaafari, J.; Mahvi, A.H. Photocatalytic Degradation of Tetracycline Using Nanosized Titanium Dioxide in Aqueous Solution. *Int. J. Environ. Sci. Technol.* **2015**, *12*, 603–616. [CrossRef]
182. Wang, P.; Yap, P.S.; Lim, T.T. C-N-S Tridoped TiO_2 for Photocatalytic Degradation of Tetracycline under Visible-Light Irradiation. *Appl. Catal. A Gen.* **2011**, *399*, 252–261. [CrossRef]
183. Mohammadi, M.; Sabbaghi, S.; Binazadeh, M.; Ghaedi, S. Chemosphere Type-1 α-Fe_2O_3/TiO_2 Photocatalytic Degradation of Tetracycline from Wastewater Using CCD-Based RSM Optimization. *Chemosphere* **2023**, *336*, 139311. [CrossRef]
184. Sharma, M.; Mandal, M.K.; Pandey, S.; Kumar, R.; Dubey, K.K. Visible-Light-Driven Photocatalytic Degradation of Tetracycline Using Heterostructured Cu_2O-TiO_2 Nanotubes, Kinetics, and Toxicity Evaluation of Degraded Products on Cell Lines. *ACS Omega* **2022**, *7*, 33572–33586. [CrossRef] [PubMed]
185. Ahmadi, M.; Ramezani Motlagh, H.; Jaafarzadeh, N.; Mostoufi, A.; Saeedi, R.; Barzegar, G.; Jorfi, S. Enhanced Photocatalytic Degradation of Tetracycline and Real Pharmaceutical Wastewater Using MWCNT/TiO_2 Nano-Composite. *J. Environ. Manag.* **2017**, *186*, 55–63. [CrossRef] [PubMed]
186. Wang, W.; Fang, J.; Shao, S.; Lai, M.; Lu, C. Compact and Uniform TiO_2@g-C_3N_4 Core-Shell Quantum Heterojunction for Photocatalytic Degradation of Tetracycline Antibiotics. *Appl. Catal. B* **2017**, *217*, 57–64. [CrossRef]
187. Phakathi, N.A.; Tichapondwa, S.M.; Chirwa, E.M.N. Photocatalytic Degradation of Tetracycline Using Visible-Light-Driven Porous g-C_3N_4 Nanosheets Catalyst. *Chem. Eng. Trans.* **2022**, *96*, 391–396. [CrossRef]

188. Gaffar, S.; Kumar, A.; Alam, J.; Riaz, U. Efficient Visible Light–Induced Photocatalytic Degradation of Tetracycline Hydrochloride Using $CuFe_2O_4$ and PANI/$CuFe_2O_4$ Nanohybrids. *Environ. Sci. Pollut. Res.* **2023**, *30*, 108878–108888. [CrossRef] [PubMed]
189. Liu, W.; Wei, C.; Peng, R.; Chu, R.; Sun, H.; Zhang, X.; Xie, F. Persulfate Assisted Photocatalytic Degradation of Tetracycline by Bismuth Titanate under Visible Light Irradiation. *New J. Chem.* **2022**, *46*, 10854–10862. [CrossRef]
190. Kaushal, S.; Kumar, A.; Bains, H.; Singh, P.P. Photocatalytic Degradation of Tetracycline Antibiotic and Organic Dyes Using Biogenic Synthesized CuO/Fe_2O_3 Nanocomposite: Pathways and Mechanism Insights. *Environ. Sci. Pollut. Res.* **2023**, *30*, 37092–37104. [CrossRef]
191. Semeraro, P.; Bettini, S.; Sawalha, S.; Pal, S.; Licciulli, A.; Marzo, F.; Lovergine, N.; Valli, L.; Giancane, G. Photocatalytic Degradation of Tetracycline by ZnO/γ-Fe_2O_3 Paramagnetic Nanocomposite Material. *Nanomaterials* **2020**, *10*, 1458. [CrossRef]
192. Nguetsa Kuate, L.J.; Chen, Z.; Lu, J.; Wen, H.; Guo, F.; Shi, W. Photothermal-Assisted Photocatalytic Degradation of Tetracycline in Seawater Based on the Black g-C_3N_4 Nanosheets with Cyano Group Defects. *Catalysts* **2023**, *13*, 1147. [CrossRef]
193. Chen, F.; Yang, Q.; Sun, J.; Yao, F.; Wang, S.; Wang, Y.; Wang, X.; Li, X.; Niu, C.; Wang, D.; et al. Enhanced Photocatalytic Degradation of Tetracycline by AgI/$BiVO_4$ Heterojunction under Visible-Light Irradiation: Mineralization Efficiency and Mechanism. *ACS Appl. Mater. Interfaces* **2016**, *8*, 32887–32900. [CrossRef]
194. Jia, X.; Wang, F.; Xu, X.; Liu, C.; Zhang, L.; Jiao, S.; Zhu, G.; Wang, X.; Yu, G. Highly Efficient Photocatalytic Degradation of Tetracycline by Modifying UiO-66 via Different Regulation Strategies. *ACS Omega* **2023**, *8*, 27375–27385. [CrossRef]
195. Oluwole, A.O.; Olatunji, O.S. Photocatalytic Degradation of Tetracycline in Aqueous Systems under Visible Light Irridiation Using Needle-like SnO_2 Nanoparticles Anchored on Exfoliated g-C_3N_4. *Environ. Sci. Eur.* **2022**, *34*, 5. [CrossRef]
196. Zhang, X.X.; Wang, X.J.; Niu, Y.Y. Photocatalytic Degradation of Tetracycline by Supramolecular Materials Constructed with Organic Cations and Silver Iodide. *Catalysts* **2022**, *12*, 1581. [CrossRef]
197. Doosti, M.; Jahanshahi, R.; Laleh, S.; Sobhani, S.; Sansano, J.M. Solar Light Induced Photocatalytic Degradation of Tetracycline in the Presence of $ZnO/NiFe_2O_4/Co_3O_4$ as a New and Highly Efficient Magnetically Separable Photocatalyst. *Front. Chem.* **2022**, *10*, 1013349. [CrossRef]
198. Li, S.; Wang, C.; Liu, Y.; Cai, M.; Wang, Y.; Zhang, H.; Guo, Y.; Zhao, W.; Wang, Z.; Chen, X. Photocatalytic Degradation of Tetracycline Antibiotic by a Novel $Bi_2Sn_2O_7/Bi_2MoO_6$ S-Scheme Heterojunction: Performance, Mechanism Insight and Toxicity Assessment. *Chem. Eng. J.* **2022**, *429*, 132519. [CrossRef]
199. Huang, J.; Xue, P.; Wang, S.; Han, S.; Lin, L.; Chen, X.; Wang, Z. Fabrication of Zirconium-Based Metal-Organic Frameworks@tungsten Trioxide (UiO-66-NH_2@WO_3) Heterostructure on Carbon Cloth for Efficient Photocatalytic Removal of Tetracycline Antibiotic under Visible Light. *J. Colloid Interface Sci.* **2022**, *606*, 1509–1523. [CrossRef]
200. Palominos, R.A.; Mondaca, M.A.; Giraldo, A.; Peñuela, G.; Pérez-Moya, M.; Mansilla, H.D. Photocatalytic Oxidation of the Antibiotic Tetracycline on TiO_2 and ZnO Suspensions. *Catal. Today* **2009**, *144*, 100–105. [CrossRef]
201. Wang, X.; Wang, A.; Ma, J. Visible-Light-Driven Photocatalytic Removal of Antibiotics by Newly Designed C_3N_4@$MnFe_2O_4$-Graphene Nanocomposites. *J. Hazard. Mater.* **2017**, *336*, 81–92. [CrossRef] [PubMed]
202. Alyani, S.J.; Pirbazari, A.E.; Khalilsaraei, F.E.; Kolur, N.A.; Gilani, N. Growing Co-Doped TiO_2 Nanosheets on Reduced Graphene Oxide for Efficient Photocatalytic Removal of Tetracycline Antibiotic from Aqueous Solution and Modeling the Process by Artificial Neural Network. *J. Alloys Compd.* **2019**, *799*, 169–182. [CrossRef]
203. Song, J.; Wu, X.; Zhang, M.; Liu, C.; Yu, J.; Sun, G.; Si, Y.; Ding, B. Highly Flexible, Core-Shell Heterostructured, and Visible-Light-Driven Titania-Based Nanofibrous Membranes for Antibiotic Removal and E. *Coil Inactivation. Chem. Eng. J.* **2020**, *379*, 122269. [CrossRef]
204. Wang, Q.; Li, P.; Zhang, Z.; Jiang, C.; Zuojiao, K.; Liu, J.; Wang, Y. Kinetics and Mechanism Insights into the Photodegradation of Tetracycline Hydrochloride and Ofloxacin Mixed Antibiotics with the Flower-like $BiOCl/TiO_2$ Heterojunction. *J. Photochem. Photobiol. A Chem.* **2019**, *378*, 114–124. [CrossRef]
205. Wierzbicka, E.; Altomare, M.; Wu, M.; Liu, N.; Yokosawa, T.; Fehn, D.; Qin, S.; Meyer, K.; Unruh, T.; Spiecker, E.; et al. Reduced Grey Brookite for Noble Metal Free Photocatalytic H2evolution. *J. Mater. Chem. A Mater.* **2021**, *9*, 1168–1179. [CrossRef]
206. Belhouchet, N.; Hamdi, B.; Chenchouni, H.; Bessekhouad, Y. Photocatalytic Degradation of Tetracycline Antibiotic Using New Calcite/Titania Nanocomposites. *J. Photochem. Photobiol. A Chem.* **2019**, *372*, 196–205. [CrossRef]
207. Lakhera, S.K.; Hafeez, H.Y.; Veluswamy, P.; Ganesh, V.; Khan, A.; Ikeda, H.; Neppolian, B. Enhanced Photocatalytic Degradation and Hydrogen Production Activity of in Situ Grown TiO_2 Coupled $NiTiO_3$ Nanocomposites. *Appl. Surf. Sci.* **2018**, *449*, 790–798. [CrossRef]
208. Yang, Q.; Wang, S.; Chen, F.; Luo, K.; Sun, J.; Gong, C.; Yao, F.; Wang, X.; Wu, J.; Li, X.; et al. Enhanced Visible-Light-Driven Photocatalytic Removal of Refractory Pollutants by Zn/Fe Mixed Metal Oxide Derived from Layered Double Hydroxide. *Catal. Commun.* **2017**, *99*, 15–19. [CrossRef]
209. Galedari, M.; Mehdipour Ghazi, M.; Rashid Mirmasoomi, S. Photocatalytic Process for the Tetracycline Removal under Visible Light: Presenting a Degradation Model and Optimization Using Response Surface Methodology (RSM). *Chem. Eng. Res. Des.* **2019**, *145*, 323–333. [CrossRef]
210. Yu, J.; Kiwi, J.; Zivkovic, I.; Rønnow, H.M.; Wang, T.; Rtimi, S. Quantification of the Local Magnetized Nanotube Domains Accelerating the Photocatalytic Removal of the Emerging Pollutant Tetracycline. *Appl. Catal. B* **2019**, *248*, 450–458. [CrossRef]

211. Hong, Y.; Li, C.; Yin, B.; Li, D.; Zhang, Z.; Mao, B.; Fan, W.; Gu, W.; Shi, W. Promoting Visible-Light-Induced Photocatalytic Degradation of Tetracycline by an Efficient and Stable Beta-Bi_2O_3@ g-C_3N_4 Core/Shell Nanocomposite. *Chem. Eng. J.* **2018**, *338*, 137–146. [CrossRef]
212. Tun, P.P.; Wang, J.; Khaing, T.T.; Wu, X.; Zhang, G. Fabrication of Functionalized Plasmonic Ag Loaded Bi_2O_3/Montmorillonite Nanocomposites for Efficient Photocatalytic Removal of Antibiotics and Organic Dyes. *J. Alloys Compd.* **2020**, *818*, 152836. [CrossRef]
213. Heidari, S.; Haghighi, M.; Shabani, M. Sono-Photodeposition of Ag over Sono-Fabricated Mesoporous $Bi_2Sn_2O_7$-Two Dimensional Carbon Nitride: Type-II Plasmonic Nano-Heterojunction with Simulated Sunlight-Driven Elimination of Drug. *Chem. Eng. J.* **2020**, *389*, 123418. [CrossRef]
214. Wang, C.Y.; Zhang, X.; Qiu, H.B.; Huang, G.X.; Yu, H.Q. $Bi_{24}O_{31}Br_{10}$ Nanosheets with Controllable Thickness for Visible–Light–Driven Catalytic Degradation of Tetracycline Hydrochloride. *Appl. Catal. B* **2017**, *205*, 615–623. [CrossRef]
215. Ha, G.H.; Mohan, H.; Oh, H.S.; Kim, G.; Seralathan, K.K.; Shin, T. Photocatalytic Degradation of Tetracycline Using Hybrid Ag/Ag_2S@BiOI Nanowires: Degradation Mechanism and Toxicity Evaluation. *Chemosphere* **2022**, *303*, 135091. [CrossRef] [PubMed]
216. Ye, J.; Liu, J.; Huang, Z.; Wu, S.; Dai, X.; Zhang, L.; Cui, L. Effect of Reduced Graphene Oxide Doping on Photocatalytic Reduction of Cr(VI) and Photocatalytic Oxidation of Tetracycline by ZnAlTi Layered Double Oxides under Visible Light. *Chemosphere* **2019**, *227*, 505–513. [CrossRef]
217. Yu, F.; Li, Y.; Han, S.; Ma, J. Adsorptive Removal of Antibiotics from Aqueous Solution Using Carbon Materials. *Chemosphere* **2016**, *153*, 365–385. [CrossRef] [PubMed]
218. Bai, X.; Wang, Y.J.; Li, Y.; Wang, X.J. Adsorption–Photocatalytical Remediation for Series of Tetracycline Contaminants with BiOCl–CdS Composite under Simulated Sunlight. *J. Taiwan Inst. Chem. Eng.* **2019**, *104*, 94–105. [CrossRef]
219. Méndez-Arriaga, F.; Torres-Palma, R.A.; Pétrier, C.; Esplugas, S.; Gimenez, J.; Pulgarin, C. Ultrasonic Treatment of Water Contaminated with Ibuprofen. *Water Res.* **2008**, *42*, 4243–4248. [CrossRef]
220. Lee, Y.; Lee, S.; Cui, M.; Ren, Y.; Park, B.; Ma, J.; Han, Z.; Khim, J. Activation of Peroxodisulfate and Peroxymonosulfate by Ultrasound with Different Frequencies: Impact on Ibuprofen Removal Efficient, Cost Estimation and Energy Analysis. *Chem. Eng. J.* **2021**, *413*, 127487. [CrossRef]
221. da Luz, V.C.; Bazoti, S.F.; Behling, L.; Dalla Rosa, C.; Pasquali, G.D.L. Enhanced UV Direct Photolysis and UV/H_2O_2 for Oxidation of Triclosan and Ibuprofen in Synthetic Effluent: An Experimental Study. *Water Air Soil Pollut.* **2022**, *233*, 126. [CrossRef]
222. Adityosulindro, S.; Julcour, C.; Barthe, L. Heterogeneous Fenton Oxidation Using Fe-ZSM5 Catalyst for Removal of Ibuprofen in Wastewater. *J. Environ. Chem. Eng.* **2018**, *6*, 5920–5928. [CrossRef]
223. Adityosulindro, S.; Julcour, C.; Riboul, D.; Barthe, L. Degradation of Ibuprofen by Photo-Based Advanced Oxidation Processes: Exploring Methods of Activation and Related Reaction Routes. *Int. J. Environ. Sci. Technol.* **2022**, *19*, 3247–3260. [CrossRef]
224. Méndez-Arriaga, F.; Esplugas, S.; Giménez, J. Degradation of the Emerging Contaminant Ibuprofen in Water by Photo-Fenton. *Water Res.* **2010**, *44*, 589–595. [CrossRef]
225. Quero-Pastor, M.J.; Garrido-Perez, M.C.; Acevedo, A.; Quiroga, J.M. Ozonation of Ibuprofen: A Degradation and Toxicity Study. *Sci. Total Environ.* **2014**, *466–467*, 957–964. [CrossRef]
226. Almeida, V.M.; Orge, C.A.; Pereira, M.F.R.; Soares, O.S.G.P. O_3 Based Advanced Oxidation for Ibuprofen Degradation. *Chin. J. Chem. Eng.* **2022**, *42*, 277–284. [CrossRef]
227. Rao, Y.F.; Xue, D.; Pan, H.; Feng, J.; Li, Y. Degradation of Ibuprofen by a Synergistic UV/Fe(III)/Oxone Process. *Chem. Eng. J.* **2016**, *283*, 65–75. [CrossRef]
228. Li, X.; Wang, Y.; Yuan, S.; Li, Z.; Wang, B.; Huang, J.; Deng, S.; Yu, G. Degradation of the Anti-Inflammatory Drug Ibuprofen by Electro-Peroxone Process. *Water Res.* **2014**, *63*, 81–93. [CrossRef] [PubMed]
229. Miranda, M.O.; Cabral Cavalcanti, W.E.; Barbosa, F.F.; Antonio De Sousa, J.; Ivan Da Silva, F.; Pergher, S.B.C.; Braga, T.P. Photocatalytic Degradation of Ibuprofen Using Titanium Oxide: Insights into the Mechanism and Preferential Attack of Radicals. *RSC Adv.* **2021**, *11*, 27720–27733. [CrossRef] [PubMed]
230. Candido, J.P.; Andrade, S.J.; Fonseca, A.L.; Silva, F.S.; Silva, M.R.A.; Kondo, M.M. Ibuprofen Removal by Heterogeneous Photocatalysis and Ecotoxicological Evaluation of the Treated Solutions. *Environ. Sci. Pollut. Res.* **2016**, *23*, 19911–19920. [CrossRef]
231. Khalaf, S.; Shoqeir, J.H.; Lelario, F.; Scrano, L.; Bufo, S.A.; Karaman, R. TiO_2 and Active Coated Glass Photodegradation of Ibuprofen. *Catalysts* **2020**, *10*, 560. [CrossRef]
232. Agócs, T.Z.; Puskás, I.; Varga, E.; Molnár, M.; Fenyvesi, É. Stabilization of Nanosized Titanium Dioxide by Cyclodextrin Polymers and Its Photocatalytic Effect on the Degradation of Wastewater Pollutants. *Beilstein J. Org. Chem.* **2016**, *12*, 2873–2882. [CrossRef]
233. Jallouli, N.; Pastrana-Martínez, L.M.; Ribeiro, A.R.; Moreira, N.F.F.; Faria, J.L.; Hentati, O.; Silva, A.M.T.; Ksibi, M. Heterogeneous Photocatalytic Degradation of Ibuprofen in Ultrapure Water, Municipal and Pharmaceutical Industry Wastewaters Using a TiO_2/UV-LED System. *Chem. Eng. J.* **2018**, *334*, 976–984. [CrossRef]
234. Jiménez-Salcedo, M.; Monge, M.; Tena, M.T. Photocatalytic Degradation of Ibuprofen in Water Using TiO_2/UV and g-C_3N_4/Visible Light: Study of Intermediate Degradation Products by Liquid Chromatography Coupled to High-Resolution Mass Spectrometry. *Chemosphere* **2019**, *215*, 605–618. [CrossRef]

235. Li, F.; Kang, Y.; Chen, M.; Liu, G.; Lv, W.; Yao, K.; Chen, P.; Huang, H. Photocatalytic Degradation and Removal Mechanism of Ibuprofen via Monoclinic $BiVO_4$ under Simulated Solar Light. *Chemosphere* **2016**, *150*, 139–144. [CrossRef] [PubMed]
236. Alhumade, H.; Akhtar, J.; Al-Shahrani, S.; Moujdin, I.A.; Tahir, M.B. Ozonation of Ibuprofen in Presence of $SrWO_4/ZnO$ Photo-Catalyst. *Emerg. Contam.* **2022**, *8*, 391–399. [CrossRef]
237. Fidelis, M.Z.; Favaro, Y.B.; dos Santos, A.S.G.G.; Pereira, M.F.R.; Brackmann, R.; Lenzi, G.G.; Soares, O.S.G.P.; Andreo, O.A.B. Enhancing Ibuprofen and 4-Isobutylacetophenone Degradation: Exploiting the Potential of Nb_2O_5 Sol-Gel Catalysts in Photocatalysis, Catalytic Ozonation, and Photocatalytic Ozonation. *J. Environ. Chem. Eng.* **2023**, *11*, 110690. [CrossRef]
238. Espíndola, J.C.; Cristóvão, R.O.; Mendes, A.; Boaventura, R.A.R.; Vilar, V.J.P. Photocatalytic Membrane Reactor Performance towards Oxytetracycline Removal from Synthetic and Real Matrices: Suspended vs Immobilized TiO_2-P25. *Chem. Eng. J.* **2019**, *378*, 122114. [CrossRef]
239. Sun, H.; Zhou, T.; Kang, J.; Zhao, Y.; Zhang, Y.; Wang, T.; Yin, X. High-Efficient Degradation of Oxytetracycline by Visible Photo-Fenton Process Using $MnFe_2O_4/g\text{-}C_3N_4$: Performance and Mechanisms. *Sep. Purif. Technol.* **2022**, *299*, 121771. [CrossRef]
240. Zhang, Z.; Du, C.; Zhang, Y.; Yu, G.; Xiong, Y.; Zhou, L.; Liu, Y.; Chi, T.; Wang, G.; Su, Y.; et al. Degradation of Oxytetracycline by Magnetic MOFs Heterojunction Photocatalyst with Persulfate: High Stability and Wide Range. *Environ. Sci. Pollut. Res.* **2022**, *29*, 30019–30029. [CrossRef]
241. Ma, Y.; Wang, R.; Gao, C.; Han, R. Carbon Nanotube-Loaded Copper-Nickel Ferrite Activated Persulfate System for Adsorption and Degradation of Oxytetracycline Hydrochloride. *J. Colloid Interface Sci.* **2023**, *640*, 761–774. [CrossRef]
242. Li, K.; Yediler, A.; Yang, M.; Schulte-Hostede, S.; Hung, M. Ozonation of Oxytetracycline and Toxicological Assessment of Its Oxidation By-Products. *Chemosphere* **2008**, *72*, 473–478. [CrossRef]
243. Do Vinh, T.; Nguyen, D.; Nguyen Trung, K.; Le, P. TiO_2 and Au-TiO_2 Nanomaterials for Rapid Photocatalytic Degradation of Antibiotic Residues in Aquaculture Wastewater. *Materiasl* **2019**, *12*, 2434. [CrossRef]
244. Augugliaro, V.; García-López, E.; Loddo, V.; Malato-Rodríguez, S.; Maldonado, I.; Marcì, G.; Molinari, R.; Palmisano, L. Degradation of Lincomycin in Aqueous Medium: Coupling of Solar Photocatalysis and Membrane Separation. *Sol. Energy* **2005**, *79*, 402–408. [CrossRef]
245. Deng, Y.; Liu, J.; Huang, Y.; Ma, M.; Liu, K.; Dou, X.; Wang, Z.; Qu, S.; Wang, Z. Engineering the Photocatalytic Behaviors of g/C_3N_4-Based Metal-Free Materials for Degradation of a Representative Antibiotic. *Adv. Funct. Mater.* **2020**, *30*, 2002353. [CrossRef]
246. Kontogiannis, A.; Evgenidou, E.; Nannou, C.; Bikiaris, D.; Lambropoulou, D. MOF-Based Photocatalytic Degradation of the Antibiotic Lincomycin Enhanced by Hydrogen Peroxide and Persulfate: Kinetics, Elucidation of Transformation Products and Toxicity Assessment. *J. Environ. Chem. Eng.* **2022**, *10*, 108112. [CrossRef]
247. Tan, J.X.; Chen, Z.Y.; Chen, C.H.; Hsieh, M.F.; Lin, A.Y.C.; Chen, S.S.; Wu, K.C.W. Efficient Adsorption and Photocatalytic Degradation of Water Emerging Contaminants through Nanoarchitectonics of Pore Sizes and Optical Properties of Zirconium-Based MOFs. *J. Hazard. Mater.* **2023**, *451*, 131113. [CrossRef]
248. Farhadian, N.; Akbarzadeh, R.; Pirsaheb, M.; Jen, T.C.; Fakhri, Y.; Asadi, A. Chitosan Modified N, S-Doped TiO_2 and N, S-Doped ZnO for Visible Light Photocatalytic Degradation of Tetracycline. *Int. J. Biol. Macromol.* **2019**, *132*, 360–373. [CrossRef]
249. Çağlar Yılmaz, H.; Akgeyik, E.; Bougarrani, S.; El Azzouzi, M.; Erdemoğlu, S. Photocatalytic Degradation of Amoxicillin Using Co-Doped TiO_2 Synthesized by Reflux Method and Monitoring of Degradation Products by LC–MS/MS. *J. Dispers. Sci. Technol.* **2020**, *41*, 414–425. [CrossRef]
250. Balarak, D.; Mengelizadeh, N.; Rajiv, P.; Chandrika, K. Photocatalytic Degradation of Amoxicillin from Aqueous Solutions by Titanium Dioxide Nanoparticles Loaded on Graphene Oxide. *Environ. Sci. Pollut. Res.* **2021**, *28*, 49743–49754. [CrossRef]
251. Aba-Guevara, C.G.; Medina-Ramírez, I.E.; Hernández-Ramírez, A.; Jáuregui-Rincón, J.; Lozano-Álvarez, J.A.; Rodríguez-López, J.L. Comparison of Two Synthesis Methods on the Preparation of Fe, N-Co-Doped TiO_2 Materials for Degradation of Pharmaceutical Compounds under Visible Light. *Ceram. Int.* **2017**, *43*, 5068–5079. [CrossRef]
252. Al-Musawi, T.J.; Yilmaz, M.; Ramírez-Coronel, A.A.; Al-Awsi, G.R.L.; Alwaily, E.R.; Asghari, A.; Balarak, D. Degradation of Amoxicillin under a UV or Visible Light Photocatalytic Treatment Process Using Fe_2O_3/Bentonite/TiO_2: Performance, Kinetic, Degradation Pathway, Energy Consumption, and Toxicology Studies. *Optik* **2023**, *272*, 170230. [CrossRef]
253. Zhou, L.; Guo, X.; Lai, C.; Wang, W. Electro-Photocatalytic Degradation of Amoxicillin Using Calcium Titanate. *Open Chem.* **2018**, *16*, 949–955. [CrossRef]
254. Elmolla, E.S.; Chaudhuri, M. Comparison of Different Advanced Oxidation Processes for Treatment of Antibiotic Aqueous Solution. *Desalination* **2010**, *256*, 43–47. [CrossRef]
255. Ashraf, A.; Liu, G.; Yousaf, B.; Arif, M.; Ahmed, R.; Rashid, A.; Riaz, L.; Rashid, M.S. *Phyto-Mediated Photocatalysis: A Critical Review of in-Depth Base to Reactive Radical Generation for Erythromycin Degradation*; Springer: Berlin/Heidelberg, Germany, 2022; Volume 29, ISBN 0123456789.
256. Chu, L.; Zhuan, R.; Chen, D.; Wang, J.; Shen, Y. Degradation of Macrolide Antibiotic Erythromycin and Reduction of Antimicrobial Activity Using Persulfate Activated by Gamma Radiation in Different Water Matrices. *Chem. Eng. J.* **2019**, *361*, 156–166. [CrossRef]
257. Albornoz, L.L.; da Silva, S.W.; Bortolozzi, J.P.; Banús, E.D.; Brussino, P.; Ulla, M.A.; Bernardes, A.M. Degradation and Mineralization of Erythromycin by Heterogeneous Photocatalysis Using SnO_2-Doped TiO_2 Structured Catalysts: Activity and Stability. *Chemosphere* **2021**, *268*, 128858. [CrossRef] [PubMed]

258. Xekoukoulotakis, N.P.; Xinidis, N.; Chroni, M.; Mantzavinos, D.; Venieri, D.; Hapeshi, E.; Fatta-Kassinos, D. UV-A/TiO_2 Photocatalytic Decomposition of Erythromycin in Water: Factors Affecting Mineralization and Antibiotic Activity. *Catal. Today* **2010**, *151*, 29–33. [CrossRef]
259. Vignesh, K.; Rajarajan, M.; Suganthi, A. Photocatalytic Degradation of Erythromycin under Visible Light by Zinc Phthalocyanine-Modified Titania Nanoparticles. *Mater. Sci. Semicond. Process* **2014**, *23*, 98–103. [CrossRef]
260. Li, G.; Wang, B.; Zhang, J.; Wang, R.; Liu, H. Rational Construction of a Direct Z-Scheme g-C_3N_4/CdS Photocatalyst with Enhanced Visible Light Photocatalytic Activity and Degradation of Erythromycin and Tetracycline. *Appl. Surf. Sci.* **2019**, *478*, 1056–1064. [CrossRef]
261. Mohsin, M.; Bhatti, I.A.; Iqbal, M.; Naz, S.; Ashar, A.; Nisar, J.; Al-Fawzan, F.F.; Alissa, S.A. Oxidative Degradation of Erythromycin Using Calcium Carbonate under UV and Solar Light Irradiation: Condition Optimized by Response Surface Methodology. *J. Water Process Eng.* **2021**, *44*, 102433. [CrossRef]
262. Fakhri, A.; Rashidi, S.; Tyagi, I.; Agarwal, S.; Gupta, V.K. Photodegradation of Erythromycin Antibiotic by γ-Fe_2O_3/SiO_2 Nanocomposite: Response Surface Methodology Modeling and Optimization. *J. Mol. Liq.* **2016**, *214*, 378–383. [CrossRef]
263. Baran, W.; Adamek, E.; Sobczak, A.; Makowski, A. Photocatalytic Degradation of Sulfa Drugs with TiO_2, Fe Salts and TiO_2/$FeCl_3$ in Aquatic Environment-Kinetics and Degradation Pathway. *Appl. Catal. B* **2009**, *90*, 516–525. [CrossRef]
264. Mendiola-Alvarez, S.Y.; Palomino-Cabello, C.; Hernández-Ramírez, A.; Turnes-Palomino, G.; Guzmán-Mar, J.L.; Hinojosa-Reyes, L. Coupled Heterogeneous Photocatalysis Using a P-TiO_2-AFe_2O_3 Catalyst and $K_2S_2O_8$ for the Efficient Degradation of a Sulfonamide Mixture. *J. Photochem. Photobiol. A Chem.* **2020**, *394*, 112485. [CrossRef]
265. Wang, N.; Li, X.; Yang, Y.; Zhou, Z.; Shang, Y.; Zhuang, X. Photocatalytic Degradation of Sulfonamides by Bi_2O_3-TiO_2/PAC Ternary Composite: Mechanism, Degradation Pathway. *J. Water Process Eng.* **2020**, *36*, 101335. [CrossRef]
266. Batista, A.P.S.; Nogueira, R.F.P. Parameters Affecting Sulfonamide Photo-Fenton Degradation—Iron Complexation and Substituent Group. *J. Photochem. Photobiol. A Chem.* **2012**, *232*, 8–13. [CrossRef]
267. Pelalak, R.; Alizadeh, R.; Ghareshabani, E.; Heidari, Z. Degradation of Sulfonamide Antibiotics Using Ozone-Based Advanced Oxidation Process: Experimental, Modeling, Transformation Mechanism and DFT Study. *Sci. Total Environ.* **2020**, *734*, 139446. [CrossRef]
268. Du, J.; Ma, S.; Yan, Y.; Li, K.; Zhao, F.; Zhou, J. Corn-Silk-Templated Synthesis of TiO_2 Nanotube Arrays with Ag_3PO_4 Nanoparticles for Efficient Oxidation of Organic Pollutants and Pathogenic Bacteria under Solar Light. *Colloids Surf. A Physicochem. Eng. Asp.* **2019**, *572*, 237–249. [CrossRef]
269. Liu, Z.; Liu, X.; Lu, Q.; Wang, Q.; Ma, Z. $TiOF_2$/TiO_2 Composite Nanosheets: Effect of Hydrothermal Synthesis Temperature on Physicochemical Properties and Photocatalytic Activity. *J. Taiwan Inst. Chem. Eng.* **2019**, *96*, 214–222. [CrossRef]
270. El-Kemary, M.; El-Shamy, H.; El-Mehasseb, I. Photocatalytic Degradation of Ciprofloxacin Drug in Water Using ZnO Nanoparticles. *J. Lumin.* **2010**, *130*, 2327–2331. [CrossRef]
271. Wen, X.J.; Niu, C.G.; Zhang, L.; Liang, C.; Guo, H.; Zeng, G.M. Photocatalytic Degradation of Ciprofloxacin by a Novel Z-Scheme CeO_2–Ag/AgBr Photocatalyst: Influencing Factors, Possible Degradation Pathways, and Mechanism Insight. *J. Catal.* **2018**, *358*, 141–154. [CrossRef]
272. Yu, X.; Zhang, J.; Zhang, J.; Niu, J.; Zhao, J.; Wei, Y.; Yao, B. Photocatalytic Degradation of Ciprofloxacin Using Zn-Doped Cu_2O Particles: Analysis of Degradation Pathways and Intermediates. *Chem. Eng. J.* **2019**, *374*, 316–327. [CrossRef]
273. Yang, L.; Xiang, Y.; Jia, F.; Xia, L.; Gao, C.; Wu, X.; Peng, L.; Liu, J.; Song, S. Photo-Thermal Synergy for Boosting Photo-Fenton Activity with RGO-$ZnFe_2O_4$: Novel Photo-Activation Process and Mechanism toward Environment Remediation. *Appl. Catal. B* **2021**, *292*, 120198. [CrossRef]
274. Ge, X.; Meng, G.; Liu, B. Ultrasound-Assisted Preparation of $LaFeO_3$/Polystyrene for Efficient Photo-Fenton Degradation of Ciprofloxacin Hydrochloride. *J. Ind. Eng. Chem.* **2022**, *115*, 390–401. [CrossRef]
275. Gupta, A.; Garg, A. Degradation of Ciprofloxacin Using Fenton's Oxidation: Effect of Operating Parameters, Identification of Oxidized by-Products and Toxicity Assessment. *Chemosphere* **2018**, *193*, 1181–1188. [CrossRef] [PubMed]

catalysts

MDPI

Article

Assessing the Degradation of Levofloxacin in Aqueous Media by Metal-Free g-C_3N_4 Photocatalyst Under Simulated Solar Light Irradiation

Truong Nguyen Xuan [1], Dien Nguyen Thi [2], Cong Le Thanh [2], Thu Mai Thi [2], Thu Le Dieu [1], Trung Nguyen Duc [1] and Ottó Horváth [3,*]

1 School of Chemistry and Life Sciences, Hanoi University of Science and Technology, No. 1 Dai Co Viet Street, Hai Ba Trung District, Hanoi 100000, Vietnam; truong.nguyenxuan@hust.edu.vn (T.N.X.); thu.ledieu@hust.edu.vn (T.L.D.); trung.nguyenduc1@hust.edu.vn (T.N.D.)

2 Viettel Aerospace Institute, Viettel Group, Hoa Lac High-Tech Park, Thach That District, Hanoi 10000, Vietnam; diennt15@viettel.com.vn (D.N.T.); conglt14@viettel.com.vn (C.L.T.); thumt6@viettel.com.vn (T.M.T.)

3 Research Group of Environmental and Inorganic Photochemistry, Center for Natural Sciences, Faculty of Engineering, University of Pannonia, P.O. Box 1158, H-8210 Veszprém, Hungary

* Correspondence: horvath.otto@mk.uni-pannon.hu; Tel.: +36-88-624-000 (ext. 6049)

Abstract: Graphitic carbon nitride (g-C_3N_4) as a fascinating conjugated polymer has attracted considerable attention due to its outstanding electronic properties, high physicochemical stability, and unique structure. In this work, we reported the characterization of g-C_3N_4, which was simply synthesized by thermal polymerization of thiourea, the photocatalytic degradation kinetics, and the pathway of levofloxacin (LEV) using the prepared g-C_3N_4. The XRD and SEM results confirmed a crystalline graphite structure with a tri-s-triazine unit and stacked sheet-like layers of g-C_3N_4. The efficacy factor (EF) was compared to different photocatalytic processes to assess the LEV removal performance. g-C_3N_4 exhibits good stability as a photocatalyst during LEV photodegradation. Radical scavenger experiments revealed that in the oxidative degradation of LEV, $^{\bullet}O_2^-$ and h^+ played the determining roles. Moreover, based on the identification of intermediates using liquid chromatography with tandem mass spectrometry (LC-MS/MS), the degradation pathway of LEV was proposed.

Keywords: antibiotics; pH-effects; adsorption; kinetics; mechanism; radical scavenging; intermediates

1. Introduction

Antibiotics have been widely used in all aspects of human life against infectious diseases. The existence of antibiotics in diverse environments is harmful to human health due to their potential toxicity, carcinogenicity, mutagenicity, and endocrine disruption. Currently, more and more cases are resistant to antibiotics, which hinders the treatment process [1,2]. This is due to many different reasons, but one of the main reasons why bacteria are resistant to specific drugs, making them ineffective when used in humans, is antibiotic pollution [3–5].

Levofloxacin (LEV) is classified as a third-generation quinolone. It is active against both Gram-positive and Gram-negative bacteria and is used in the treatment of many diseases, including urinary, respiratory, and gastrointestinal tract infections [6–8]. LEV is excreted as unchanged drug mainly in the urine (>80% of the dose) [9]. With its complex molecular structure, high stability, and low biodegradability, LEV removal has received increasing attention. To date, photocatalysis has become a promising process for the removal of LEV in water [10–18]. This method utilizes abundant solar energy, and avoiding secondary pollution is considered a friendly environmental solution.

g-C_3N_4 is a popular metal-free semiconductor with an appealing electronic band structure and high physicochemical stability. Moreover, g-C_3N_4 has many advantages, such as an average band gap energy (E_g) value of ≈2.7 eV to aid the visible light absorption, low cost, and facile preparation, promising for photocatalytic applications. These properties make it conspicuous among other metal oxides, for example TiO_2, ZnO and SnO_2. g-C_3N_4 is synthesized from nitrogen-rich precursors such as urea [19], melamine [20], cyanamide [21], and dicyandiamide [22]. The common methods used for the preparation of g-C_3N_4 are thermal polymerization [23], hydrothermal [24], and sol-gel [25]. In this study, g-C_3N_4 was synthesized by thermal polymerization of thiourea calcining at 550 °C. The morphology, structure, and optical properties of the prepared photocatalysts were described and analyzed in detail. The kinetics and optimization of experimental parameters (initial pH value, catalyst content and initial LEV concentration) for photocatalytic degradation of LEV and the reusability of the photocatalyst were also reported. To assess the LEV decomposition efficiency by the g-C_3N_4 photocatalyst, the efficacy factor was calculated. Furthermore, the identification of intermediates was analyzed to propose the LEV degradation mechanism. Although g-C_3N_4 modified with various dopants, metal depositions or combined with other photocatalysts forming composites or Z-schemes were applied for the degradation of LEV without additional oxidants or in the presence of peroxomonosulfate [26–30], unmodified g-C_3N_4 has not been used so far for this purpose. Additionally, a systematic study involving pH-effects, adsorption, kinetics, and mechanistic investigation has not been made with this photocatalyst–pollutant combination; hence, these results serve useful comparisons, too.

2. Results and Discussion

2.1. Characterization

2.1.1. XRD Analysis

The XRD spectroscopy of g-C_3N_4 is shown in Figure 1. It can be seen the presence of two peaks corresponds to the graphite structure with a tri-s-triazine unit [3,31,32]. The sharp peak at 27.3° is attributed to the interplanar stacking of aromatic systems, which can be indexed to the (002) crystal plane [19,20]. The small peak at 13.2° is related to the cyclic arrangement of the condensed tri-s-triazine unit in the structure, which could be indexed to the (001) plane (JCPDS No. 87-1526) [21–23]. Broad and weak reflections at approximately 2θ = 17.7° and 2θ = 21.9° are typical of melamine. It can be reasonably hypothesized that a network of hydrogen-bonded melamine/triazine units is present inside the heptazine framework [33].

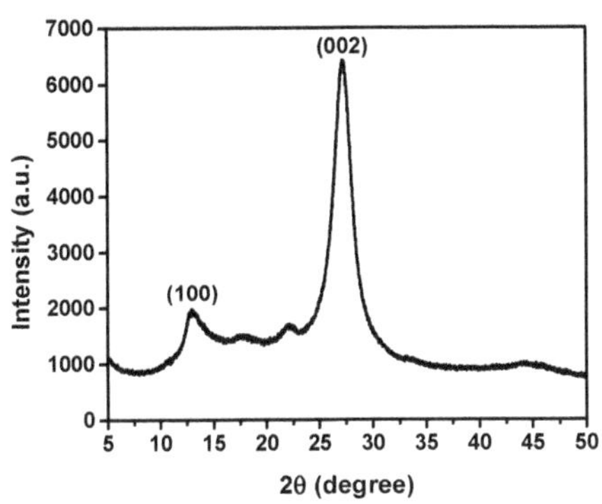

Figure 1. XRD pattern of g-C_3N_4 catalyst.

2.1.2. SEM Analysis

As can be seen in Figure 2, the morphology of g-C_3N_4 consists of stacked sheet-like layers. The sizes of the plates are not uniform and agglomerate together into a block.

Figure 2. SEM images of g-C_3N_4 catalyst with different resolutions. (**a**) 50,000× magnification, (**b**) 20,000× magnification.

2.1.3. Optical Measurements

Absorption spectroscopy was applied to study the optical properties of the synthesized sample, which can play important roles in the photocatalytic behavior. Figure S1 shows the UV-Vis diffuse reflectance spectra and the Tauc plot [34] of the g-C_3N_4 sample. The latter one indicates the band gap (E_g), which was determined to be 2.60 eV for the prepared g-C_3N_4. This is in good accordance with the value measured in our previous work [35,36]. Using the band-gap energy, the edge potential of the conduction band (CB) and valence band (VB) can be estimated by application of the following equations (Equations (1) and (2)) [17]:

$$E_{CB} = E_{VB} - E_g \tag{1}$$

$$E_{VB} = \chi - E_e + 0.5E_g \tag{2}$$

where χ designates the Mulliken electronegative symbol of g-C_3N_4 (4.73 eV [37]), while E_e represents the energy of free electrons on the hydrogen scale ($E_e \approx 4.50$ eV, [38]). Thus, E_{VB} estimated for this semiconductor is 1.53 eV, while E_{CB} is −1.07 eV. The latter is far more negative than the potential needed for the reduction of O_2 to $^\bullet O_2^-$ (−0.16 eV [39]). However, the edge potential of VB seems to be rather low for the oxidation of H_2O or OH^- to $^\bullet OH$, considering the corresponding redox potential (2.27 eV, [27]). On the basis of these results, one would expect that in the degradation of LEV in this work, hydroxyl radicals cannot play determining roles, while superoxide is a good candidate for that, and the valence-band holes can also be taken into consideration in this respect.

2.2. Optimization of the Experimental Conditions for LEV Photodegradation in Solution

2.2.1. Effect of Initial Solution pH Value

The initial pH value of the solution plays an important role in the photodegradation of antibiotics. According to previous studies, pH can affect the LEV removal ability of the catalyst [40–42]. The influence of pH media on LEV treatment with g-C_3N_4 was studied, and the results are illustrated in Figure 3.

When the initial solution pH value increased from 3 to 5, the LEV removal efficiency (H%) exhibited an extreme increase from 36.3% to 80.1%, and the corresponding k value increased from 0.006 min^{-1} to 0.021 min^{-1}. As the pH values changed from 5 to 10, the LEV degradation efficiencies decreased from 80.1% to 41.9% and the corresponding k-values decreased from 0.021 min^{-1} to 0.008 min^{-1}. These results suggest that neutral or weakly acidic media are better for the LEV removal, while strong acid or strong basic conditions hinder the catalytic degradation performance. In accordance with the degradation rates, the efficiency of the adsorption of LEV on the surface of the catalyst particles shows a similar tendency upon changing the initial pH (see the "dark" range of Figure 3a).

This confirms that adsorption of is strongly correlated with the rate and efficacy of the photocatalytic degradation.

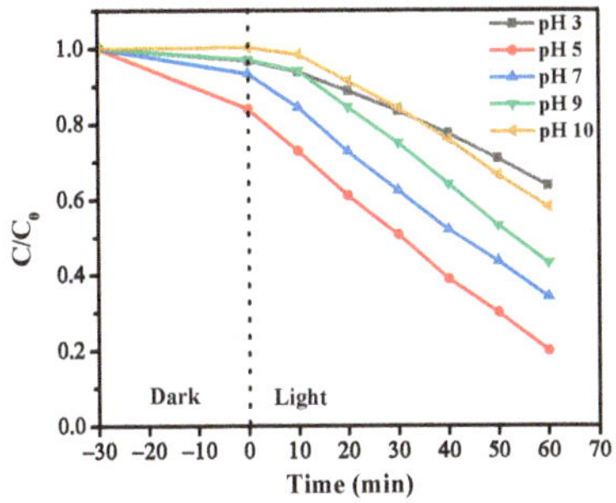

(a)

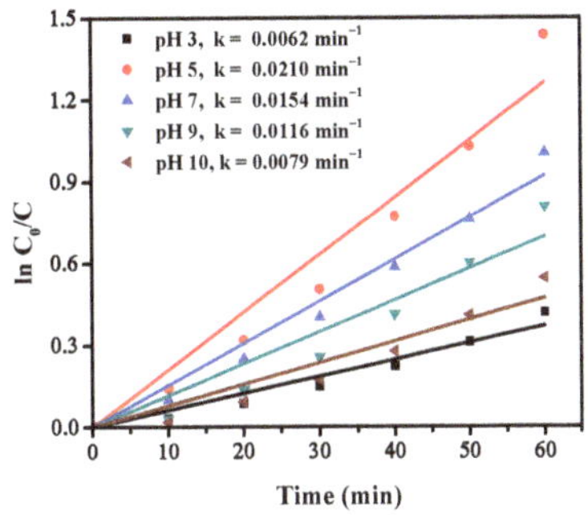

(b)

Figure 3. (**a**) Effect of initial solution pH on the photocatalytic activity and (**b**) first-order kinetic curves.

The above results can be explained by the following aspects. First, the dissociation state of the pollutant and the surface charge of the catalyst vary with pH. As is known, levofloxacin has two dissociation constants: 6.02 (pK_{a1}) and 8.15 (pK_{a2}), which means that levofloxacin exists as LEV^+ cation (pH < 6.02), $LEV^\pm$ neutral state (in the pH range 6.02–8.15) and LEV^- anion (pH > 8.15) [10,41]. In addition, g-C_3N_4 exhibited a zero charge point (pH_{pzc}) around 6.97 (Figure 4), indicating that the catalyst surface is positively charged at pH < 6.97 and negatively charged at pH > 6.97. Hence, electrostatic repulsion between LEV and g-C_3N_4 may occur due to their uniform charge characteristics under extreme acidic or alkaline conditions, which restrained the contact between the catalysts and the contaminants, thus reducing the catalytic efficiency [11,41].

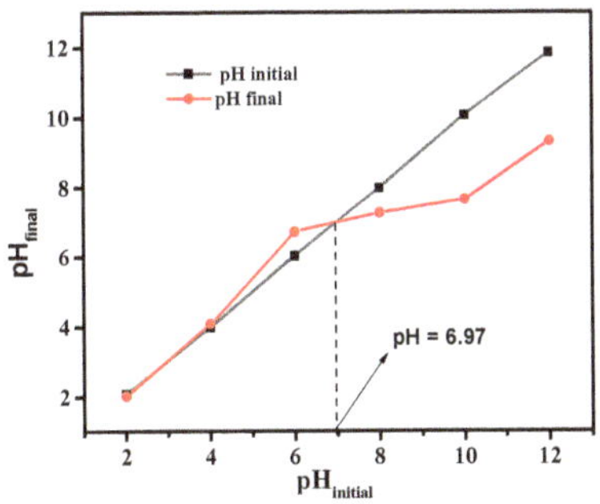

Figure 4. pH_{pzc} determination for g-C_3N_4.

The generation of reactive oxygen species, which include superoxide anion radicals ($^\bullet O_2^-$), hydroxyl radicals ($^\bullet OH$), hydroperoxyl radicals ($^\bullet HO_2$), and hydrogen peroxide (H_2O_2), during the photocatalytic process may be another important factor that accounts for the change of LEV removal efficiency with pH. At higher pH, a large amount of OH^- ions are present on the surface of the catalyst, which facilitates the formation of $^\bullet OH$. When reaching strong alkalinity (pH 10), excess $^\bullet OH$ and LEV molecules will compete for adsorption on the surface of g-C_3N_4, thus reducing the adsorption amount of LEV molecules on the surface of g-C_3N_4, resulting in a reduction in photocatalytic degradation efficiency [41]. With a pH environment near pH_{pzc}, the surface of the material is almost uncharged and practically unaffected by the electrostatic repulsion that occurred between g-C_3N_4 and LEV. LEV primarily appeared as a zwitterion as the pH value increased to neutral, making active species attack easier [12]. Therefore, pH 5 with the highest catalytic treatment efficiency was selected as the optimal pH in this experiment.

The results of the effect of photolysis on the degradation of LEV pollutants at pH 5, are depicted in Figure 5. Without the use of a photocatalyst, the photolysis experiment was carried out under the same reaction conditions (pH 5, and 10 mg/L concentration of LEV). The results showed that solar light irradiation did not significantly affect the degradation of LEV (H = 6.5%), indicating that its absorption in the UV range results in only a slight decomposition. An experiment using g-C_3N_4 photocatalyst was carried out in the dark with optimal reaction conditions to investigate the impact of adsorption on the removal of the LEV drug. In the dark, a small amount of LEV adsorption was observed, indicating that adsorption was not a main factor for the removal of LEV (H = 16.0%). A noticeable degradation of LEV was seen when a solution of LEV-containing photocatalyst was exposed to solar light. It is evident from Figure 5 that photocatalysis played a major role in the degradation of LEV (H = 80.1%).

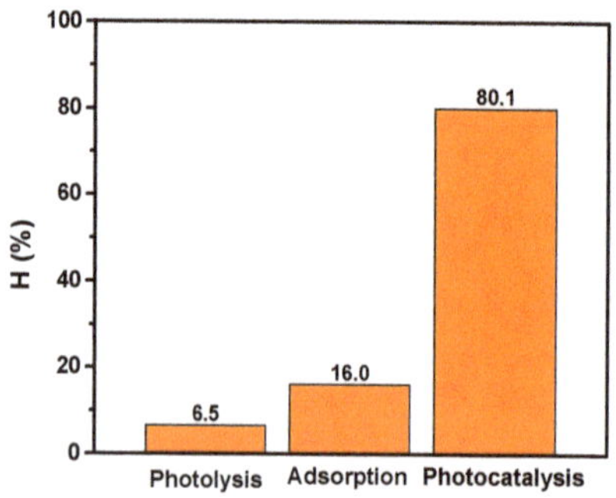

Figure 5. Comparison between photolysis, adsorption and photocatalysis by g-C_3N_4.

2.2.2. Effect of Photocatalyst Concentration

The effect of catalyst dose on the degradation of LEV was investigated using five different doses of g-C_3N_4 composite (i.e., 0.1, 0.2, 0.5, 0.7 or 1.0 g/L) in 100 mL LEV (10 mg/L) at pH 5. The results are shown in Figure 6, which shows that the degradation efficiency of LEV improved with increasing the catalyst dose. As shown in Figure 6, the corresponding removal rate constants at different catalyst concentrations were 0.0074, 0.0140, 0.0210, 0.0283 and 0.0251 min^{-1}. The effect of catalyst dosage can be explained by the following reasons. Increasing the amount of catalyst results in a higher number of active sites and a higher density of catalyst particles in the illuminated area. Therefore, the photocatalytic capacity of the material is improved, resulting in faster decomposition [43]. However, if the catalyst content is continuously increased, it may lead to an excessive density of particles suspended in solution, which hinders light penetration and promotes the light scattering effect (i.e., prevents light absorption during the reaction). Consequently, further increasing the catalyst dosage (1.0 g/L) did not lead to an improvement in photocatalytic performance. Therefore, a dose of 0.5 g/L was optimal for photodegradation of LEV in 60 min.

2.2.3. Effect of Initial LEV Concentration

The influence of the original LEV concentration on the photocatalytic efficiency is displayed in Figure 7. With the increase of LEV concentration from 5 mg L^{-1} to 20 mg L^{-1}, the removal efficiency of LEV gradually declined from 83.62% to 43.17% within 60 min irradiation, and the corresponding k value decreased from 0.0305 min^{-1} to 0.0073 min^{-1}. Clearly, the higher initial LEV content had a negative impact on the catalytic behavior of the g-C_3N_4 photocatalyst. The reduction in removal efficiency might be derived from the fact that the limited number of reactive species produced at the catalytic system was insufficient to attack the excess pollutants. When the LEV concentration is increased, the photons are blocked before reaching the catalyst surface, causing a lower amount of •OH and $•O_2^-$ radicals (the photogenerated electrons react with the dissolved O_2 on the surface of g-C_3N_4 and transform to $•O_2^-$), meaning the removal efficiency is decreased [44]. Besides, the higher LEV content would block the penetration of visible light and cover the active site of the catalyst, subsequently inhibiting the formation of the active species [42]. Simultaneously, as the initial LEV concentration increased, more and more complex intermediates might be generated in the reaction system, which would also compete with the target pollutant for constant active species, resulting in a decrease of LEV photodegradation. Nevertheless, if we take the absolute rate into account, multiplying the rate constant with the corresponding initial concentration, the highest value is obtained for the 10 mg L^{-1} concentration (0.210 mg L^{-1} min^{-1}). The absolute rate moderately decreased at higher concentrations, but even at 20 mg L^{-1}, it just slightly dropped below that observed for the 5 mg L^{-1} concentration (0.146 vs. 0.153 mg L^{-1} min^{-1}). It means that the hindering effects described above are overcompensated by the increase in the rate due to the higher concentration in most of the range studied.

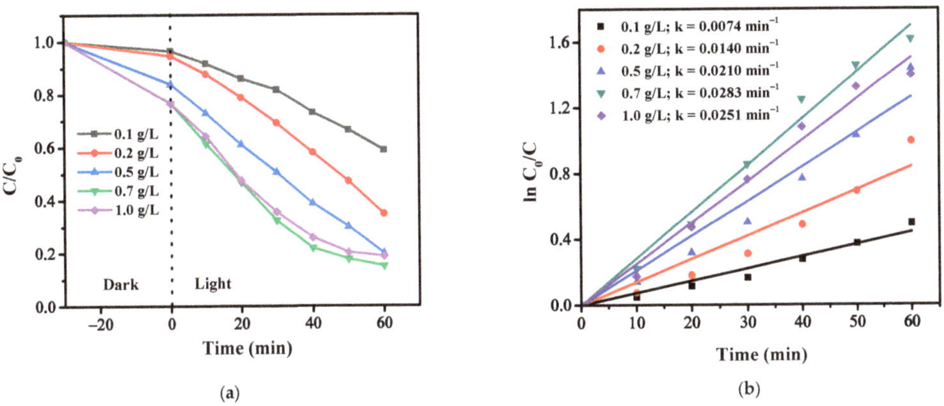

Figure 6. (**a**) Effect of photocatalyst content on the photocatalytic activity and (**b**) first-order kinetic curves.

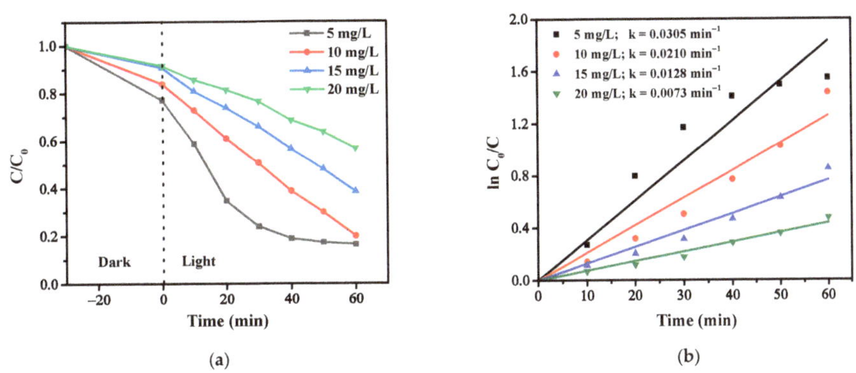

Figure 7. (**a**) Effect of initial LEV concentration on the photocatalytic activity and (**b**) first-order kinetic curves.

2.3. Efficacy Factor of the Photocatalyst

The efficacy factor (EF), which describes the number of contaminants eliminated per concentration of catalyst per treatment time, is used to assess the LEV removal performance by different photocatalytic processes. The EF calculation formula is given as Equation (3) [45]:

$$\text{EF} = \frac{\text{Concentration of LEV (mg/L)} \times \text{\% of its reduction}}{\text{Catalyst concentration (mg/L)} \times \text{time (min)}} \quad (3)$$

From the results of optimized experimental conditions, the performance factor was calculated to compare the photoactivity of reported catalysts for LEV removal. The calculated results are shown in Table 1.

Table 1. Comparison of different photocatalytic processes for degradation of LEV.

Photocatalyst	Catalyst (g/L)	LVFX (mg/L)	Removal (%)	Time (min)	EF	Ref.
g-C_3N_4	0.5	10	80.1	60	2.67×10^{-4}	This study
$CaTiO_3$/g-C_3N_4	0.2	20	87.7	120	7.31×10^{-4}	[13]
Co/Fe-CN-PMS	0.4	20	92.1	50	9.21×10^{-4}	[11]
TOB CN (1-3)	0.2	10	99.2	50	1.98×10^{-4}	[46]
BiOCl/g-C_3N_4	0.25	10	94.2	150	2.51×10^{-4}	[47]
Cu (0)/PCN	1.0	15	98.2	60	2.45×10^{-4}	[14]

The results in Table 1 show that the metal-free g-C_3N_4 photocatalyst has a relatively good EF coefficient and can be used as a potential catalyst for treatment of LEV. On the other side, if the pure g-C_3N_4 is modified by doping metal oxides (ex. $CaTiO_3$/g-C_3N_4), co-doping metals, and adding a strong oxidizer (ex. Co/Fe-CN-PMS), its catalytic activity will be improved.

2.4. Reusability of the Photocatalyst

The reusability of the g-C_3N_4 photocatalyst was evaluated over four consecutive cycles. It can be seen from Figure 8 that a lower degradation rate was obtained after recycling. Within 60 min of irradiation, the photocatalytic performance was reduced from 80.1% to 49.0% after three reuses. This can be explained by a slow recovery of the active sites on the catalyst. It may require a longer time for the g-C_3N_4 photocatalyst to reach its overall original efficiency.

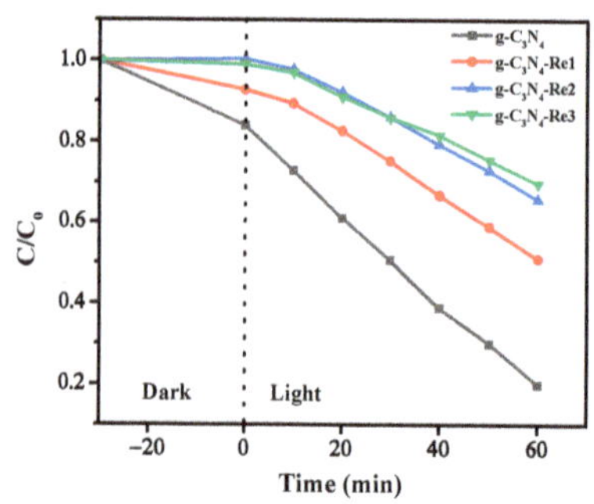

Figure 8. Recycle experiments of the LEV degradation using g-C_3N_4 photocatalyst.

2.5. Radical Trapping

Oxidizing agents such as $^{\bullet}O_2^-$, $^{\bullet}OH$, h_{VB}^+ formed during photocatalysis are considered to play important roles in levofloxacin degradation. To elucidate the photodegradation mechanism, the effects of various reactive oxidative species (ROS) on the decomposition of levofloxacin over g-C_3N_4 were investigated. In this photocatalytic study, isopropyl alcohol (IPA), benzoquinone (BQ), and ammonium oxalate (AO), were employed to capture hydroxyl radicals ($^{\bullet}OH$), superoxide ($^{\bullet}O_2^-$,) and holes (h^+) as reactive species in photocatalytic process, respectively. The photocatalytic decomposition of levofloxacin in the presence of the radical trapping agents is shown in Figure 9.

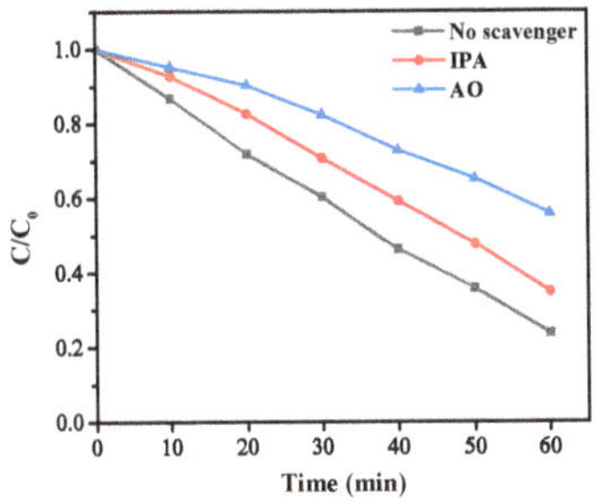

(**a**)

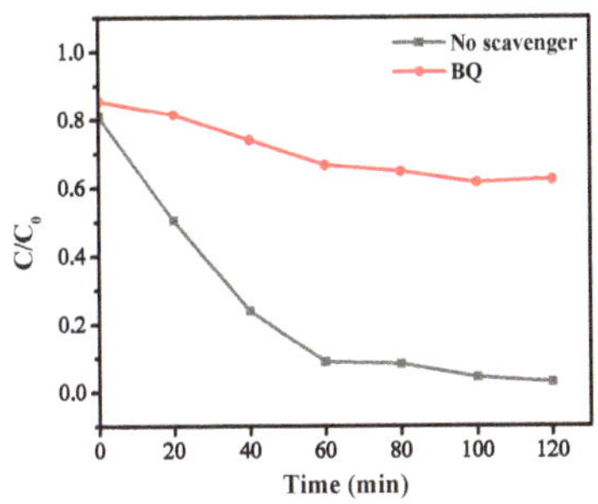

(**b**)

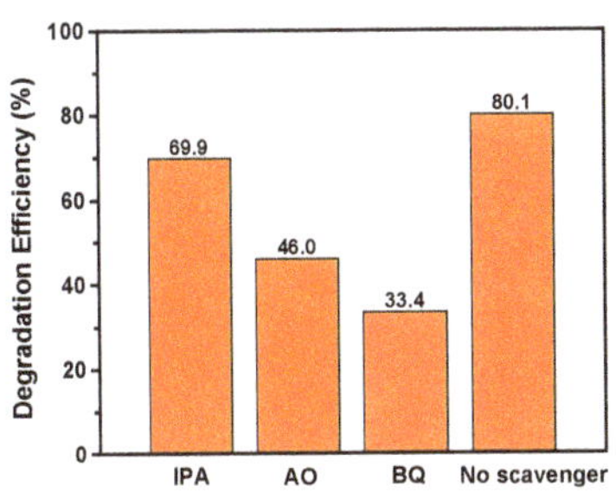

(**c**)

Figure 9. Reactive species trapping experiments: (**a**) effect of quenching agent IPA and AO (observed by UV-Vis spectrophotometry); (**b**) effect of quenching agent BQ (obtained by photoluminescence measurements); (**c**) degradation efficiency.

Various changes in photocatalytic efficiency were observed for the different scavengers. As shown in Figure 9a, in the presence of IPA, the degradation rate moderately decreased compared to the case with no scavenger. The degradation efficiency was reduced from 80.1% to 69.9% (Figure 9c). This observation suggests that $^{\bullet}OH$ radicals only slightly contribute to the photodegradation of levofloxacin. It is not surprising on the basis of the determined edge potential of the valence band of the prepared g-C_3N_4. Moreover, it is in accordance with the results obtained for modified photocatalysts with much higher EVB values [15,16]. The presence of AO resulted in a more appreciable decrease in the degradation rate (Figure 9a) and the efficiency dropped by almost a quarter of the original performance (to H = 46.0%, Figure 9c). This result indicates that the role of valence-band holes is important in the degradation process. This observation confirms that the adsorption of LEV is indispensable for its interaction with h_{VB}^+. The most significant inhibition effect was observed in the presence of BQ. As shown in Figure 9b, the decrease in the degradation rate was more remarkable, leading to a stronger reduction in the photocatalytic efficiency (to H = 33.4%, Figure 9c). This result unambiguously suggests that $^\bullet O_2^-$ plays the most important role in the photocatalytic degradation of LEV. The participation of valence-band holes in the degradation process is less significant but still considerably, while hydroxyl radicals play a minor role in this mechanism.

2.6. Proposed Removal Mechanisms of Levofloxacin by the g-C_3N_4 Photocatalyst

Possible photocatalytic mechanisms for the degradation of LEV have been proposed based on the above quenching experiments and literature reports [17,48–50]. As illustrated in Equations (4)–(11), under sunlight irradiation, g-C_3N_4 was excited to form electron-hole pairs by absorbing light. The potential of the conduction band edge of g-C_3N_4 ($E_{CB} = -1.07$ V vs. NHE) is more negative than the standard reduction potential of $O_2/^\bullet O_2^-$ pair (−0.16 V vs. NHE, pH 7) [39], indicating that the accumulated electrons can reduce O_2 to $^\bullet O_2^-$ [51]. The results of the radical trapping experiment showed that $^\bullet OH$ played a negligible role in the photocatalytic process, suggesting that degradation of LEV in the valence band is mainly due to the direct reaction with h^+ rather than $^\bullet OH$. Furthermore, $^1O_2$ can be generated by the reaction of $^\bullet O_2^-$ and h^+, Equations (10)–(12) [18,52]. Finally, ROS ($^\bullet O_2^-$, h^+, $^1O_2$, $^\bullet OH$) reacted with LEV molecules on the g-C_3N_4 surface, resulting in the decomposition and eventual complete mineralization of LEV, Equation (15). The photocatalytic mechanism of LEV decomposition by g-C_3N_4 catalyst was proposed and is shown in Figure 10.

$$\text{g-}C_3N_4 + h\upsilon \rightarrow \text{g-}C_3N_4\ (e^- + h_{VB}^+) \quad (4)$$

$$O_2 + e^- \rightarrow {}^\bullet O_2^- \quad (5)$$

$$O_2 + H^+ + e^- \rightarrow {}^\bullet HO_2 \quad (6)$$

$${}^\bullet HO_2 + {}^\bullet HO_2 \rightarrow H_2O_2 \quad (7)$$

$${}^\bullet O_2^- + 2H^+ + e^- \rightarrow H_2O_2 \quad (8)$$

$${}^\bullet HO_2 + H^+ + e^- \rightarrow H_2O_2 \quad (9)$$

$$H_2O_2 + h\upsilon \rightarrow 2{}^\bullet OH \quad (10)$$

$$H_2O_2 + H^+ + e^- \rightarrow {}^\bullet OH + 2H_2O \quad (11)$$

$$h_{VB}^+ + {}^\bullet O_2^- \rightarrow {}^1O_2 \quad (12)$$

$$h_{VB}^+ + H_2O \rightarrow {}^\bullet OH + H^+ \quad (13)$$

$$h_{VB}^+ + OH^- \rightarrow {}^\bullet OH \quad (14)$$

$${}^\bullet O_2^-/{}^\bullet HO_2/{}^\bullet OH/{}^1O_2 + \text{LEV} \rightarrow \text{degradation products} \quad (15)$$

LC-MS/MS was used to identify LEV degradation products to further analyze the degradation path of LEV by g-C_3N_4 photocatalyst. Mass spectra of levofloxacin in the photodegradation experiments (i) light-off for 30 min, (ii) light-on for 40 min, and (iii) light-on for 120 min were recorded. With the levofloxacin solution in the dark, molecular ion peaks $[M+H]^+$ (m/z = 362) were initially produced (Figure S2a). During the degradation process, intermediates were formed (Figure S2b). Molecular ion peaks undergo N-methyl piperazine ring breaking reaction to form P1 (m/z = 265) and an intermediate at m/z = 99. P2 (m/z = 218) and P3 (m/z = 217) intermediates appeared as products of decarboxylation and quinolone ring opening reactions. Meanwhile, the loss of $-C_3OH_2$ and hydroxylation occurs on P3, producing P4 (m/z = 170). P4 was further defluorinated and demethylated to form P5 (m/z = 122). Subsequently, intermediate P5 product might continue to be broken into small molecules as well as mineralization product (main peaks at $m/z < 50$ region as seen from Figure S2c).

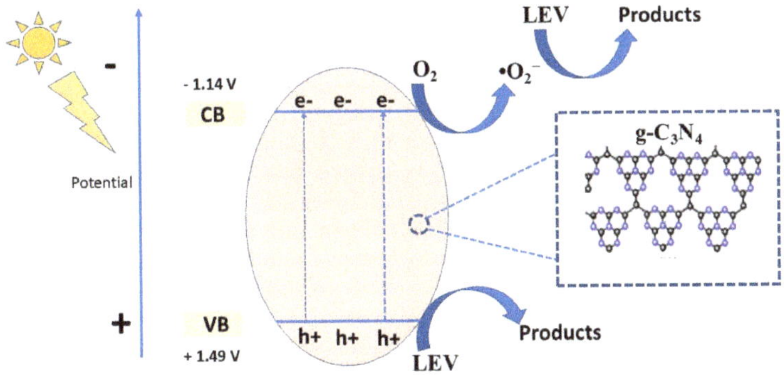

Figure 10. Photocatalytic mechanism of levofloxacin by g-C3N4 photocatalyst.

The intermediate and degradation paths are proposed as shown in Scheme 1. The degradation of LEV is mainly by piperazine ring breaking and opening, quinolone ring opening, defluorination, demethylation, decarboxylation/carboxylation, and hydroxylation processes as detailed above [40].

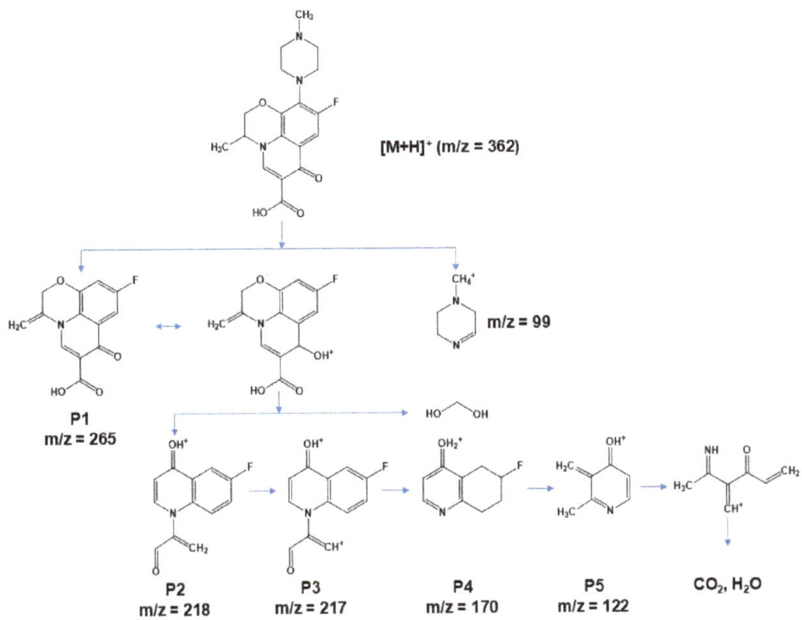

Scheme 1. The proposed degradation pathway of LEV.

3. Materials and Methods

3.1. Reagents

Thiourea (99%) and ethanol (99%) were obtained from Merck, Darmstadt, Germany. p-benzoquinone (99.0%, Xiya Reagent (Xiya Chemicals Co., Ltd., Shandong, China)), ammoniumoxalate monohydrate (99.5%), and isopropyl alcohol (99.7%) were purchased from Xilong, Shantou, China. Sodium hydroxide (NaOH, 99.8%) and hydrochloric acid (HCl, 36–38%) were obtained from Sigma-Aldrich (Burlington, MA, USA). Levofloxacin

($C_{18}H_{20}FN_3O_4$, LEV, 97.0%) was purchased from Vietnam National Institute of Drug Quality Control. All chemicals have been used without purification.

3.2. Preparation of g-C_3N_4

g-C_3N_4 was prepared by thermal polymerization of thiourea, an abundant nitrogen-rich precursor [1–3]. Typically, 3 g of thiourea was put into a porcelain crucible with a cover, then calcinated at 550 °C for 4 h at a heating rate of 2 °C min^{-1} under air atmosphere. After calcination, the samples were washed with deionized water and ethanol several times and dried at 60 °C. g-C_3N_4 was obtained, ground, and collected for further experiments. Since thiourea contains sulfur, the elemental composition of the product was checked by EDS measurements. The results showed that the whole sulfur content of the precursor had been removed during the preparation process (Figure S3).

3.3. Sample Characterization

The samples prepared were characterized by X-ray diffraction (XRD, Malvern PANalytical, Aeris, Almelo, The Netherlands) and scanning electron microscopy (SEM, NanoSEM 450 FEI, Eindhoven, The Netherlands) combined with a TEAM Apollo XL energy dispersive spectroscopy (EDS, EDAX, Cambridge, UK). UV-Vis spectra were recorded by an Agilent 8453 spectrophotometer (Santa Clara, CA, USA), while diffuse reflectance spectra were measured on a Carry 5000 UV-Vis-NIR equipment (Santa Clara, CA, USA). Luminescence was measured by a F-4700 spectrofluorometer (Hitachi, Tokyo, Japan). The reaction intermediates were analyzed by SCIEX Exion LC 20AD LC-MS/MS system (SCIEX, Framingham, MA, USA): column Agilent Eclipse Plus C18 (2.1 × 50 mm, ID 3.5 μm) and detector AB SCIEX Triple Quad 6500+.

3.4. Point-Zero Charge and Surface Charge Determination

It has been reported that the pH-drift method can be used to determine the pH of zero charge and surface charge [4,5]. In this study, it was measured on the basis of the work by Mahmood et. al. [53]. 100 mL LEV solution (10 mg/L) was adjusted to pH (2, 4, 6, 8, 10, 12) in test tubes at room temperature using NaOH and HCl (0.1 M) and the initial pH was recorded. Then 50 mg of g-C_3N_4 photocatalyst was added and stirred for 24 h under dark conditions. The suspension was removed from the tube, the pH was measured, and the final pH was recorded.

3.5. Photocatalytic Degradation Experiments

The catalytic performance of g-C_3N_4 was evaluated by the degradation of LEV under the irradiation of a 500-W Hg-lamp, as a simulated solar source. The emission spectrum of the light source is shown in Figure S4. The intensity of radiation was 26 lm W^{-1}. The experimental procedure was as follows: 50 mg of g-C_3N_4 was dispersed in a 100 mL LEV (10 mg L^{-1}) solution, and subsequently, the suspension was sonicated for about 3 min. Prior to the irradiation, the mixture was kept for 30 min in the dark with stirring at room temperature to obtain an adsorption-desorption equilibrium. At each given time interval (10 min), 3 mL of the solution was collected and filtered with a filter (0.22 μm PTFE membrane) to remove the catalyst. The concentration of LEV was examined by UV-Vis spectrophotometry in the wavelength range from 200 to 400 nm. The photocatalytic degradation efficiency was calculated according to the following equation (Equation (16)).

$$H = \frac{C_0 - C_t}{C_0} = \frac{A_0 - A_t}{A_0} \quad (16)$$

where, C_t and A_t—concentration and absorbance of the solution at time t; C_0 and A_0— adsorption/desorption equilibrium concentration and absorbance of the solution at time t_0.

A simplified pseudo-first-order kinetic model of Langmuir–Hinshewood (Equation (17)) was used to explore the kinetic of the photocatalysis process.

$$In\frac{C_0}{C_t} = k_{app}t = kKt \quad (17)$$

where C—concentration of the antibiotic (mg L^{-1}), t—time for degradation (min), k_{app}—apparent rate constant (min^{-1}), k—reaction rate constant (min^{-1}), and K—adsorption coefficient of the antibiotic over catalyst particles.

3.6. Radical Trapping Experiments

Isopropyl alcohol (IPA, 5.0×10^{-3} M), benzoquinone (BQ, 2.5×10^{-4} M) and ammonium oxalate (5.0×10^{-3} M) were used as scavengers for trapping the photogenerated $^{\bullet}OH$, $^{\bullet}O_2^-$, and holes radicals during the degradation of LEV.

3.7. Intermediates Identification

The intermediates generated during the photocatalytic degradation of levofloxacin solution were further investigated by LC-MS/MS measurement. The mass spectra of the photodegradation products are shown in Figure S2. They were used to propose the photodegradation route of levofloxacin.

3.8. Recycling Experiments

To investigate the reusability of the g-C_3N_4, consecutive experiments for degradation of LEV were carried out in four cycles. After each cycle, the catalyst was recovered by filtration and washed several times with distilled water and ethanol then dried at 60 °C for 10 h.

4. Conclusions

The g-C_3N_4 photocatalyst was successfully synthesized from thiourea precursor at a calcination temperature of 550 °C. The prepared catalyst was characterized by XRD and SEM. The experimental parameters, including the initial pH solution, catalyst dose, and initial LEV concentration, greatly influence the photodegradation of LEV antibiotics. The disappearance of LEV followed the Langmuir-Hinshelwood (L-H) kinetic model. Radical scavenger experiments confirmed that $^{\bullet}O_2^-$ and h^+ played significant roles during the photocatalytic degradation of LEV. Optimized reaction conditions lead to the conclusion that a g-C_3N_4 and a use of 0.5 g/L photocatalyst concentration and 10 mg L^{-1} LEV solution at pH 5 eliminate 80.1% LEV within a 60-min irradiation with solar-simulated light. Based on the identification of intermediates using LC-MS/MS, the photocatalytic degradation pathways of LEV were proposed. The present catalyst exhibits good stability even after four cycles. The EF factor indicates that g-C_3N_4 can be used as a potential photocatalyst for antibiotic degradation.

Supplementary Materials: The following supporting information can be downloaded at: https://www.mdpi.com/article/10.3390/catal14110837/s1, Figure S1. (a) DR/UV-Vis spectra; (b) energy bandgap determination by the Tauc plots. Figure S2. Mass spectra of levofloxacin in the photodegradation experimental: (a) light-off for 30 minutes, (b) light-on for 40 minutess and (c) light-on for 120 minutes. Figure S3. EDS measurement of the prepared g-C_3N_4. Figure S4. Profile of the 500-W Hg-lamp (Philips Lighting product data).

Author Contributions: Conceptualization, T.N.X.; methodology, T.N.X. and O.H.; software, T.N.X.; validation, D.N.T., C.L.T., T.M.T., T.L.D. and T.N.D.; formal analysis, T.N.X. and O.H.; investigation, D.N.T., C.L.T., T.M.T., T.L.D. and T.N.D.; resources, T.N.X. and O.H.; data curation, D.N.T., C.L.T., T.M.T., T.L.D. and T.N.D.; writing—original draft preparation, T.N.X. and O.H.; writing—review and editing, T.N.X. and O.H.; visualization, T.N.X. and O.H.; supervision, T.N.X. and O.H.; project administration, T.N.X. and O.H.; funding acquisition, T.N.X. and O.H. All authors have read and agreed to the published version of the manuscript.

Funding: This work was supported by the National Research, Development, and Innovation Office of Hungary in the frame of the bilateral Hungarian-Vietnamese S&T Co-operation Program (project code 2019-2.1.12-TÉT_VN-2020-00009 and NĐT/HU/22/21) and by the Ministry for Innovation and Technology of Hungary from the National Research, Development and Innovation Fund, financed under the 2021 Thematic Excellence Program funding scheme (grant number TKP2021-NKTA-21).

Data Availability Statement: The data presented in this study are available on request from the corresponding author. The data are not publicly available due to privacy.

Conflicts of Interest: The authors declare no conflicts of interest.

References

1. Hong, Y.; Liu, E.; Shi, J.; Lin, X.; Sheng, L.; Zhang, M.; Wang, L.; Chen, J. A direct one-step synthesis of ultrathin g-C_3N_4 nanosheets from thiourea for boosting solar photocatalytic H_2 evolution. *Int. J. Hydrog. Energy* **2019**, *44*, 7194–7204. [CrossRef]
2. Wang, H.; Sun, Z.; Li, Q.; Tang, Q.; Wu, Z. Surprisingly advanced CO_2 photocatalytic conversion over thiourea derived g-C_3N_4 with water vapor while introducing 200–420 nm UV light. *J. CO2 Util.* **2016**, *14*, 143–151. [CrossRef]
3. Paul, D.R.; Sharma, R.; Nehra, S.P.; Sharma, A. Effect of calcination temperature, pH and catalyst loading on photodegradation efficiency of urea derived graphitic carbon nitride towards methylene blue dye solution. *RSC Adv.* **2019**, *9*, 15381–15391. [CrossRef] [PubMed]
4. Noh, J.S.; Schwarz, J.A. Estimation of the point of zero charge of simple oxides by mass titration. *J. Colloid Interface Sci.* **1989**, *130*, 157–164. [CrossRef]
5. Zhu, P.; Luo, D.; Liu, M.; Duan, M.; Lin, J.; Wu, X. Flower-globular BiOI/$BiVO_4$/g-C_3N_4 with a dual Z-scheme heterojunction for highly efficient degradation of antibiotics under visible light. *Sep. Purif. Technol.* **2022**, *297*, 121503. [CrossRef]
6. Czyrski, A. Analytical Methods for Determining Third and Fourth Generation Fluoroquinolones: A Review. *Chromatographia* **2017**, *80*, 181–200. [CrossRef]
7. Sousa, J.; Alves, G.; Fortuna, A.; Falcão, A. Analytical methods for determination of new fluoroquinolones in biological matrices and pharmaceutical formulations by liquid chromatography: A review. *Anal. Bioanal. Chem.* **2012**, *403*, 93–129. [CrossRef]
8. Van, T.T.; Minejima, E.; Chiu, C.A.; Butler-Wu, S.M. Don't Get Wound Up: Revised Fluoroquinolone Breakpoints for Enterobacteriaceae and Pseudomonas aeruginosa. *J. Clin. Microbiol.* **2019**, *57*, e02072-18. [CrossRef]
9. Fish, D.N.; Chow, A.T. The Clinical Pharmacokinetics of Levofloxacin. *Clin. Pharmacokinet.* **1997**, *32*, 101–119. [CrossRef]
10. Goulart, L.A.; Moratalla, A.; Lanza, M.R.V.; Saez, C.; Rodrigo, M.A. Photocatalytic performance of Ti/MMO/ZnO at degradation of levofloxacin: Effect of pH and chloride anions. *J. Electroanal. Chem.* **2021**, *880*, 114894. [CrossRef]
11. Guo, P.; Hu, X. Co, Fe co-doped g-C_3N_4 composites as peroxymonosulfate activators under visible light irradiation for levofloxacin degradation: Characterization, performance and synergy mechanism. *Colloids Surf. A Physicochem. Eng. Asp.* **2022**, *648*, 129423. [CrossRef]
12. Guo, P.; Hu, X. ZIF-derived $CoFe_2O_4$/Fe_2O_3 combined with g-C_3N_4 as high-efficient photocatalysts for enhanced degradation of levofloxacin in the presence of peroxymonosulfate. *J. Alloys Compd.* **2022**, *914*, 165338. [CrossRef]
13. Zhao, J.; Cao, X.; Bai, Y.; Chen, J.; Zhang, C. Simple synthesis of $CaTiO_3$/g-C_3N_4 heterojunction for efficient photodegradation of methylene blue and levofloxacin. *Opt. Mater.* **2023**, *135*, 113239. [CrossRef]
14. Li, F.; Zhu, P.; Wang, S.; Xu, X.; Zhou, Z.; Wu, C. One-pot construction of Cu and O co-doped porous g-C3N4 with enhanced photocatalytic performance towards the degradation of levofloxacin. *RSC Adv.* **2019**, *9*, 20633–20642. [CrossRef]
15. Gupta, G.; Kaur, A.; Sinha, A.S.K.; Kansal, S.K. Photocatalytic degradation of levofloxacin in aqueous phase using Ag/AgBr/BiOBr microplates under visible light. *Mater. Res. Bull.* **2017**, *88*, 148–155. [CrossRef]
16. Kaur, A.; Kansal, S.K. Bi_2WO_6 nanocuboids: An efficient visible light active photocatalyst for the degradation of levofloxacin drug in aqueous phase. *Chem. Eng. J.* **2016**, *302*, 194–203. [CrossRef]
17. Prabavathi, S.L.; Saravanakumar, K.; Park, C.M.; Muthuraj, V. Photocatalytic degradation of levofloxacin by a novel Sm_6WO_{12}/g-C_3N_4 heterojunction: Performance, mechanism and degradation pathways. *Sep. Purif. Technol.* **2021**, *257*, 117985. [CrossRef]
18. Wang, F.; Feng, Y.; Chen, P.; Wang, Y.; Su, Y.; Zhang, Q.; Zeng, Y.; Xie, Z.; Liu, H.; Liu, Y.; et al. Photocatalytic degradation of fluoroquinolone antibiotics using ordered mesoporous g-C_3N_4 under simulated sunlight irradiation: Kinetics, mechanism, and antibacterial activity elimination. *Appl. Catal. B Environ.* **2018**, *227*, 114–122. [CrossRef]
19. Martin, D.J.; Qiu, K.; Shevlin, S.A.; Handoko, A.D.; Chen, X.; Guo, Z.; Tang, J. Highly Efficient Photocatalytic H_2 Evolution from Water using Visible Light and Structure-Controlled Graphitic Carbon Nitride. *Angew. Chem. Int. Ed.* **2014**, *53*, 9240–9245. [CrossRef]
20. Zhang, Y.; Liu, J.; Wu, G.; Chen, W. Porous graphitic carbon nitride synthesized via direct polymerization of urea for efficient sunlight-driven photocatalytic hydrogen production. *Nanoscale* **2012**, *4*, 5300. [CrossRef]
21. Dong, F.; Wu, L.; Sun, Y.; Fu, M.; Wu, Z.; Lee, S.C. Efficient synthesis of polymeric g-C_3N_4 layered materials as novel efficient visible light driven photocatalysts. *J. Mater. Chem.* **2011**, *21*, 15171. [CrossRef]
22. Yang, Y.; Zhang, Q.; Zhang, R.; Ran, T.; Wan, W.; Zhou, Y. Compressible and Recyclable Monolithic g-C_3N_4/Melamine Sponge: A Facile Ultrasonic-Coating Approach and Enhanced Visible-Light Photocatalytic Activity. *Front. Chem.* **2018**, *6*, 156. [CrossRef]

23. Das, D.; Banerjee, D.; Das, B.; Das, N.S.; Chattopadhyay, K.K. Effect of cobalt doping into graphitic carbon nitride on photo induced removal of dye from water. *Mater. Res. Bull.* **2017**, *89*, 170–179. [CrossRef]
24. Komatsu, T. Prototype carbon nitrides similar to the symmetric triangular form of melon. *J. Mater. Chem.* **2001**, *11*, 802–803. [CrossRef]
25. Komatsu, T. The First Synthesis and Characterization of Cyameluric High Polymers. *Macromol. Chem. Phys.* **2001**, *202*, 19–25. [CrossRef]
26. Machín, A.; Cotto, M.; Duconge, J.; Arango, J.C.; Morant, C.; Pinilla, S.; Soto-Vázquez, L.; Resto, E.; Márquez, F. Hydrogen production via water splitting using different Au@ZnO catalysts under UV–vis irradiation. *J. Photochem. Photobiol. A Chem.* **2018**, *353*, 385–394. [CrossRef]
27. Rameel, M.I.; Wali, M.; Al-Humaidi, J.Y.; Liaqat, F.; Khan, M.A. Enhanced photocatalytic degradation of levofloxacin over heterostructured C_3N_4/Nb_2O_5 system under visible light. *Heliyon* **2023**, *9*, e20479. [CrossRef]
28. Luo, Y.; Zhu, Y.; Han, Y.; Ye, H.; Liu, R.; Lan, Y.; Xue, M.; Xie, X.; Yu, S.; Zhang, L.; et al. g-C_3N_4-based photocatalysts for organic pollutant removal: A critical review. *Carbon Res.* **2023**, *2*, 14. [CrossRef]
29. Zhong, X.; Ji, M.; Wu, W.; Lu, C.; Liu, W.; Jiang, F. Enhanced Degradation of Levofloxacin through Visible-Light-Driven Peroxymonosulfate Activation over $CuInS_2$/g-C_3N_4 Heterojunctions. *Nanomaterials* **2023**, *14*, 74. [CrossRef]
30. Zhang, X.; Tian, Y.; Zhou, L.; Wang, L.; Zhang, J.; Liu, Y.; Lei, J. Efficient degradation of levofloxacin using a g-C_3N_4@glucose-derived carbon catalyst with adjustable N content via peroxymonosulfate activation. *Chemosphere* **2023**, *314*, 137684. [CrossRef]
31. Yan, S.C.; Li, Z.S.; Zou, Z.G. Photodegradation Performance of g-C_3N_4 Fabricated by Directly Heating Melamine. *Langmuir* **2009**, *25*, 10397–10401. [CrossRef] [PubMed]
32. Wang, X.; Chen, X.; Thomas, A.; Fu, X.; Antonietti, M. Metal-Containing Carbon Nitride Compounds: A New Functional Organic–Metal Hybrid Material. *Adv. Mater.* **2009**, *21*, 1609–1612. [CrossRef]
33. Negro, P.; Cesano, F.; Casassa, S.; Scarano, D. Combined DFT-D3 Computational and Experimental Studies on g-C_3N_4: New Insight into Structure, Optical, and Vibrational Properties. *Materials* **2023**, *16*, 3644. [CrossRef] [PubMed]
34. Pattnaik, S.; Rai, V.K. Insight into the spectroscopic and thermometric properties of titanate phosphors via a novel co-excited laser system. *Mater. Sci. Eng. B* **2021**, *272*, 115318. [CrossRef]
35. Xuan, T.N.; Thi, D.N.; Ngoc, T.N.; Quoc, K.D.; Németh, M.; Mukhtar, S.; Horváth, O. Effect of Ruthenium Modification of g-C3N4 in the Visible-Light-Driven Photocatalytic Reduction of Cr(VI). *Catalysts* **2023**, *13*, 964. [CrossRef]
36. Nguyen Xuan, T.; Nguyen Thi, D.; Tran Thuong, Q.; Nguyen Ngoc, T.; Dang Quoc, K.; Molnár, Z.; Mukhtar, S.; Szabó-Bárdos, E.; Horváth, O. Effect of Copper-Modification of g-C_3N_4 on the Visible-Light-Driven Photocatalytic Oxidation of Nitrophenols. *Molecules* **2023**, *28*, 7810. [CrossRef]
37. Machín, A.; Fontánez, K.; Duconge, J.; Cotto, M.C.; Petrescu, F.I.; Morant, C.; Márquez, F. Photocatalytic Degradation of Fluoroquinolone Antibiotics in Solution by Au@ZnO-rGO-gC_3N_4 Composites. *Catalysts* **2022**, *12*, 166. [CrossRef]
38. Cao, J.; Han, F.; Wang, L.; Huang, X.; Cao, Y.; He, P.; Yang, H.; Chen, J.; Li, H. Ru/g-C_3N_4 as an efficient catalyst for selective hydrogenation of aromatic diamines to alicyclic diamines. *RSC Adv.* **2020**, *10*, 16515–16525. [CrossRef]
39. Li, J.; Jiang, M.; Zhou, H.; Jin, P.; Cheung, K.M.C.; Chu, P.K.; Yeung, K.W.K. Vanadium Dioxide Nanocoating Induces Tumor Cell Death through Mitochondrial Electron Transport Chain Interruption. *Glob. Chall.* **2019**, *3*, 1800058. [CrossRef]
40. Lu, X.; Wu, L.; Liang, L.; Liu, D.; Chen, Y.; Zeng, Y.; Zhong, M.; Jia, B. Levofloxacin degradation by porous Co_x/CN activated peroxymonosulfate: Investigation of efficiency, mechanism, and degradation pathways. *J. Water Process Eng.* **2023**, *56*, 104427. [CrossRef]
41. Xing, Z.; Wang, Z.; Chen, W.; Zhang, M.; Fu, X.; Gao, Y. Degradation of levofloxacin in wastewater by photoelectric and ultrasonic synergy with TiO_2/g-C_3N_4@AC combined electrode. *J. Environ. Manag.* **2023**, *330*, 117168. [CrossRef] [PubMed]
42. Li, X.; Chen, T.; Qiu, Y.; Zhu, Z.; Zhang, H.; Yin, D. Magnetic dual Z-scheme g-$C_3N_4/BiVO_4/CuFe_2O_4$ heterojunction as an efficient visible-light-driven peroxymonosulfate activator for levofloxacin degradation. *Chem. Eng. J.* **2023**, *452*, 139659. [CrossRef]
43. Ahmed, S.; Rasul, M.G.; Martens, W.N.; Brown, R.; Hashib, M.A. Advances in Heterogeneous Photocatalytic Degradation of Phenols and Dyes in Wastewater: A Review. *Water Air Soil Pollut.* **2011**, *215*, 3–29. [CrossRef]
44. Yan, H.; Luo, M.; Chen, Q.; Jeong, T.; Zhang, J.; Wang, L. Efficacy and mechanism of chemical-free VUV/UV process for oxytetracycline degradation: Continuous-flow experiment and CFD modeling. *Chem. Eng. J. Adv.* **2020**, *4*, 100059. [CrossRef]
45. Kohantorabi, M.; Moussavi, G.; Oulego, P.; Giannakis, S. Synthesis of a novel, ternary AgI/CeO_2@g-C_3N_4 nanocomposite with exceptional stability and reusability for visible light-assisted photocatalytic reduction of hexavalent chromium. *Appl. Surf. Sci.* **2021**, *555*, 149692. [CrossRef]
46. Balakumar, V.; Selvarajan, S.; Baishnisha, A.; Kathiresan, S. In-situ growth of TiO_2@B-doped g-C_3N_4 core-shell nanospheres for boosts the photocatalytic detoxification of emerging pollutants with mechanistic insight. *Appl. Surf. Sci.* **2022**, *577*, 151924. [CrossRef]
47. Singh, S.; Sharma, N.; Sehrawat, P.; Kansal, S.K. Solar-light-driven photocatalytic degradation of pharmaceutical pollutants utilizing 2D g-C3N4/BiOCl composite. *Environ. Toxicol. Pharmacol.* **2023**, *99*, 104110. [CrossRef]
48. Liu, W.; Li, Z.; Kang, Q.; Wen, L. Efficient photocatalytic degradation of doxycycline by coupling α-Bi_2O_3/g-C_3N_4 composite and H_2O_2 under visible light. *Environ. Res.* **2021**, *197*, 110925. [CrossRef]
49. Chen, C.; Xie, M.; Kong, L.; Lu, W.; Feng, Z.; Zhan, J. Mn_3O_4 nanodots loaded g-C_3N_4 nanosheets for catalytic membrane degradation of organic contaminants. *J. Hazard. Mater.* **2020**, *390*, 122146. [CrossRef]

50. He, Y.; Ma, Z.; Junior, L.B. Distinctive binary g-C_3N_4/MoS_2 heterojunctions with highly efficient ultrasonic catalytic degradation for levofloxacin and methylene blue. *Ceram. Int.* **2020**, *46*, 12364–12372. [CrossRef]
51. Zhang, C.; Ouyang, Z.; Yang, Y.; Long, X.; Qin, L.; Wang, W.; Zhou, Y.; Qin, D.; Qin, F.; Lai, C. Molecular engineering of donor-acceptor structured g-C_3N_4 for superior photocatalytic oxytetracycline degradation. *Chem. Eng. J.* **2022**, *448*, 137370. [CrossRef]
52. Wang, F.; Wang, Y.; Feng, Y.; Zeng, Y.; Xie, Z.; Zhang, Q.; Su, Y.; Chen, P.; Liu, Y.; Yao, K.; et al. Novel ternary photocatalyst of single atom-dispersed silver and carbon quantum dots co-loaded with ultrathin g-C_3N_4 for broad spectrum photocatalytic degradation of naproxen. *Appl. Catal. B Environ.* **2018**, *221*, 510–520. [CrossRef]
53. Mahmood, T.; Saddique, M.T.; Naeem, A.; Westerhoff, P.; Mustafa, S.; Alum, A. Comparison of Different Methods for the Point of Zero Charge Determination of NiO. *Ind. Eng. Chem. Res.* **2011**, *50*, 10017–10023. [CrossRef]

MDPI AG
Grosspeteranlage 5
4052 Basel
Switzerland
Tel.: +41 61 683 77 34

Catalysts Editorial Office
E-mail: catalysts@mdpi.com
www.mdpi.com/journal/catalysts

www.ingramcontent.com/pod-product-compliance
Lightning Source LLC
Chambersburg PA
CBHW040836250926
51850CB00007BA/1331
* 9 7 8 3 7 2 5 8 6 2 2 8 3 *